BIG DATA AND
SOCIAL SCIENCE

Chapman & Hall/CRC
Statistics in the Social and Behavioral Sciences Series

Series Editors
Jeff Gill, Steven Heeringa, Wim J. van der Linden, Tom Snijders

Recently Published Titles

Multilevel Modelling Using Mplus
Holmes Finch and Jocelyn Bolin

Applied Survey Data Analysis, Second Edition
Steven G. Heering, Brady T. West, and Patricia A. Berglund

Adaptive Survey Design
Barry Schouten, Andy Peytchev, and James Wagner

Handbook of Item Response Theory, Volume One: Models
Wim J. van der Linden

Handbook of Item Response Theory, Volume Two: Statistical Tools
Wim J. van der Linden

Handbook of Item Response Theory, Volume Three: Applications
Wim J. van der Linden

Bayesian Demographic Estimation and Forecasting
John Bryant and Junni L. Zhang

Multivariate Analysis in the Behavioral Sciences, Second Edition
Kimmo Vehkalahti and Brian S. Everitt

Analysis of Integrated Data
Li-Chun Zhang and Raymond L. Chambers

Multilevel Modeling Using R, Second Edition
W. Holmes Finch, Joselyn E. Bolin, and Ken Kelley

Modelling Spatial and Spatial-Temporal Data: A Bayesian Approach
Robert Haining and Guangquan Li

Measurement Models for Psychological Attributes
Klaas Sijtsma and Andries van der Ark

Handbook of Automated Scoring: Theory into Practice
Duanli Yan, André A. Rupp, and Peter W. Foltz

Interviewer Effects from a Total Survey Error Perspective
Kristen Olson, Jolene D. Smyth, Jennifer Dykema, Allyson Holbrook, Frauke Kreuter, and Brady T. West

Statistics and Elections: Polling, Prediction, and Testing
Ole J. Forsberg

Big Data and Social Science: Data Science Methods and Tools for Research and Practice, Second Edition
Ian Foster, Rayid Ghani, Ron S. Jarmin, Frauke Kreuter and Julia Lane

Analyzing Spatial Models of Choice and Judgment, Second Edition
David A. Armstrong II, Ryan Bakker, Royce Carroll, Christopher Hare, Keith T. Poole and Howard Rosenthal

For more information about this series, please visit: https://www.routledge.com/Chapman--HallCRC-Statistics-in-the-Social-and-Behavioral-Sciences/book-series/CHSTSOBESCI

Chapman & Hall/CRC
Statistics in the Social and Behavioral Sciences Series

BIG DATA AND SOCIAL SCIENCE

Data Science Methods and Tools for Research and Practice

Second Edition

Edited by

Ian Foster

University of Chicago
Argonne National Laboratory

Rayid Ghani

University of Chicago

Ron S. Jarmin

U.S. Census Bureau

Frauke Kreuter

University of Maryland
University of Manheim
Institute for Employment Research

Julia Lane

New York University
American Institutes for Research

CRC Press
Taylor & Francis Group
Boca Raton London New York

CRC Press is an imprint of the
Taylor & Francis Group, an **informa** business

A CHAPMAN & HALL BOOK

Second edition published 2021
by CRC Press
6000 Broken Sound Parkway NW, Suite 300, Boca Raton, FL 33487-2742

and by CRC Press
2 Park Square, Milton Park, Abingdon, Oxon, OX14 4RN

First edition published by CRC Press 2016

CRC Press is an imprint of Taylor & Francis Group, LLC

ISBN: 9780367341879 (hbk)
ISBN: 9780367568597 (pbk)
ISBN: 9780429324383 (ebk)

Typeset in Kerkis
by Nova Techset Private Limited, Bengaluru & Chennai, India

Contents

7   Machine Learning

Rayid Ghani and Malte Schierholz

8 Text Analysis

Preface

The class on which this book is based was created in response to a very real challenge: how to introduce new ideas and methodologies about economic and social measurement into a workplace focused on producing high-quality statistics. Since the publication of the first edition, we have been fortunate to train more than 450 participants in the Applied Data Analytics classes, resulting in increased data analytics capacity, in terms of both human and technical resources. What we have learned in delivering these classes has greatly influenced the second edition. We also have added a new chapter on Bias and Fairness in Machine Learning as well as reorganized some of the chapters.

As with any book, there are many people to be thanked. The Coleridge Initiative team at New York University, the University of Maryland, and the University of Chicago have been critical in shaping the format and structure—we are particularly grateful to Clayton Hunter, Jody Derezinski Williams, Graham Henke, Jonathan Morgan, Drew Gordon, Avishek Kumar, Brian Kim, Christoph Kern, and all of the book chapter authors for their contributions to the second edition.

We also thank the critical reviewers solicited from CRC Press and everyone from whom we received revision suggestions online, in particular Stas Kolenikov, who carefully examined the first edition and suggested updates. We owe a great debt of gratitude to the copyeditor, Anna Stamm; the project manager, Arun Kumar; the editorial assistant, Vaishali Singh; the project editor, Iris Fahrer; and the publisher, Rob Calver, for their hard work and dedication.

Editors

Ian Foster, PhD, is a professor of computer science at the University of Chicago as well as a senior scientist and distinguished fellow at Argonne National Laboratory. His research addresses innovative applications of distributed, parallel, and data-intensive computing technologies to scientific problems in such domains as climate change and biomedicine. Methods and software developed under his leadership underpin many large national and international cyberinfrastructures. He is a fellow of the American Association for the Advancement of Science, the Association for Computing Machinery, and the British Computer Society. He earned a PhD in computer science from Imperial College London.

Prof. Rayid Ghani is a professor in the Machine Learning Department (in the School of Computer Science) and the Heinz College of Information Systems and Public Policy at Carnegie Mellon University. His research focuses on developing and using Machine Learning, AI, and Data Science methods for solving high impact social good and public policy problems in a fair and equitable way across criminal justice, education, healthcare, energy, transportation, economic development, workforce development and public safety. He is also the founder and director of the "Data Science for Social Good" summer program for aspiring data scientists to work on data mining, machine learning, big data, and data science projects with social impact. Previously Prof. Ghani was a faculty member at the University of Chicago, and prior to that, served as the Chief Scientist for Obama for America (Obama 2012 Campaign).

Ron S. Jarmin, PhD, is the deputy director at the U.S. Census Bureau. He earned a PhD in economics from the University of Oregon and has published in the areas of industrial organization, business dynamics, entrepreneurship, technology and firm performance, urban economics, Big Data, data access and statistical disclosure avoidance. He oversees the Census Bureau's large portfolio of data collection, research and dissemination activities for critical economic and social statistics including the 2020 Decennial Census of Population and Housing.

Frauke Kreuter, PhD, is a professor at the University of Maryland in the Joint Program in Survey Methodology, professor of Statistics and Methodology at the University of Mannheim and head of the Statistical Methods group at the Institute for Employment Research in Nuremberg, Germany. She is the founder of the International Program in Survey and Data Science, co-founder of the Coleridge Initiative, fellow of the American Statistical Association (ASA), and recipient of the WSS Cox and the ASA Links Lecture Awards. Her research focuses on data quality, privacy, and the effects of bias in data collection on statistical estimates and algorithmic fairness.

Julia Lane, PhD, is a professor at the NYU Wagner Graduate School of Public Service. She is also an NYU Provostial Fellow for Innovation Analytics. She co-founded the Coleridge Initiative as well as UMETRICS and STAR METRICS programs at the National Science Foundation, established a data enclave at NORC/University of Chicago, and co-founded the Longitudinal Employer-Household Dynamics Program at the U.S. Census Bureau and the Linked Employer Employee Database at Statistics New Zealand. She is the author/editor of 10 books and the author of more than 70 articles in leading journals, including *Nature and Science*. She is an elected fellow of the American Association for the Advancement of Science and a fellow of the American Statistical Association.

Contributors

Stefan Bender
Deutsche Bundesbank
Frankfurt, Germany

Paul P. Biemer
RTI International
Raleigh, NC, USA
University of North Carolina
Chapel Hill, NC, USA

Jordan Boyd-Graber
University of Maryland
College Park, MD, USA

Pascal Heus
Metadata Technology North America
Knoxville, TN, USA

Clayton Hunter
New York University
New York, NY, USA

Christoph Kern
University of Mannheim
Mannheim, Germany

Brian Kim
University of Maryland
College Park, MD, USA

Evgeny Klochikhin
Parkofon Inc.
Washington, DC, USA

Avishek Kumar
Intuit AI

Cameron Neylon
Curtin University
Perth, Australia

Jason Owen-Smith
University of Michigan
Ann Arbor, MI, USA

Catherine Plaisant
University of Maryland
College Park, MD, USA

Kit T. Rodolfa
Carnegie Mellon University
Pittsburgh, PA, USA

Pedro Saleiro
Feedzai
California, USA

Malte Schierholz
Institute for Employment Research (IAB)
Nuremberg, Germany

Jonathan Scott Morgan
University of Mannheim
Mannheim, Germany

Claudio Silva
New York University
New York, NY, USA

Joshua Tokle
Amazon
Seattle, WA, USA

Huy Vo
City University of New York
New York, NY, USA

M. Adil Yalçın
Keshif
Washington, DC, USA

Chapter 1

Introduction

1.1   Why this book?

The world has changed for empirical social scientists. The new types of "big data" have generated an entire new research field—that of data science. That world is dominated by computer scientists who have generated new ways of creating and collecting data, developed new analytical techniques, and provided new ways of visualizing and presenting information. The results have been to change the nature of the work that social scientists do.

Social scientists have been enthusiastic in responding to the new opportunity. Python and R are becoming as, and hopefully more, well-known as SAS and Stata—indeed, the 2018 Nobel Laureate in Economics, Paul Romer, is a Python convert (Kopf, 2018). Research also has changed. Researchers draw on data that are "found" rather than "made" by federal agencies; those publishing in leading academic journals are much less likely today to draw on preprocessed survey data (Figure 1.1). Social science workflows can become more automated, replicable, and reproducible (Yarkoni et al., 2019).

Policy also has changed. The Foundations of Evidence-based Policy Act, which was signed into law in 2019, requires agencies to utilize evidence and data in making policy decisions (Hart, 2019). The Act, together with the Federal Data Strategy (Office of Management and Budget, 2019), establishes both Chief Data Officers to oversee the collection, use of, and access to many new types of data and a learning agenda to build the data science capacity of agency staff.

In addition, the jobs have changed. The new job title of "data scientist" is highlighted in job advertisements on CareerBuilder.com and Burningglass—supplanting the demand for statisticians, economists, and other quantitative social scientists if starting salaries are useful indicators. At the federal level, the Office of Personnel Management has created a new data scientist job title.

The goal of this book is to provide social scientists with an understanding of the key elements of this new science, the value of the

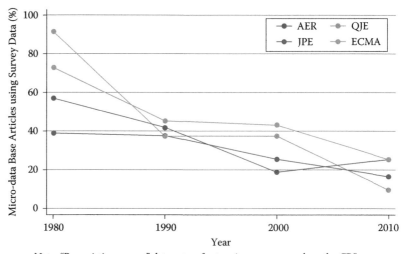

Note: "Pre-existing survey" data sets refer to micro surveys such as the CPS or
SIPP and do not include surveys designed by researchers for their study.
Sample excludes studies whose primary data source is from developing countries.

Figure 1.1. Use of pre-existing survey data in publications in leading journals,
1980–2010 (Chetty, 2012)

tools, and the opportunities for doing better work. The goal is also
to identify the many ways in which the analytical toolkits possessed
by social scientists can enhance the generalizability and usefulness
of the work done by computer scientists.

We take a pragmatic approach, drawn on our experience of working with data to tackle a wide variety of policy problems. Most
social scientists set out to solve a real world social or economic
problem: they frame the problem, identify the data, conduct the
analysis, and then draw inferences. At all points, of course, the
social scientist needs to consider the ethical ramifications of their
work, particularly respecting privacy and confidentiality. The book
follows the same structure. We chose a particular problem—the
link between research investments and innovation—because that is
a major social science policy issue, and one in which social scientists have been addressing the use of big data techniques.

1.2   Defining big data and its value

There are almost as many definitions of big data as there are new
types of data. One approach is to define big data as anything too big

to fit onto your computer. **Another approach is to define it as data with high volume, high velocity, and great variety.** We choose the description adopted by the American Association of Public Opinion Research: "The term 'Big Data' is an imprecise description of a rich and complicated set of characteristics, practices, techniques, ethical issues, and outcomes all associated with data" (Japec et al., 2015).

> ▶ This topic will be discussed in more detail in Chapter 5.

The value of the new types of data for social science is quite substantial. Personal data have been hailed as the "new oil" of the 21st century (Greenwood et al., 2014). Policymakers have found that detailed data on human beings can be used to reduce crime (Lynch, 2018), improve health delivery (Pan et al., 2017), and better manage cities (Glaeser, 2019). Society can gain as well—much cited work shows data-driven businesses are 5% more productive and 6% more profitable than their competitors (Brynjolfsson et al., 2011). Henry Brady provides a succinct overview when he says, "Burgeoning data and innovative methods facilitate answering previously hard-to-tackle questions about society by offering new ways to form concepts from data, to do descriptive inference, to make causal inferences, and to generate predictions. They also pose challenges as social scientists must grasp the meaning of concepts and predictions generated by convoluted algorithms, weigh the relative value of prediction versus causal inference, and cope with ethical challenges as their methods, such as algorithms for mobilizing voters or determining bail, are adopted by policy makers" (Brady, 2019).

Example: New potential for social science

The billion prices project is a great example of how researchers can use new web-scraping techniques to obtain online prices from hundreds of websites and thousands of webpages to build datasets customized to fit specific measurement and research needs in ways that were unimaginable 20 years ago (Cavallo and Rigobon, 2016); other great examples include the way in which researchers use text analysis of political speeches to study political polarization (Peterson and Spirling, 2018) or of Airbnb postings to obtain new insights into racial discrimination (Edelman et al., 2017).

Of course, these new sources come with their own caveats and biases that need to be considered when drawing inferences. We will cover this later in the book in more detail.

But most interestingly, the new data can change the way we think about behavior. For example, in a study of environmental effects on health, researchers combine information on public school cafeteria deliveries with children's school health records to show that simply putting water jets in cafeterias reduced milk consumption and also reduced childhood obesity (Schwartz et al., 2016). Another study which sheds new light into the role of peers on productivity finds that the productivity of a cashier increases if they are within eyesight of a highly productive cashier but not otherwise (Mas and Moretti, 2009). Studies such as these show ways in which clever use of data can lead to greater understanding of the effects of complex environmental inputs on human behavior.

New types of data also can enable us to study and examine small groups—the tails of a distribution—in a way that is not possible with small data. Much of the interest in human behavior is driven by those tails, such as health care costs by small numbers of ill people (Stanton and Rutherford, 2006) or economic activity and employment by a small number of firms (Evans, 1987; Jovanovic, 1982).

Our excitement about the value of new types of data must be accompanied by a recognition of the lessons learned by statisticians and social scientists from their past experience with surveys and small scale data collection. The next sections provide a brief overview.

1.3 The importance of inference

It is critically important to be able to use data to generalize from the data source to the population. That requirement exists, regardless of the data source. Statisticians and social scientists have developed methodologies for survey data to overcome problems in the data-generating process. A guiding principle for survey methodologists is the total survey error framework, and statistical methods for weighting, calibration, and other forms of adjustment are commonly used to mitigate errors in the survey process. Likewise for "broken" experimental data, techniques such as propensity score adjustment and principal stratification are widely used to fix flaws in the data-generating process.

If we take a look across the social sciences, including economics, public policy, sociology, management, (parts of) psychology, and the like, their scientific activities can be grouped into three categories with three different inferential goals: Description, Causation, and Prediction.

1.3.1   Description

The job of many social scientists is to provide descriptive statements about the population of interest. These could be univariate, bivariate, or even multivariate statements.

Usually, descriptive statistics are created based on census data or sample surveys to create some summary statistics such as a mean, a median, or a graphical distribution to describe the population of interest. In the case of a census, the work ends there. With sample surveys, the point estimates come with measures of uncertainties (standard errors). The estimation of standard errors has been established for most descriptive statistics and common survey designs, even complex ones that include multiple layers of sampling and disproportional selection probabilities (Hansen et al., 1993; Valliant et al., 2018).

Example: Descriptive statistics

The Census Bureau's American Community Survey (ACS) "helps local officials, community leaders, and businesses understand the changes taking place in their communities. It is the premier source for detailed population and housing information about our nation" (https://www.census.gov/programs-surveys/acs). The summary statistics are used by planners to allocate resources—but it's important to pay attention to the standard errors, particularly for small samples. For example, in one county (Autauga) in Alabama, with a total population of about 55,000, the ACS estimates that 139 children under age 5 live in poverty—plus or minus 178! So the plausible range is somewhere between 0 and 317 (Spielman and Singleton, 2015).

Proper inference from a sample survey to the population usually depends on (1) knowing that everyone from the target population has had the chance to be included in the survey and (2) calculating the selection probability for each element in the population. The latter does not necessarily need to be known prior to sampling, but eventually a probability is assigned for each case. Obtaining correct selection probabilities is particularly important when reporting totals (Lohr, 2009). Unfortunately in practice, samples that begin as probability samples can suffer from a high rate of nonresponse. Because the survey designer cannot completely control which units respond, the set of units that ultimately respond cannot be considered to be a probability sample. Nevertheless, starting with a probability sample provides some degree of assurance that a sample will have limited coverage errors (nonzero probability of being in the sample).

1.3.2 Causation

Identifying causal relationships is another common goal for social science researchers (Varian, 2014). Ideally, such explanations stem from data that allow causal inference: typically randomized experiments or strong non-experimental study designs. When examining the effect of X on Y, knowing how cases have been selected into the sample or dataset is much less important for estimating causal effects than they are for descriptive studies, e.g., population means. What is important is that all elements of the inferential population have a chance to be selected for the treatment (Imbens and Rubin, 2015). In the debate about probability and non-probability surveys, this distinction often is overlooked. Medical researchers have operated with unknown study selection mechanisms for years, e.g, randomized trials that enroll very select samples.

Example: New data and causal inference

If the data-generating process is not understood, resources can be badly misallocated. Overreliance on, for example, Twitter data, in targeting resources after hurricanes can lead to the misallocation of resources towards young internet-savvy people with cell phones and away from elderly or impoverished neighborhoods (Shelton et al., 2014). Of course, all data collection approaches have had similar risks. Bad survey methodology is what led the *Literary Digest* to incorrectly call the 1936 election for Landon, not Roosevelt (Squire, 1988). Inadequate understanding of coverage, incentive, and quality issues, together with the lack of a comparison group, has hampered the use of administrative records—famously in the case of using administrative records on crime to make inferences about the role of death penalty policy in crime reduction (Donohue and Wolfers, 2006).

In practice, regardless of how much data are available, researchers must consider at least two things: (1) how well the results generalize to other populations (Athey and Imbens, 2017) and (2) whether the treatment effect on the treated population is different than the treatment effect on the full population of interest (Stuart, 2010). New methods to address generalizability are under development (DuGoff et al., 2014). While unknown study selection probabilities usually make it difficult to estimate population causal effects, as long as we are able to model the selection process there is no reason not to do causal inference from so-called non-probability data.

1.3.3 Prediction

Forecasting or prediction tasks. The potential for massive amounts of data to improve prediction is undeniable. However, just like the causal inference setting, it is of the utmost importance that we know the process that has generated the data, so that biases due to unknown or unobserved systematic selection can be minimized. Predictive policing is a good example of the challenges. The criminal justice system generates massive amounts of data that can be used to better allocate police resources—but if the data do not represent the population at large, the predictions could be biased and, more importantly, the interventions assigned using those predictions could harm society.

Example: Learning from the flu

"Five years ago [in 2009], a team of researchers from Google announced a remarkable achievement in one of the world's top scientific journals, *Nature*. Without needing the results of a single medical check-up, they were nevertheless able to track the spread of influenza across the US. What's more, they could do it more quickly than the Centers for Disease Control and Prevention (CDC). Google's tracking had only a day's delay, compared with the week or more it took for the CDC to assemble a picture based on reports from doctors' surgeries. Google was faster because it was tracking the outbreak by finding a correlation between what people searched for online and whether they had flu symptoms." ...

"Four years after the original *Nature* paper was published, *Nature News* had sad tidings to convey: the latest flu outbreak had claimed an unexpected victim: Google Flu Trends. After reliably providing a swift and accurate account of flu outbreaks for several winters, the theory-free, data-rich model had lost its nose for where flu was going. Google's model pointed to a severe outbreak but when the slow-and-steady data from the CDC arrived, they showed that Google's estimates of the spread of flu-like illnesses were overstated by almost a factor of two.

The problem was that Google did not know—could not begin to know—what linked the search terms with the spread of flu. Google's engineers weren't trying to figure out what caused what. They were merely finding statistical patterns in the data. They cared about correlation rather than causation" (Harford, 2014).

1.4 The importance of understanding how data are generated

The costs of realizing the benefits of the new types of data are nontrivial. Even if data collection is cheap, the costs of cleaning, curating, standardizing, integrating, and using the new types of data are substantial. In essence, just as with data from surveys, data still need to be processed—cleaned, normalized, and variables coded—but this needs to be done at scale. But even after all of these tasks are completed, social scientists have a key role in describing the quality of the data. This role is important, because most data in the real world are noisy, inconsistent, and exhibit missing values. Data quality can be characterized in multiple ways (see Christen, 2012a; National Academies of Sciences, Engineering, and Medicine and others [2018]), such as:

▶ This topic will be discussed in more detail in Section 1.5.

- Accuracy: How accurate are the attribute values in the data?

- Completeness: Are the data complete?

- Consistency: How consistent are the values in and between different database(s)?

- Timeliness: How timely are the data?

- Accessibility: Are all variables available for analysis?

In the social science world, the assessment of data quality has been integral to the production of the resultant statistics. That has not necessarily been easy when assessing new types of data. A good example of the importance of understanding how data are generated arose in one of our classes a few years ago, when class participants were asked to develop employment measures for ex-offenders in the period after they were released from prison (Kreuter et al., 2019).

For people working with surveys, the definition was already pre-constructed: in the Current Population Survey (CPS), respondents were asked about their work activity in the week covering the 12th of the month. Individuals were counted as employed if they had at least one hour of paid work in that week (with some exceptions for family and farm work). But the class participants were working with administrative records from the Illinois Department of Employment Security and the Illinois Department of Corrections (Kreuter et al., 2019). Those records provided a report of all jobs in every quarter that each individual held in the state; when matched to data about formerly incarcerated individuals, it could provide rich information about their employment patterns. A group of class participants produced Figure 1.2—the white boxes represent quarters in which an individual does not have a job and the blue boxes represent quarters in which an individual does have a job.

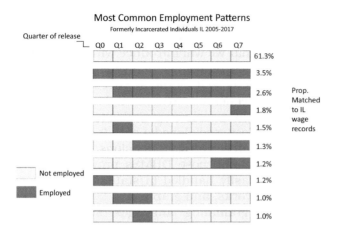

Figure 1.2. Most common employment patterns, formerly incarcerated individuals in Illinois, 2005–2017

A quick look at the results is very interesting. First, the participants present an entirely new dynamic way of looking at employment—not just the relatively static CPS measure. Second, the results are a bit shocking. More than 61% of Illinois exoffenders do not have a job in any of the eight quarters after their release. Only 3.5% have a job in all of the quarters. This is where social scientists and government analysts can contribute—because they know how the data are generated. The matches between the two agencies have been conducted on (deidentified) Social Security numbers (SSNs). It is likely that there are several gaps in those matches. First, agency staff know that the quality of SSNs in prisons is quite low, so that may be one reason for the low match rate. Second, the matches are only to Illinois jobs, and many formerly incarcerated individuals could be working across state lines (if allowed). Third, the individuals may be attending community college, or accepting social assistance, or reincarcerated. More data can be used to examine these different possibilities—but we believe this illustrates the value that social scientists and subject matter experts provide to measuring the quality issues we highlight at the beginning of this section.

1.5 New tools for new data

The new data sources that we have discussed frequently require working at scales for which the social scientist's familiar tools are not designed. Fortunately, the wider research and data analytics community has developed a wide variety of often more scalable and flexible tools—tools that we will introduce within this book.

▶ This topic will be discussed in more detail in Chapter 4.

Relational database management systems (DBMSs) are used throughout business as well as the sciences to organize, process, and search large collections of structured data. NoSQL DBMSs are used for data that is extremely large and/or unstructured, such as collections of web pages, social media data (e.g., Twitter messages), sensor data, and clinical notes. Extensions to these systems and also specialized single-purpose DBMSs provide support for data types that are not easily handled in statistical packages such as geospatial data, networks, and graphs.

Open source programming languages, such as Python (used extensively throughout this book) and R, provide high-quality implementations of numerous data analysis and visualization methods, from regression to machine learning, text analysis, network analysis, and much more. Finally, parallel computing platforms, such as Hadoop and Spark, can be used to harness parallel computing clusters for extremely large data sets and computationally intensive analyses.

These various components may not always work together as smoothly as do integrated packages, such as SAS, SPSS, and Stata, but they allow researchers to address problems of greater scale and complexity. Furthermore, they are developing at a tremendous rate as the result of work by thousands of people worldwide. For these reasons, the modern social scientist needs to be familiar with their capabilities.

1.6   The book's "use case"

This book is about the uses of new types of computational and data analysis methods in social science. Our focus is on working through the use of data as a social scientist normally approaches research. That involves thinking through how to use such data to address a question from beginning to end, and thereby learning about the associated tools—rather than simply engaging in coding exercises and then thinking about how to apply them to a potpourri of social science examples.

There are many examples of the use of big data in social science research. Our challenge in designing the book has been to find a use case that is interesting, that does not require access to confidential microdata, that utilizes all of the methods and tools of interest to a typical researcher, and that could be applied to most of the many other use cases that may be relevant to other instructors (such as criminal justice, health care, welfare, education, or economic development). We have chosen to make use of the great surge of interest in examining the impact of investments in research and development on economic and social outcomes, and which constitutes one of the first large-scale big data social science data infrastructures. Many of the data sources are public, and they are the sources used in this book.

We believe the question also should be of broad interest to many potential users, even if they are not subject matter specialists. The surge of interest has been in response to a call from the President's Science Advisor, Jack Marburger, for a *science of science policy* (Marburger, 2005), which reflects a desire to have a scientific response to the questions that he has been asked about the impact of investments in science.

Example: The science of science policy

During his tenure as Science Advisor, Marburger made some very insightful points. He was skeptical of the calls for more and more investment in science, particularly of the European push for 3% of GDP to be allocated to research and development. He wanted to both understand and be able to explain what would be the returns on that kind of expenditure. In a very famous editorial, he asked: "How much should a nation spend on science? What kind of science? How much from private versus public sectors? Does demand for funding by potential science performers imply a shortage of funding or a surfeit of performers? These and related science policy questions tend to be asked and answered today in a highly visible advocacy context that makes assumptions that are deserving of closer scrutiny. A new 'science of science policy' is emerging, and it may offer more compelling guidance for policy decisions and for more credible advocacy. . . .

"Relating R&D to innovation in any but a general way is a tall order, but not a hopeless one. We need econometric models that encompass enough variables in a sufficient number of countries to produce reasonable simulations of the effect of specific policy choices. This need won't be satisfied by a few grants or workshops, but demands the attention of a specialist scholarly community. As more economists and social scientists turn to these issues, the effectiveness of science policy will grow, and of science advocacy too" (Marburger, 2005).

In order to answer his question, an entire research field has developed that pulls together relevant data from a wide variety of different sources using widely differing methodologies and approaches. This effort addresses challenges often faced by social science and computer science researchers trying to use new data to answer important questions—namely, that inputs, outputs, and outcomes are not generated or combined in a systematic fashion, even though producing consistent and reliable answers to stakeholder requests requires the use of common data sources and standardized methodologies. They have been able to pull together new digital sources of data and apply modern technologies to analyze them. In this book, we use three primary examples to show how this has been accomplished. The first is to describe what research is being done, using data produced from multiple agencies on grand funding. The second is to use award and patent administrative records to describe who is doing the research (and with whom). The third is to use patent data to describe what results the funding has generated (Lane et al., 2018; Weinberg et al., 2014).

Showing how those challenges can be addressed coordinates with the goal of this book. Our focus is highlighting how to use new digital technologies to capture the data needed to understand and address a set of questions, with an illustrative focus on the broad results of Federal Science and Technology investments. We are able to draw on the public availability of a wide variety of research inputs, such as federal grant information, and some outputs, particularly patent data. We also are able to draw on new and more accurate methods for reliably attributing research products to researchers, a nontrivial task due to considerable ambiguity in author names (Han et al., 2004; Kim et al., 2016a; Li et al., 2014; Smalheiser and Torvik, 2009). Figure 1.3 provides an abstract representation of the empirical approach that is needed: data about grants, the people who are funded on grants, and the subsequent scientific and economic activities and shows.

First, data must be captured on what is funded and, because the data are in text format, computational linguistics tools must be applied (Chapter 8). Second, data must be captured on who is funded and how they interact in teams, so network tools and analysis must be used (Chapter 9). Third, information about the type of results must be gleaned from the web and other sources (Chapter 2).

Finally, the disparate complex data sets need to be stored in databases (Chapter 4), integrated (Chapter 3), analyzed (Chapter 7), and used to make inferences (Chapter 10).

The use case serves as the thread that ties many of the ideas together. Rather than asking the reader to learn how to code "hello, world," we build on data that have been put together to answer

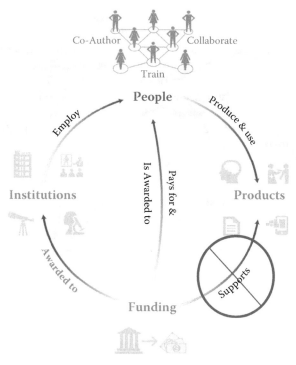

Figure 1.3. A visualization of the complex links between what and who is funded as well as the results; note that tracing a direct link between funding and results is incorrect and misleading

a real-world question, and we provide explicit examples based on that data. Then, we provide examples that show how the approach generalizes.

For example, Chapter 8 shows how to use natural language processing to describe *what* research is being done, using proposal and award text to identify the research topics in a portfolio (Evans and Foster, 2011; Talley et al., 2011). But then, the chapter also shows how the approach can be used to address a problem that is not limited to science policy only—the conversion of massive amounts of knowledge, stored in text, to usable information.

Similarly, Chapter 9 provides specific examples utilizing UMET-RICS data and shows how such data can be used to create new units of analysis—the networks of researchers who do science and the networks of vendors who supply research inputs. This chapter also shows how networks can be used to study a wide variety of other social science questions.

★ Application programming
interfaces

In another example, we use APIs[*] provided by publishers to describe the results generated by research funding in terms of publications and other measures of scientific impact, and we also provide code that can be repurposed for many similar APIs.

And, of course, because all of these new types of data are provided in a variety of different formats, some of which are quite large (or voluminous), and with a variety of different timestamps (or velocity), we discuss how to store the data in different types of data formats.

Box 1.1: The IRIS data infrastructure

Much more information is available at the Institute for Research on Innovation and Science (IRIS, https://iris.isr. umich.edu/) at the University of Michigan. The Institute has extended the data infrastructure to bring together both confidential and public data in a secure environment. Working with universities interested in documenting the results of their grant funding, it has been able to trace the spending of almost 400,000 grants to more than 600,000 individuals and 820,000 vendors—and it has shown the direct effects of that funding on their subsequent scientific and economic activity (Institute for Research on Innovation and Science, IRIS). To date, more than 100 researchers from dozens of institutions have worked with the new data infrastructure to provide an empirical response to Marburger's call.

Although we focus on one particular use case, the methods covered in this book are broadly applicable across a variety of policy areas—indeed, we have used this book to teach classes in such fields as education, criminal justice, workforce and economic development, and social services (https://coleridgeinitiative.org/training).

The methods have been used to answer questions such as:

- "What are the earnings and employment outcomes of individuals graduating from two and four year colleges?"

- "How does placement in different types of firms change the likelihood that the formerly incarcerated will recidivate?"

- "What factors increase the likelihood that welfare recipients on TANF (Temporary Assistance to Needy Families) will leave welfare?"

and

- "How do regulatory agencies move from reactive, complaint-based, health and safety inspections for workplaces and housing to a more proactive approach that focuses on prevention?"

Box 1.2: Data science skills and the provision of human services

Operating under the Department of Health and Human Services' Administration for Children and Families (ACF), the Office of Family Assistance (OFA) and Office of Planning, Research, and Evaluation (OPRE) have sponsored the TANF Data Collaborative (TDC) (https://www.tanfdata.org/) to help professionals working with Temporary Assistance for Needy Families (TANF) and related human services agencies develop key data science skills. The focus is on teaching participants how to scope a problem, conduct record linkage, apply machine learning and visualization tools, and learn about privacy issues when working with confidential data.

1.7 The structure of the book

We organize the book in three parts, based around the way social scientists approach doing research (see Figure 1.4 for a prototypical data science workflow). The first set of chapters addresses the new ways to capture, curate, and store data. The second set of chapters describes what tools are available to process and analyze data. The last set considers the appropriate handling of data on individuals and organizations as well as what inferences can be drawn from the data and the analysis that has been performed. Of course, we assume that, before starting with the data and analysis, time has been spent formulating the problem or question that is being addressed. We do not include that in this book but refer readers to resources such as "Data Science Project Scoping" for more information.

> ▶ http://www.dssgfellow-ship.org/2016/10/27/scoping-data-science-for-social-good-projects/

1.7.1 Part I: Capture and curation

The four chapters in Part I (see Figure 1.5) tell you how to collect, store, link, and manage data.

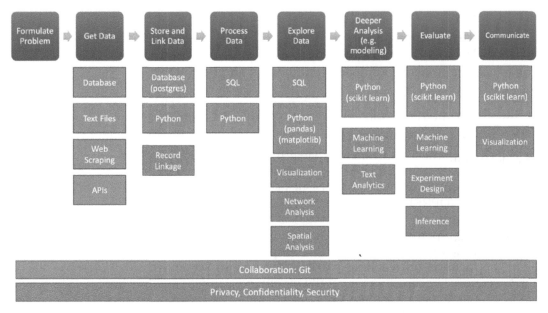

Figure 1.4. The data science project workflow. Blue represents each step in the project, orange represents the tools used in that step, and green represents the methods for analysis

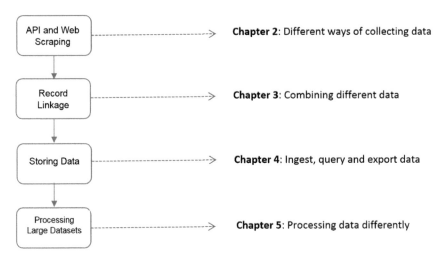

Figure 1.5. The four chapters of Part I focus on *data capture* and *curation*

Chapter 2 describes how to extract information from data sources on the web, including social media. The particular application will be to develop links to authors' articles on Twitter using PLOS articles and to pull information about authors and articles from web sources by using an API. You will learn how to retrieve link data from book-marking services, citations from Crossref, links from Facebook, and information from news coverage. In keeping with the social science grounding that is a core feature of this book, the chapter discusses what data can be captured from online sources, what is potentially reliable, and how to manage data quality issues.

This data differs from survey data in that typically we must combine data from multiple sources to obtain a complete picture of the activities of interest. Although computer scientists sometimes may simply "mash" data sets together, social scientists are rightfully concerned about issues of missing links, duplicative links, and erroneous links. Chapter 3 provides an overview of traditional rule-based and probabilistic approaches to data linkage, as well as machine learning approaches that are more adaptive and tunable.

After data have been collected and linked, it is necessary to store and organize it. Social scientists are accustomed to working with one analytical file, often in statistical software tools such as SAS or Stata. Chapter 4 describes different approaches to storing data in ways that facilitate rapid, scalable, and reliable exploration and analysis.

Big data is sometimes defined as data that are too big to fit onto the analyst's computer. Chapter 5 provides an overview of programming techniques that facilitate the scalable use of data (often using parallel computing). While the focus is on one of the most widely used big data programming paradigms and its most popular implementation, Apache Hadoop, the goal of the chapter is to provide a conceptual framework to the key challenges that the approach is designed to address.

1.7.2   Part II: Modeling and analysis

The four chapters in Part II (see Figure 1.6) introduce four of the most important tools that can be used by social scientists to conduct new and exciting research: information visualization, machine learning, text analysis, and social network analysis.

Chapter 6 introduces information visualization methods and describes how you can use those methods to explore data and communicate results so that data can be turned into interpretable, actionable information. There are many ways of presenting statistical information that convey content in a rigorous manner. The goal of this chapter is to explore different approaches and examine the information content and analytical validity of the different approaches. It provides an overview of effective visualizations; in fact, using visualization even in early analysis stages is key

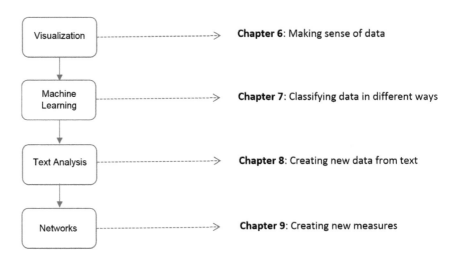

Figure 1.6. The four chapters in Part II focus on data *modeling* and *analysis*

to obtaining a good understanding of data quality and potential pitfalls.

Chapter 7 introduces machine learning methods. It shows the power of machine learning in a variety of different contexts, particularly focusing on clustering, classification, and prediction. It will provide an overview of basic approaches and how those approaches are applied. The chapter builds from a conceptual framework on how to formulate social science problems as machine learning problems, how to perform machine learning analysis, and how to evaluate the analysis. Then, these concepts are translated into code to ensure that the analysis can be put into practical use by social science researchers and practitioners.

Chapter 8 describes how social scientists can make use of text data through text analysis and natural language processing methods. Dealing with text and analyzing text is not new to social scientists. What is different these days is that the vast amounts of data that are stored in documents now can be analyzed and searched at scale, so that different types of information can be retrieved. Documents (and the underlying activities of the entities that generated the documents) can be categorized into topics or fields as well as summarized. In addition, machine translation can be used to compare documents in different languages.

Typically, social scientists are interested in describing the activities of individuals and organizations (such as households and firms) in a variety of economic and social contexts. The frames within which data are collected typically have been generated from tax or other programmatic sources. The new types of data permit new units of analysis—particularly network analysis—largely enabled by advances in mathematical graph theory. Thus, Chapter 9 describes how social scientists can use network theory to generate measurable representations of patterns of relationships connecting entities. As the author notes, the value of the new framework is not only in constructing different right-side variables but also in studying an entirely new unit of analysis that lies somewhere between the largely atomistic actors that occupy the markets of neo-classical theory and the tightly managed hierarchies that are the traditional object of inquiry of sociologists and organizational theorists.

1.7.3 Part III: Inference and ethics

The three chapters in Part III (see Figure 1.7) cover three advanced topics relating to data inference and ethics—errors and inference, bias, and privacy and confidentiality—and introduce the workbooks that provide access to the practical exercises associated with the text.

Chapter 10 addresses inference and the errors associated with big data. Social scientists know only too well the cost associated with bad data—as highlighted in the classic *Literary Digest* example in the introduction to this chapter, as well as the more recent Google Flu Trends. Although the consequences are well understood,

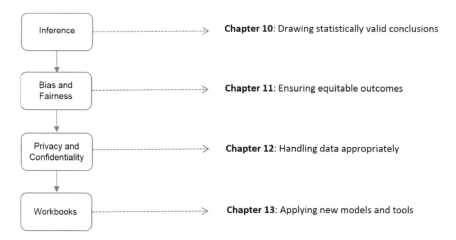

Figure 1.7. The four chapters in Part III focus on *inference* and *ethics*

the new types of data are so large and complex that their properties often cannot be studied in traditional ways. In addition, the data-generating function is such that the data are often selective, incomplete, and erroneous. Without proper data hygiene, errors can compound quickly. This chapter provides a systematic way to think about the error framework in a big data setting.

Interest in algorithmic fairness and bias has been growing recently, but it is easy to get lost in the large number of definitions and metrics. There are many different, often competing, ways to measure whether a given model and the resulting system is "fair." In Chapter 11, we provide an overview of these metrics along with some specific examples to help navigate these concepts and understand the tradeoffs involved in choosing to optimize one metric over others.

Chapter 12 addresses the issue that sits at the core of any study of human beings—privacy and confidentiality. In a new field, such as the one covered in this book, it is critical that many researchers have access to the data so that work can be replicated and built upon—that there be a scientific basis to data science. Yet, the rules that social scientists traditionally have used for survey data, namely anonymity and informed consent, no longer apply when the data are collected "in the wild." This concluding chapter identifies the issues that must be addressed for the practice of responsible and ethical research.

▶ See https://jupyter.org/.

Finally, Chapter 13 provides an overview of the practical work that accompanies each chapter—the workbooks that are designed, using *Jupyter notebooks*, to enable students and interested practitioners to apply the new techniques and approaches in selected chapters. This final chapter gives a broad overview of the tools needed to work with these workbooks and some instructions on how to use the workbooks if you decide to teach a class using this content. The chapter also informs broadly about the data and problems these workbooks tackle, and about the general structure of the workbooks. We are constantly expanding and updating the set of available workbooks, so check GitHub regularly if you would like to see the latest version. We hope you have a lot of fun with them.

1.8   Resources

For more information on the science of science policy, see Husbands Fealing et al.'s book for a full discussion of many issues (Husbands Fealing et al., 2011) and peruse the awards made by the National Science Foundation's Science of Science: Discovery, Communication, and Impact program (https://www.nsf.gov/funding/pgm_summ.jsp?pims_id=505730).

This book is, above all, a *practical* introduction to the methods and tools that the social scientist can use to make sense of big data, and thus programming resources are also important. We make extensive use of the Python programming language and databases in both the book and its supporting workbooks. We recommend that any social scientist who aspires to work with large data sets become proficient in the use of these two systems and GitHub. All three, fortunately, are quite accessible and are supported by excellent online resources. Time spent mastering them will be repaid many times over in more productive research.

For Python, Alex Bell's *Python for Economists* (available online: Bell [2012]) provides a wonderful 30-page introduction to the use of Python in the social sciences, complete with XKCD cartoons. Economists Tom Sargent and John Stachurski provide a very useful set of lectures and examples at https://quantecon.org/. For more detail, we recommend Charles Severance's *Python for Informatics: Exploring Information* (Severance, 2013), which not only covers basic Python but also provides material relevant to web data (the subject of Chapter 2) and MySQL (the subject of Chapter 4). This book is also freely available online and is supported by excellent online lectures and exercises.

> ► Read this: http://alexbell.net/pyseminar.html

For SQL, Chapter 4 provides introductory material and pointers to additional resources, so we will not say more here.

We also recommend that you master GitHub. A version control system is a tool for keeping track of changes that have been made to a document over time; GitHub is a hosting service for projects that use the Git version control system. As Strasser (2014) explains, Git/GitHub makes it straightforward for researchers to create digital lab notebooks that record the data files, programs, papers, and other resources associated with a project, with automatic tracking of the changes that are made to those resources over time. GitHub also makes it easy for collaborators to work together on a project, whether a program or a paper: changes made by each contributor are recorded and can be easily reconciled. For example, we have used GitHub to create this book, with authors and editors contributing changes and comments at different times and from many time zones. We also use GitHub to provide access to the supporting workbooks. Ram (2013) provides a nice description of how Git/GitHub can be used to promote reproducibility and transparency in research.

One more resource that is outside the scope of this book but that you may well want to master is the cloud (Armbrust et al., 2010; Lifka et al., 2013). Previously, when your data and computations became too large to analyze on your laptop, you were "out of luck" unless your employer (or a friend) had a larger computer; now, with the emergence of cloud storage and computing services from Amazon Web Services, Google, and Microsoft, powerful computers are available to anyone with a credit card. We and many others have had positive experiences using such systems for the analysis of urban (Catlett et al., 2014), environmental (Elliott et al., 2014), and genomic (Bhuvaneshwar et al., 2015) data analysis and modeling, for example.

Part I
Capture and Curation

Chapter 2

Working with Web Data and APIs

Cameron Neylon

In many social science problems, we have to augment our primary data with external data sources. Often, the external data are available on the web, either on web pages directly or accessible through application programming interfaces (APIs). Gathering this data requires understanding how to scrape web pages or calling the APIs with parameters about the information we need. One common example of this is augmenting our primary data with data from the American Community Survey (ACS) or from Open Data Portals maintained by local, state, and federal agencies. These data sources can be either downloaded in bulk or used dynamically through APIs. The same is true for data from social media sources, such as Twitter, Instagram, and Facebook. In this chapter, we will cover tools (specifically using Python) that can be used by social science researchers to programmatically gather this type of external data from web pages and APIs.

2.1 Introduction

The Internet is an excellent resource for vast amounts of data on businesses, people, and their activity on social media. But how can we capture the information and make use of it as we might make use of more traditional data sources?

In social science, we often explore information on people, organizations, or locations. The web can be a rich source of additional information when performing this type of analysis, pointing to new sources of information, allowing a pivot from one perspective to another, or from one kind of query to another. Sometimes, this data from the web is completely unstructured, existing in web pages

> ▶ Chapter 12 will discuss ethical issues when addressing and using "publicly" available data for research and policy purposes.

spread across a site, and sometimes they are provided in a machine-readable form. In order to manage this variety, we need a sufficiently diverse toolkit to bring all of this information together.

Using the example of data on researchers and research outputs, this chapter will focus on obtaining information directly from web pages (*web scraping*) as well as explore the uses of APIs—web services that allow programmatic retrieval of data. In both this chapter and the next, you will see how the crucial pieces of integration often lie in making connections between disparate data sets and how, in turn, making those connections requires careful quality control. The emphasis throughout this chapter is on the importance of focusing on the purpose for which the data will be used as a guide for data collection. While much of this is specific to data about research and researchers, the ideas are generalizable to wider issues of data and public policy. Although we use Python as the programming language in this chapter, data

> ▶ If you have examples from your own research using the methods we describe in this chapter, please submit a link to the paper (and/or code) here: https://textbook.coleridgeinitiative.org/submitexamples.

Box 2.1: Web data and API applications

In addition to the examples that will be shown in this chapter, here are three papers that show a wide variety of projects using data from web pages or APIs.

Kim et al. (2016b) use social media data about e-cigarettes from Twitter for public health research.

Göbel and Munzert (2018) use the online encyclopedia, *Wikipedia*, to study how politicians enhance and change their appearance over time. They trace changes to biographies derived from the German parliament using data that cover the entire edit histories of biographies on all parliament members for the three last legislative periods. The authors have workshop material and code on GitHub demonstrating how they have performed the web scraping and API used for this project (https://github.com/simonmunzert/political-wikipedia-workshop).

King et al. (2013) investigate how censorship in China allows government criticism but silences collective expression using a system to locate, download, and analyze the content of millions of social media posts originating from nearly 1,400 different social media services across China before the Chinese government is able to find, evaluate, and censor (i.e., remove from the Internet) the subset they deem objectionable.

collection through web scraping and APIs can be accomplished in most modern programming languages as well as using software designed specifically for that purpose.

2.2 Scraping information from the web

With the range of information available on the web, our first task is to learn how to access it. The simplest approach is often to manually go to the web and look for data files or other information. For instance, on the NSF website, it is possible to obtain data downloads of all grant information. Sometimes, data are available through web pages, or our interest is in only a subset of this information. In this case, web scraping is often a viable approach.

▶ https://nsf.gov/award search/download.jsp

Web scraping involves writing code to download and process web pages programmatically. We need to look at the website, identify how to obtain the information we want from it, and then write code to do it. Many websites deliberately make this difficult to prevent easy access to their underlying data, while some websites explicitly prohibit this type of activity in their terms of use. Another challenge when scraping data from websites is that the structure of the websites changes often, requiring researchers to continually update their code. In fact, this aspect applies when using the code in this chapter, too. Although the code accurately captures the data from a website at the time of this writing, it may not be valid in the future as the structure and content of the website changes.

2.2.1 Obtaining data from websites

Let us suppose we are interested in obtaining information on those investigators who are funded by the Howard Hughes Medical Institute (HHMI). The HHMI has a website that includes a search function for funded researchers, including the ability to filter by field, state, and role. But, there does not appear to be a downloadable data set of this information. However, we can automate the process with code to create a data set that you might compare with other data.

```
https://www.hhmi.org/scientists/browse?
    sort_by=field_scientist_last_name&
    sort_order=ASC&items_per_page=24
```

Obtaining information from this web page programmatically requires us to follow these steps. 1. Construct a URL that will produce the desired results. 2. Retrieve the contents of the page using that URL. 3. Process the HTML response to extract the targeted pieces of information (such as names and specialties of the scientists).

2.2.1.1   Constructing the URL

This process involves first understanding how to construct a URL that perform the desired search. This is most easily done by playing with search functionality and investigating the URL structures that are returned.

With HHMI, if we perform a general search and make changes to the structure of the URL, we can see some of the elements of the URL that we can think of as a query. As we want to see *all* investigators, we do not need to limit the search; with some "fiddling," we produce a URL similar to the following [note that we have broken the one-line URL into three lines for ease of presentation].

```
http://www.hhmi.org/scientists/browse?
   kw=&sort_by=field_scientist_last_name&
   sort_order=ASC&items_per_page=24&page=0
```

We can click on different links on the page to modify part of this URL and see how the search results change. For example, if we click on "Sort by Institution," the URL changes to:

```
https://www.hhmi.org/scientists/browse?
   sort_by=field_scientist_academic_institu&
   sort_order=ASC&items_per_page=24&page=0.
```

If we click on "next" at the bottom, the URL changes to:

```
https://www.hhmi.org/scientists/browse?
   sort_by=field_scientist_academic_institu&
   sort_order=ASC&items_per_page=24&page=1.
```

This allows us to see that the URL is constructed using a few parameters, such as sort_by, sort_order, items_per_page, and page, that can be programmatically modified to yield the desired search results.

2.2.1.2   Obtaining the contents of the page from the URL

The `requests` module, available natively in Jupyter Python notebooks, is a useful set of tools for handling interactions with websites.

It lets us construct the request that we just presented in terms of a base URL and query terms, as follows.

```
>> BASE_URL = "http://www.hhmi.org/scientists/browse"
>> query = {
          "kw" : "",
          "sort_by" : "field_scientist_last_name",
          "sort_order" : "ASC",
          "items_per_page" : 24,
          "page" : None
          }
```

With our request constructed, we can make the call to the web page to obtain a response.

```
>> import requests
>> response = requests.get(BASE_URL, params=query)
```

The first thing to do when building a script that hits a web page is to make sure that your call has been successful. This can be checked by looking at the response code that the web server sent—and, obviously, by checking the actual HTML that has been returned. A 200 code means success and that everything should be okay. Other codes may mean that the URL has been constructed incorrectly or that a server error has occurred.

```
>> response.status_code
200
```

2.2.1.3 Processing the HTML response

With the page successfully returned, now we need to process the text it contains into the data we want. This is not a trivial exercise. Web pages typically are written in a "markup" language (i.e., Hyptertext Markup Language [HTML]). This language tells the web browser how to display the content on that web page, such as making a piece of text bold or italic, creating numbered lists, or showing images. When we use Python to retrieve a webpage, running the code yields the HTML text. Then, we have to process this text to extract the content that we seek. There are a range of tools in Python that can be helpful when processing HTML data. One of the most popular is a module, BeautifulSoup (Richardson, nd), which provides a number of useful functions for this kind of processing. The module documentation provides more details.

```
<div class="view-content">
    <div class="views-row views-row-1 views-row-odd views-row-first">

    <div class="views-field views-field-field-scientist-image-thumbnail">      <span class="field-content"><a href="/scientists/laurence-f-abbott"><img src="http://www.hhmi.org/sites/default/files/Our%20
    <div class="views-field views-field-field-scientist-classification">       <span class="field-content"><a href="/scientists/laurence-f-abbott">Laurence  Abbott, PhD</a> <br /> Janelia Senior Fellow
</span>  </div>
    <div class="views-field views-field-field-scientist-research-abs-nod">     <span class="field-content"><a href="/research/computational-and-mathematical-modeling-neurons-and-neural-networks">Computa
    <div class="views-field views-field-field-scientist-academic-institu">    <span class="field-content">Janelia Research Campus</span>  </div>
    <div class="views-field views-field-field-scientist-institutionstate">    <span class="field-content">Ashburn, VA</span>  </div>  </div>
    <div class="views-row views-row-2 views-row-even">

    <div class="views-field views-field-field-scientist-image-thumbnail">      <span class="field-content"><a href="/scientists/susan-l-ackerman"><img src="http://www.hhmi.org/sites/default/files/Our%20S
    <div class="views-field views-field-field-scientist-classification">       <span class="field-content"><a href="/scientists/susan-l-ackerman">Susan  Ackerman, PhD</a> <br /> Investigator
</span>  </div>
    <div class="views-field views-field-field-scientist-research-abs-nod">     <span class="field-content"><a href="/research/identification-molecular-mechanisms-underlying-neurodegeneration">Identifica
    <div class="views-field views-field-field-scientist-academic-institu">    <span class="field-content">The Jackson Laboratory</span>  </div>
    <div class="views-field views-field-field-scientist-institutionstate">    <span class="field-content">Bar Harbor, ME</span>  </div>  </div>
    <div class="views-row views-row-3 views-row-odd">
```

Figure 2.1. Source HTML from the portion of an HHMI results page containing information on HHMI investigators; note that the web scraping has produced poorly formatted HTML which is difficult to read

We need to check the details of the page source to find where the information we are seeking is kept (see, for example, Figure 2.1). Here, all of the details on HHMI investigators can be found in a `'div'` element with the class attribute `view-content`. This structure is not something that can be determined in advance. It requires knowledge of the structure of the page itself. Nested inside this `'div'` element is another series of divs, each of which corresponds to one investigator. These have the class attribute `view-rows`. Again, there is nothing obvious about finding these; the process requires a close examination of the page HTML itself for any specific case you are investigating.

First, we process the page using the BeautifulSoup module (into the variable `soup`), and then we find the `div` element that holds the information on investigators (`investigator_list`). As this element is unique on the page (I checked using my web browser), we can use the `find` method. Then, we process that `div` (using `find_all`) to create an iterator object that contains each of the page segments detailing a single investigator (`investigators`).

```
>> from bs4 import BeautifulSoup
>> soup = BeautifulSoup(response.text, "html5lib")
>> investigator_list = soup.find('div', class_ = "view-content")
>> investigators = investigator_list.find_all("div", class_ = "
   views-row")
```

As our query parameters specify that we want 24 results per page, we should check whether our list of page sections has the correct length.

```
>> len(investigators)
24
```

```python
# Given a request response object, parse for HHMI investigators
def scrape(page_response):
    # Obtain response HTML and the correct <div> from the page
    soup = BeautifulSoup(response.text, "html5lib")
    inv_list = soup.find('div', class_ = "view-content")

    # Create a list of all the investigators on the page
    investigators = inv_list.find_all("div", class_ = "views-row")

    data = [] # Make the data object to store scraping results

    # Scrape needed elements from investigator list
    for investigator in investigators:
        inv = {} # Create a dictionary to store results

        # Name and role are in same HTML element; this code
        # separates them into two data elements
        name_role_tag = investigator.find("div",
            class_ = "views-field-field-scientist-classification")
        strings = name_role_tag.stripped_strings
        for string,a in zip(strings, ["name", "role"]):
            inv[a] = string

        # Extract other elements from text of specific divs or from
        # class attributes of tags in the page (e.g., URLs)
        research_tag = investigator.find("div",
            class_ = "views-field-field-scientist-research-abs-nod")
        inv["research"] = research_tag.text.lstrip()
        inv["research_url"] = "http://hhmi.org"
            + research_tag.find("a").get("href")
        institution_tag = investigator.find("div",
            class_ = "views-field-field-scientist-academic-institu")
        inv["institute"] = institution_tag.text.lstrip()
        town_state_tag = investigator.find("div",
            class_ = "views-field-field-scientist-institutionstate"
    )
        inv["town"], inv["state"] = town_state_tag.text.split(",")
        inv["town"] = inv.get("town").lstrip()
        inv["state"] = inv.get("state").lstrip()

        thumbnail_tag = investigator.find("div",
            class_ = "views-field-field-scientist-image-thumbnail")
        inv["thumbnail_url"] = thumbnail_tag.find("img")["src"]
        inv["url"] = "http://hhmi.org"
            + thumbnail_tag.find("a").get("href")

        # Add the new data to the list
        data.append(inv)
    return data
```

Listing 2.1. Python code to parse for HHMI investigators

Finally, we need to process each of these segments to obtain the data we seek. This is the actual "scraping" of the page to retrieve the information we want. Again, this involves looking closely at the HTML itself, identifying where the information is held as well as what tags can be used to find it, and often performing some post-processing to "clean it up" (removing spaces, splitting different elements, etc.).

Listing 2.1 provides a function to manage all of this. The function accepts the response object from the requests module as its input, processes the page text to soup, and then finds the `investigator_list` as above and processes it into an actual list of the investigators. For each investigator, it then processes the HTML to find and clean up the information required, converting it to a dictionary and adding it to our growing list of data.

Let us check what the first two elements of our data set now look like. You can see two dictionaries, one relating to Laurence Abbott, who is a senior fellow at the HHMI Janelia Farm Campus, and one for Susan Ackerman, an HHMI investigator based at the Jackson Laboratory in Bar Harbor, Maine. Note that we also have obtained URLs that provide more details on the researcher and their research program (`research_url` and `url` keys in the dictionary) that could provide an input useful to textual analysis or topic modeling (see Chapter 8).

```
>> data = scrape(response)
>> data[0:2]
[{'institute': u'Janelia Research Campus ',
  'name': u'Laurence Abbott, PhD',
  'research': u'Computational and Mathematical Modeling of Neurons
      and Neural... ',
  'research_url': u'http://hhmi.org/research/computational-and-
    mathematical-modeling-neurons-and-neural-networks',
  'role': u'Janelia Senior Fellow',
  'state': u'VA ',
  'thumbnail_url': u'http://www.hhmi.org/sites/default/files/Our
    %20Scientists/Janelia/Abbott-112x112.jpg',
  'town': u'Ashburn',
  'url': u'http://hhmi.org/scientists/laurence-f-abbott'},
 {'institute': u'The Jackson Laboratory ',
  'name': u'Susan Ackerman, PhD',
  'research': u'Identification of the Molecular Mechanisms
    Underlying... ',
  'research_url': u'http://hhmi.org/research/identification-
    molecular-mechanisms-underlying-neurodegeneration',
  'role': u'Investigator',
  'state': u'ME ',
  'thumbnail_url':
```

```
u'http://www.hhmi.org/sites/default/files/Our%20Scientists/
    Investigators/Ackerman-112x112.jpg',
  'town': u'Bar Harbor',
  'url': u'http://hhmi.org/scientists/susan-l-ackerman'}]
```

2.2.2 Programmatically iterating over the search results

Now, we know we can process a page from a website to generate useful structured data. However, this is only the first page of results.
We need to do this for each page of results if we want to capture
all of the HHMI investigators. We could simply look at the number
of pages that our search returned manually, but to make this more
general we can actually scrape the page to find that piece of information and use that to calculate how many pages through which
we need to iterate.

The number of results is found in a `div` with the class "view-
headers" as a piece of free text ("Showing 1–24 of 517 results"). We
need to grab the text, split it up (we do so based on spaces), find the
right number (the one that is before the word "results"), and convert
that to an integer. Then, we can divide by the number of items we
requested per page (24 in our case) to find how many pages we need
to work process. A quick mental calculation confirms that, if page
0 has results 1–24, page 22 would have results 505–517.

```
>> # Check total number of investigators returned
>> view_header = soup.find("div", class_ = "view-header")
>> words = view_header.text.split(" ")
>> count_index = words.index("results.") - 1
>> count = int(words[count_index])

>> # Calculate number of pages, given count & items_per_page
>> num_pages = count/query.get("items_per_page")
>> num_pages
22
```

Then, it is a matter of putting the function we constructed earlier into a loop to work through the correct number of pages. As we
start to hit the website repeatedly, we need to consider whether we
are being polite. Most websites have a file in the root directory, called
robots.txt, that contains guidance on using programs to interact with
the website. In the case of http://hhmi.org, the file states first that we
are allowed (or, more properly, not forbidden) to query http://www.
hhmi.org/scientists/ programmatically. Thus, you may pull down
all of the more detailed biographical or research information, if you
so desire. The file also states that there is a requested "Crawl-delay"

of 10. This means that, if you are making repeated queries (as we will be in retrieving the 22 pages), you should wait for 10 seconds between each query. This request is easily accommodated by adding a timed delay between each page request.

```
>> for page_num in range(num_pages):
>> # We already have page zero and we need to go to 22:
>> # range(22) is [0,1,...,21]
>>     query["items_per_page"] = page_num + 1
>>     page = requests.get(BASE_URL, params=query)
>> # We use extend to add list for each page to existing list
>>     data.extend(scrape(page))
>> print ("Retrieved and scraped page number:", query.get("
     items_per_page"))
>> time.sleep(10) # robots.txt at hhmi.org specifies a crawl delay
     of 10 seconds
Retrieved and scraped page number: 1
Retrieved and scraped page number: 2
...
Retrieved and scraped page number: 22
```

Finally, we can check that we have the right number of results after our scraping. This should correspond to the 517 records that the website reports.

```
>> len(data)
493
```

2.2.3 Limits of scraping

While scraping websites is often necessary, it can be a fragile and messy way of working. It is problematic for a number of reasons: for example, many websites are designed in ways that make scraping difficult or impossible, and other sites explicitly prohibit this kind of scripted analysis. (Both reasons apply in the case of the NSF and Grants.gov websites, which is why we use the HHMI website in our example.) The structure of websites also changes frequently, forcing you to continuously modify your code to keep up with the structure.

In many cases, a better choice is to process a data download from an organization. For example, the NSF and Wellcome Trust both provide data sets for each year that include structured data on all of their awarded grants. In practice, integrating data is a continual challenge of determining what is the easiest way to proceed, what is allowed, and what is practical and useful. Often, the selection of data will be driven by pragmatic rather than theoretical concerns.

Increasingly, organizations are providing APIs to enable scripted and programmatic access to the data they hold. These tools are much easier and generally more effective, and they will be the focus of much of the rest of this chapter.

2.3 Application programming interfaces

An API is simply a tool that allows a program to interface with a service. The APIs can take many different forms and be of varying quality and usefulness. In this section, we will focus on one common type of API and examples of important publicly available APIs relevant to research communications. In addition, we will discuss combining APIs and the benefits and challenges of bringing multiple data sources together.

2.3.1 Relevant APIs and resources

There is a wide range of other sources of information that can be used in combination with the APIs featured above to develop an overview of research outputs and of where and how they are being used. There are also other tools that can allow deeper analysis of the outputs themselves. Table 2.1 shows a partial list of key data sources and APIs that are relevant to the analysis of research outputs.

2.3.2 RESTful APIs, returned data, and Python wrappers

The APIs we will focus on here are all examples of RESTful services. Although REST stands for Representational State Transfer (Fielding and Taylor, 2002; Wikipedia, nd), for our purposes it is most easily understood as a means of transferring data using web protocols. Other forms of APIs require additional tools or systems for working with them, but RESTful APIs work directly over the web. This has the advantage that a human user can, with relative ease, play with the API to understand how it works. Indeed, some websites work simply by formatting the results of API calls.

As an example, let us look at the Crossref API. This provides a range of information associated with Digital Object Identifiers (DOIs) registered with Crossref. The DOIs uniquely identify an object, and Crossref DOIs refer to research objects, primarily (but not entirely) research articles. If you use a web browser to navigate to http://api.crossref.org/works/10.1093/nar/gni170, you should receive a webpage that looks something like the following. (We have laid it out nicely to make it more readable.)

Table 2.1. Popular sources of data relevant to the analysis of research outputs

Source	Description	API	Free
Bibliographic data			
PubMed	Online index combining bibliographic data from Medline and PubMed Central; PubMed Central and Europe PubMed Central also provide information.	Y	Y
Web of Science	Bibliographic database provided by Thomson Reuters; ISI Citation Index is also available.	Y	N
Scopus	Bibliographic database provided by Elsevier; it also provides citation information.	Y	N
Crossref	Provides a range of bibliographic metadata and information obtained from members registering DOIs.	Y	Y
Google Scholar	Provides a search index for scholarly objects and aggregates citation information.	N	Y
Microsoft Academic Search	Provides a search index for scholarly objects and aggregates citation information; not as complete as Google Scholar but has an API.	Y	Y
Social media			
Altmetric.com	Provides aggregated data on social media and mainstream media attention of research outputs; most comprehensive source of information across different social media and mainstream media conversations.	Y	N
Twitter	Provides an API that allows a user to search for recent tweets and obtain some information on specific accounts.	Y	Y
Facebook	The Facebook API gives information on the number of pages, likes, and posts associated with specific web pages.	Y	Y
Author profiles			
ORCID	Unique identifiers for research authors; profiles include information on publication lists, grants, and affiliations.	Y	Y
LinkedIn	CV-based profiles, projects, and publications.	Y	*
Funder information			
Gateway to Research	Database of funding decisions and related outputs from Research Councils UK.	Y	Y
NIH Reporter	Online search for information on National Institutes of Health grants; does not provide an API but a downloadable data set is available.	N	Y
NSF Award Search	Online search for information on NSF grants; does not provide an API but downloadable data sets by year are available.	N	Y

*The data are restricted: sometimes fee-based, other times not.

```
{ "status" : "ok",
  "message-type" : "work",
  "message-version" : "1.0.0",
  "message" :
  { "subtitle": [],
    "subject" : ["Genetics"],
    "issued" : { "date-parts" : [[2005,10,24]] },
```

```
    "score" : 1.0,
    "prefix" : "http://id.crossref.org/prefix/10.1093",
    "author" : [ "affiliation" : [],
                 "family" : "Whiteford",
                 "given" : "N."}],
    "container-title" : ["Nucleic Acids Research"],
    "reference-count" : 0,
    "page" : "e171-e171",
    "deposited" : {"date-parts" : [[2013,8,8]],
                   "timestamp" : 1375920000000},
    "issue" : "19",
    "title" :
      ["An analysis of the feasibility of short read sequencing"]
    ,
    "type" : "journal-article",
    "DOI" : "10.1093/nar/gni170",
    "ISSN" : ["0305-1048","1362-4962"],
    "URL" : "http://dx.doi.org/10.1093/nar/gni170",
    "source" : "Crossref",
    "publisher" : "Oxford University Press (OUP)",
    "indexed" : {"date-parts" : [[2015,6,8]],
                 "timestamp" : 1433777291246},
    "volume" : "33",
    "member" : "http://id.crossref.org/member/286"
  }
}
```

This is a package of JavaScript Object Notation (JSON)[*] data returned in response to a query. The query is contained entirely in the URL, which can be broken up into pieces: the root URL (http://api.crossref.org) and a data "query," in this case composed of a "field" (`works`) and an identifier (the DOI `10.1093/nar/gni170`). The Crossref API provides information about the article identified with this specific DOI.

> ★ JSON is an open standard way of storing and exchanging data.

2.4 Using an API

Similar to what we have demonstrated with web scraping, using an API involves (1) constructing HTTP requests and (2) processing the data that are returned. Here, we use the Crossref API to illustrate how this is done. Crossref is the provider of DOIs used by many publishers to uniquely identify scholarly works. Crossref is not the only organization to provide DOIs, however; the scholarly communication space DataCite is another important provider. The documentation is available at the Crossref website.

> ▶ http://api.crossref.org

Once again, the `requests` Python library provides a series of convenience functions that make it easier to make HTTP calls and to process returned JSON. Our first step is to import the module and set a base URL variable.

```
>> import requests
>> BASE_URL = "http://api.crossref.org/."
```

A simple example is to obtain metadata for an article associated with a specific DOI. This is a straightforward call to the Crossref API, similar to what we displayed previously.

```
>> doi = "10.1093/nar/gni170"
>> query = "works/"
>> url = BASE_URL + query + doi
>> response = requests.get(url)
>> url
http://api.crossref.org/works/10.1093/nar/gni170
>> response.status_code
200
```

The `response` object that the `requests` library has created has a range of useful information, including the URL called and the response code from the web server (in this case 200, which means everything is okay). We need the JSON body from the response object (which is currently text from the perspective of our script) converted to a Python dictionary. The `requests` module provides a convenient function for performing this conversion, as the following code shows. (All strings in the output are in Unicode, hence the u′ notation.)

```
>> response_dict = response.json()
>> response_dict
{ u'message' :
  { u'DOI' : u'10.1093/nar/gni170',
    u'ISSN' : [ u'0305-1048', u'1362-4962' ],
    u'URL' : u'http://dx.doi.org/10.1093/nar/gni170',
    u'author' : [ {u'affiliation' : [],
                   u'family' : u'Whiteford',
                   u'given' : u'N.'} ],
    u'container-title' : [ u'Nucleic Acids Research' ],
    u'deposited' : { u'date-parts' : [[2013, 8, 8]],
                     u'timestamp' : 1375920000000 },
    u'indexed' : { u'date-parts' : [[2015, 6, 8]],
                   u'timestamp' : 1433777291246 },
    u'issue' : u'19',
    u'issued' : { u'date-parts' : [[2005, 10, 24]] },
    u'member' : u'http://id.crossref.org/member/286',
    u'page' : u'e171-e171',
    u'prefix' : u'http://id.crossref.org/prefix/10.1093',
    u'publisher' : u'Oxford University Press (OUP)',
    u'reference-count' : 0,
    u'score' : 1.0,
```

```
      u'source' : u'Crossref',
      u'subject' : [u'Genetics'],
      u'subtitle' : [],
      u'title' : [u'An analysis of the feasibility of short read
      sequencing'],
      u'type' : u'journal-article',
      u'volume' : u'33'
    },
  u'message-type' : u'work',
  u'message-version' : u'1.0.0',
  u'status' : u'ok'
}
```

Now, this data object can be processed in whatever way the user wishes, using standard manipulation techniques.

The Crossref API can, of course, do much more than simply look up article metadata. It is also valuable as a search resource and for cross-referencing information by journal, funder, publisher, and other criteria. More details can be found at the Crossref website.

2.5 Another example: Using the ORCID API via a wrapper

The ORCID, which stands for Open Research and Contributor Identifier (see orcid.org; see also Haak et al., 2012), is a service that provides unique identifiers for researchers. Researchers can claim an ORCID profile and populate it with references to their research works, funding, and affiliations. The ORCID provides an API for interacting with this information. For many APIs, there is a convenient Python wrapper that can be used. The ORCID–Python wrapper works with the ORCID v1.2 API to make various API calls straightforward. This wrapper only works with the public ORCID API and can therefore only access publicly available data.

Using the API and wrapper together provides a convenient means of obtaining this information. For instance, given an ORCID, it is straightforward to retrieve profile information. Here, we show a list of publications associated with my ORCID and look at the the first item on the list.

```
>> import orcid
>> cn = orcid.get("0000-0002-0068-716X")
>> cn
<Author Cameron Neylon, ORCID 0000-0002-0068-716X>
>> cn.publications[0]
<Publication "Principles for Open Scholarly Infrastructures-v1">
```

The wrapper has created Python objects that make it easier to work with and manipulate the data. It is common to take the return from an API and create objects that behave as would be expected in Python. For instance, the `publications` object is a list populated with publications (which are also Python-like objects). Each publication in the list has its own attributes, which then can be examined individually. In this case, the external IDs attribute is a list of further objects that include a DOI for the article and the ISSN of the journal in which the article has been published.

```
>> len(cn.publications)
70
>> cn.publications[12].external_ids
[<ExternalID DOI:10.1371/journal.pbio.1001677>, <ExternalID ISSN
    :1545-7885>]
```

As a simple example of data processing, we can iterate over the list of publications to identify those for which a DOI has been provided. In this case, we can see that, of the 70 publications listed in this ORCID profile (at the time of testing), 66 have DOIs.

```
>> exids = []
>> for pub in cn.publications:
       if pub.external_ids:
       exids = exids + pub.external_ids
>> DOIs = [exid.id for exid in exids if exid.type == "DOI"]
>> len(DOIs)
66
```

Wrappers generally make operating with an API simpler and cleaner by abstracting away the details of making HTTP requests. Achieving the same by directly interacting with the ORCID API would require constructing the appropriate URLs and parsing the returned data into a usable form. Where a wrapper is available, it is generally much easier to use. However, wrappers may not be actively developed and may lag the development of the API. Where possible, use a wrapper that is directly supported or recommended by the API provider.

2.6 Integrating data from multiple sources

We often must work across multiple data sources to gather the information needed to answer a research question. A common pattern is to search in one location to create a list of identifiers and then use those identifiers to query another API. In the ORCID example above, we create a list of DOIs from a single ORCID profile. We could use those DOIs to obtain further information from the Crossref API and other sources. This models a common path for analysis of research outputs: identifying a corpus and then seeking information on its performance.

One task we often want to do is to analyze relationships between people. As an exercise, we suggest writing code that is able to generate data about relationships between researchers working in similar areas. This could involve using data sources related to researchers, publications, citations, and tweets about those publications, and researchers who are citing or tweeting about them. One way of generating this data for further analysis is to use APIs that give you different pieces of this information and connect them programmatically. We could take the following steps to accomplish this.

Given a twitter handle, obtain the ORCID for that twitter handle. From the ORCID, obtain a list of DOIs. For each DOI, retrieve citations, citing articles, tweets (and twitter handles) associated.

The result is a list of related twitter handles that can be analyzed to look for communities and networks.

The goal of this example is to use ORCID and Crossref to collect a set of identifiers and use a range of APIs to gather metadata and information on articles' performance. Our example is using the PLOS Lagotto API. Lagotto is the software that has been built to support the Article Level Metrics program at PLOS, the open access publisher, and its API provides information on various metrics of PLOS articles. A range of other publishers and service providers, including Crossref, also provide an instance of this API, meaning the same tools can be used to collect information on articles from a range of sources.

2.7 Summary

This chapter focuses on approaches to augment our data with external data sources on the web. We provide steps and code to gather data web pages directly or through Application Programming Interfaces (APIs). While scraping websites is often necessary, it can be fragile because (1) many websites are designed in ways that make scraping difficult or impossible (or explicitly prohibit it), and (2) the structure of websites also changes frequently, forcing you to continuously modify your code to match their structure. Increasingly, organizations are providing APIs to enable scripted and programmatic access to the data they hold. There are many good introductions to web scraping using BeautifulSoup and other libraries as well as API usage in general. In addition, the *APIs* notebook in Chapter 13 will provide a practical introduction to some of these techniques. Given the pace at which APIs and Python libraries change, however, the best and most up to date source of information is likely to be a web search.

▶ See https://workbooks. coleridgeinitiative.org.

As we collect data through scraping and APIs, we then have to understand how to effectively integrate it with our primary data because we may not have access to unique and reliable identifiers. The next chapter, Chapter 3, will address issues of data cleaning, disambiguation, and linking different types of data sources to perform further analysis and research.

Chapter 3

Record Linkage

Joshua Tokle and Stefan Bender

As we mention in the last chapter, it is often necessary to combine data from multiple sources to obtain a complete picture of the activities of interest. In addition to just linking data to obtain additional information, we also are concerned about issues of missing links, duplicative links, and erroneous links. This chapter provides an overview of traditional rule-based and probabilistic approaches, as well as the modern approaches to record linkage using machine learning.

3.1 Motivation

New sources of data offer social scientists great opportunities to bring together many different types of data, from many different sources. Merging different data sets provides new ways of creating population frames that are generated from the digital traces of human activity rather than, for example, tax records. These opportunities, however, create different kinds of challenges from those posed by survey data. Combining information from different sources about an individual, business, or geographic entity means that the social scientist must determine whether or not two entities in two different data sources are the same. This determination is not always easy.

We regularly encounter situations where we need to combine data from different agencies about the same people to understand future employment or health outcomes for people on social service benefits or those who have recently been released from prison. In the UMETRICS data, for example, if data are to be used to measure the impact of research grants, is David A. Miller from Stanford, CA the same as David Andrew Miller from Fairhaven, NJ in a list of inventors? Is IBM the same as Big Blue if the productivity and growth of R&D-intensive firms is to be studied? Or, more generally, is individual A the same person as the one who appears on a list that has been compiled? Does the product that a customer is searching for match the products that business B has for sale?

The consequences of poor record linkage decisions can be substantial. In the business arena, Christen reports that as much as 12% of business revenues are lost due to bad linkages (Christen, 2012a). In the security arena, failure to match travelers to a "known terrorist" list may result in those individuals entering the country, while overzealous matching could lead to numbers of innocent citizens being detained. In finance, incorrectly detecting a legitimate purchase as a fraudulent one annoys the customer, but failing to identify a thief will lead to credit card losses. Less dramatically, in the scientific arena when studying patenting behavior, if it is decided that two inventors are the same person, when in fact they are not, then records will be incorrectly grouped together and one researcher's productivity will be overstated. Conversely, if the records for one inventor are believed to correspond to multiple individuals, then that inventor's productivity will be understated.

This chapter discusses current approaches to joining multiple data sets together—commonly called *record linkage*.[*]

We draw heavily here on work by Winkler, Scheuren, and Christen (in particular, Christen, 2012a,b; Herzog et al., 2007). To ground ideas, we use examples from a recent paper examining the effects of different algorithms on studies of patent productivity (Ventura et al., 2015).

★ Other names associated with record linkage are entity disambiguation, entity resolution, co-reference resolution, matching, and data fusion, meaning that records which are linked or co-referent can be thought of as corresponding to the same underlying entity. The number of names is reflective of a vast literature in social science, statistics, computer science, and information sciences.

3.2   Introduction to record linkage

There are many reasons to link data sets. Linking to existing data sources to solve a measurement need instead of implementing a new survey results in cost savings (and almost certainly time savings as well) and reduces the burden on potential survey respondents. For some research questions (e.g., a survey of the reasons for death of a longitudinal cohort of individuals), a new survey may not be possible. In the case of administrative data or other automatically generated data, the sample size is much greater than would be possible from a survey.

Record linkage can be used to compensate for data quality issues. If a large number of observations for a particular field are missing, it may be possible to link to another data source to fill in the missing values. For example, survey respondents might not want to share a

---

Box 3.1: Record linkage examples

In addition to our examples in this chapter, here are a few papers that show the wide variety of projects that use combining records from different sources.

Glennon (2019) uses a unique matched firm-level dataset of H-1B visas and multinational firm activity to show that restrictions on H-1B immigration has caused increases in foreign affiliate employment. Restrictions also have caused increases in foreign patenting, suggesting that there is a change in the location of innovative activity as well.

▶ If you have examples from your own research using the methods we describe in this chapter, please submit a link to the paper (and/or code) here: https://textbook.coleridgeini tiative.org/submitexamples.

Rodolfa et al. (2020) use machine learning-based record linkage to link data about the same individuals together from a criminal justice case management system to help the Los Angeles City Attorney's office develop individually-tailored social service interventions in a fair and equitable manner. Because the system lacks a global unique person-level identifier, case-level defendant data are used to link cases belonging to the same person using first and last name, date of birth, address, driver's license number (where available), and California Information and Identification (CII) number (where available).

The National Center for Health Statistics (NCHS) (2019) links the data from the National Health Interview Survey (NHIS) to records from the Social Security Administration, the Centers for Medicare & Medicaid Services, and the National Death Index to investigate the relationship between health and sociodemographic information reported in the surveys and medical care costs, future use of medical services, and mortality.

---

sensitive datum such as income. If the researcher has access to an official administrative list with income data, then those values can be used to supplement the survey (Abowd et al., 2006).

Often, record linkage is used to create new longitudinal data sets by linking the same entities over time (Jarmin and Miranda, 2002). More generally, linking separate data sources makes it possible to create a combined data set that is richer in coverage and measurement than any of the individual data sources (Abowd et al., 2004).

Linking is straightforward if each entity has a corresponding unique identifier that appears in the data sets to be linked. For example, two lists of US employees may both contain Social Security numbers. When a unique identifier exists in the data or can be created, no special techniques are necessary to join the data sets.

Example: The Administrative Data Research Network

★ "Administrative data" typically refers to data generated by the administration
of a government program,
as distinct from deliberate
survey collection.

The UK's Administrative Data Research Network* (ADRN) is a major investment by
the United Kingdom to "improve our knowledge and understanding of the society
we live in . . . [and] provide a sound base for policymakers to decide how to tackle a
range of complex social, economic and environmental issues" by linking administrative data from a variety of sources, such as health agencies, court records, and
tax records in a confidential environment for approved researchers. The linkages
are done by trusted third-party providers (Economic and Social Research Council,
2016).

If there is no unique identifier available, then the task of identifying unique entities is challenging. One instead relies on fields that
only partially identify the entity, such as names, addresses, or dates
of birth. The problem is further complicated by poor data quality
and duplicate records, issues well attested in the record linkage literature (Christen, 2012b) and certain to become more important in
the context of big data. Data quality issues include input errors
(typographical errors [typos], misspellings, truncation, extraneous
letters, abbreviations, and missing values) as well as differences in
the way variables are coded between the two data sets (age versus
date of birth, for example). In addition to record linkage algorithms,
we will discuss different data preprocessing steps that are necessary
first steps for the best results in record linkage.

To find all possible links between two data sets, it would be necessary to compare each record of the first data set with each record
of the second data set. The computational complexity of this grows
quadratically with the size of the data—an important consideration,
especially with large amounts of data. To compensate for this complexity, the standard second step in record linkage, after preprocessing, is indexing or blocking, which uses some set of heuristics
to create subsets of similar records and reduces the total number
of comparisons.

The outcome of the matching step is a set of predicted links—
record pairs that are likely to correspond to the same entity. After
these are produced, the final stage of the record linkage process is
to evaluate the result and estimate the resulting error rates. Unlike
other areas of application for predictive algorithms, ground truth
or gold standard data sets are rarely available. The only way to
create a reliable truth data set sometimes is through an expensive
human review process that may not be viable for a given application.
Instead, error rates must be estimated.

An input data set may contribute to the linked data in a variety of ways, such as increasing coverage, expanding understanding of the measurement or mismeasurement of underlying latent variables, or adding new variables to the combined data set. Therefore, it is important to develop a well-specified reason for linking the data sets, and to specify a loss function to proxy the cost of false negative matches versus false positive matches that can be used to guide match decisions. It is also important to understand the coverage of the different data sets being linked because differences in coverage may result in bias in the linked data. For example, consider the problem of linking Twitter data to a sample-based survey—elderly adults and very young children are unlikely to use Twitter and so the set of records in the linked data set will have a youth bias, even if the original sample was representative of the population. It is also essential to engage in critical thinking about what latent variables are being captured by the measures in the different data sets—an "occupational classification" in a survey data set may be very different from a "job title" in an administrative record or a "current position" in LinkedIn data.

▶ This topic will be discussed in more detail in Chapter 10.

Example: Employment and earnings outcomes of doctoral recipients

A recent paper in *Science* matched UMETRICS data on doctoral recipients to census data on earnings and employment outcomes. The authors note that some 20% of the doctoral recipients are not matched for several reasons: (1) the recipient does not have a job in the US, either for family reasons or because he/she has returned to his/her home country; (2) he/she starts up a business rather than choosing employment; or (3) it is not possible to uniquely match him/her to a Census Bureau record. They correctly note that there may be biases introduced in case (3), because Asian names are more likely duplicated and more difficult to uniquely match (Zolas et al., 2015). Improving the linkage algorithm would increase the estimate of the effects of investments in research and the result would be more accurate.

Comparing the kinds of heterogeneous records associated with big data is a new challenge for social scientists, who have traditionally used a technique first developed in the 1960s to apply computers to the problem of medical record linkage. There is a reason why this approach has survived: it has been highly successful in linking survey data to administrative data, and efficient implementations of this algorithm can be applied at the big data scale. However, the approach is most effective when the two files

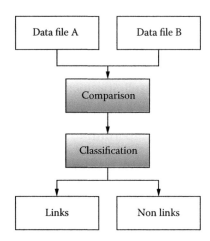

Figure 3.1. The preprocessing pipeline

being linked have a number of fields in common. In the new land-scape of big data, there is a greater need to link files that have few fields in common but whose noncommon fields provide additional predictive power to determine which records should be linked. In some cases, when sufficient training data can be produced, more modern machine learning techniques may be applied.

The canonical record linkage workflow process is shown in Figure 3.1 for two data files, A and B. The goal is to identify all pairs of records in the two data sets that correspond to the same underlying individual. One approach is to compare all data units from file A with all units in file B and classify all of the comparison outcomes to decide whether or not the records match. In a perfect statistical world, the comparison would end with a clear determination of links and nonlinks.

Alas, a perfect world does not exist, and there is likely to be noise in the variables that are common to both data sets and that will be the main identifiers for the record linkage. Although the original files A and B are the starting point, the identifiers must be preprocessed before they can be compared. Determining identifiers for the linkage and deciding on the associated cleaning steps are extremely important, as they result in a necessary reduction of the possible search space.

In the next section, we begin our overview of the record link-age process with a discussion of the main steps in data prepro-cessing. This is followed by a section on approaches to record linkage that includes rule-based, probabilistic, and machine learning algorithms. Then, we cover classification and evaluation of links, and we conclude with a discussion of data privacy in record linkage.

3.3 Preprocessing data for record linkage

As noted in the introductory chapter, all data work involves preprocessing, and data that need to be linked is no exception. Preprocessing refers to a workflow that transforms messy and noisy data into a well-defined, clearly structured, and quality-tested data set. Elsewhere in this book, we discuss general strategies for data preprocessing. In this section, we focus specifically on preprocessing steps relating to the choice of input fields for the record linkage algorithm. Preprocessing for any kind of a new data set is a complex and time-consuming process because it is "hands-on": it requires judgment and cannot be effectively automated. It may be tempting to minimize this demanding work under the assumption that the record linkage algorithm will account for issues in the data, but it is difficult to overstate the value of preprocessing for record linkage quality. As Winkler notes: "In situations of reasonably high-quality data, preprocessing can yield a greater improvement in matching efficiency than string comparators and 'optimized' parameters. In some situations, 90% of the improvement in matching efficiency may be due to preprocessing" (Winkler, 2009).

> ▶ This topic (quality of data, preprocessing issues) is discussed in more detail in Chapter 1.

The first step in record linkage is to develop link keys, which are the record fields that will be used to estimate if there is a link between two records. These can include common identifiers such as first and last name. Survey and administrative data sets may include a number of clearly identifying variables such as address, birth date, and sex. Other data sets, such as transaction records or social media data, often will not include address or birth date but may still include other identifying fields, such as occupation, a list of interests, or connections on a social network. Consider this chapter's illustrative example of the US Patent and Trademark Office (USPTO) data (Ventura et al., 2015):

> USPTO maintains an online database of all patents issued in the United States. In addition to identifying information about the patent, the database contains each patent's list of inventors and assignees, the companies, organizations, individuals, or government agencies to which the

patent is assigned.... However, inventors and assignees in the USPTO database are not given unique identification numbers, making it difficult to track inventors and assignees across their patents or link their information to other data sources.

There are some basic precepts that are useful when considering identifying fields. The more different values a field can take, the less likely it is that two randomly chosen individuals in the population will agree on those values. Therefore, fields that exhibit a wider range of values are more powerful as link keys: names are much better link keys than sex or year of birth.

Example: Link keys in practice

"A Harvard professor has re-identified the names of more than 40% of a sample of anonymous participants in a high-profile DNA study, highlighting the dangers that ever greater amounts of personal data available in the Internet era could unravel personal secrets. ... Of the 1,130 volunteers Sweeney and her team reviewed, about 579 provided zip code, date of birth and gender, the three key pieces of information she needs to identify anonymous people combined with information from voter rolls or other public records. Of these, Sweeney succeeded in naming 241, or 42% of the total. The Personal Genome Project confirmed that 97% of the names matched those in its database if nicknames and first name variations were included" (Tanner, 2013).

Complex link keys such as addresses can be broken down into components so that the components can be compared independently of one another. In this way, errors due to data quality can be further isolated. For example, assigning a single comparison value to the complex fields "1600 Pennsylvania" and "160 Pennsylvania Ave" is less informative than assigning separate comparison values to the street number and street name portions of those fields. A record linkage algorithm that uses the decomposed field can make more nuanced distinctions by assigning different weights to errors in each component.

Sometimes a data set can include different variants of a field, such as legal first name and nickname. In these cases, match rates can be improved by including all variants of the field in the record comparison. For example, if only the first list includes both variants, and the second list has a single "first name" field that could be either a legal first name or a nickname, then match rates can be improved by comparing both variants and then keeping the better of the two comparison outcomes. It is important to remember, however, that some record linkage algorithms expect field comparisons to be somewhat independent. In our example, using the outcome from both comparisons as separate inputs into the probabilistic model we describe below may result in a higher rate of false negatives. If a record has the same value in the legal name and nickname fields, and if that value happens to agree with the first name field in the second file, then the agreement is being double-counted. Similarly, if a person in the first list has a nickname that differs significantly from their legal first name, then a comparison of that record to the corresponding record will unfairly penalize the outcome because at least one of those name comparisons will show a low level of agreement.

Preprocessing serves two purposes in record linkage. First, it can correct for issues in data quality such as those we have described. Second, it can account for the different ways in which the input files have been generated, which may result in the same underlying data being recorded on different scales or according to different conventions.

Once preprocessing is finished, it is possible to start linking the records in the different data sets. In the next section, we describe a technique to improve the efficiency of the matching step.

3.4 Indexing and blocking

There is a practical challenge to consider when comparing the records in two files. If both files are roughly the same size, e.g., 100 records in the first and 100 records in the second file, then there are 10,000 possible comparisons, because the number of pairs is the product of the number of records in each file. More generally, if the number of records in each file is approximately n, then the total number of possible record comparisons is approximately n^2. Assuming that there are no duplicate records in the input files, the proportion of record comparisons that correspond to a link is only $1/n$. If we naively proceed with all n^2 possible comparisons, the linkage algorithm will spend the bulk of its time comparing records that are not matches. Thus, it is possible to speed up record linkage significantly by skipping comparisons between record pairs that are not likely to be linked.

Indexing refers to techniques that determine which of the possible comparisons will be made in a record linkage application. The most used technique for indexing is blocking. In this approach, you construct a "blocking key" for each record by concatenating fields or parts of fields. Two records with identical blocking keys are said to be in the same block, and only records in the same block are compared. This technique is effective because performing an exact comparison of two blocking keys is a relatively quick operation compared to a full record comparison, which may involve multiple applications of a fuzzy string comparator.

Example: Blocking in practice

Given two lists of individuals, one might construct the blocking key by concatenating the first letter of the last name and the postal code and then "blocking" on first character of last name and postal code. This reduces the total number of comparisons by only comparing those individuals in the two files who live in the same locality and whose last names begin with the same letter.

There are important considerations when choosing the blocking key. First, the choice of blocking key creates a potential bias in the linked data because true matches that do not share the same blocking key will not be found. In the example, the blocking strategy could fail to match records for individuals whose last name changed or who moved. Second, because blocking keys are compared exactly, there is an implicit assumption that the included fields will not have typos or other data entry errors. In practice, however, the blocking fields will exhibit typos. If those typos are not uniformly distributed over the population, then there is again the possibility of bias in the linked data set. One simple strategy for managing imperfect blocking keys is to implement multiple rounds of blocking and matching. After the first set of matches is produced, a new blocking strategy is deployed to search for additional matches in the remaining record pairs.

> ▶ This topic will be discussed in more detail in Chapter 10.

Blocking based on exact field agreements is common in practice, but there are other approaches to indexing that attempt to be more error tolerant. For example, one may use clustering algorithms to identify sets of similar records. In this approach, an index key, which is analogous to the blocking key above, is generated for both data sets and then the keys are combined into a single list. A distance function must be chosen and pairwise distances computed for all keys. Then, the clustering algorithm is applied to the combined list, and only record pairs that are assigned to the same cluster are compared. This is a theoretically appealing approach but it has the drawback that the similarity metric has to be computed for all pairs of records. Even so, computing the similarity measure for a pair of blocking keys is likely to be cheaper than computing the full record comparison, so there is still a gain in efficiency. Whang et al. (2009) provide a nice review of indexing approaches.

In addition to reducing the computational burden of record linkage, indexing plays an important secondary role. Once implemented, the fraction of comparisons made that correspond to true links will be significantly higher. For some record linkage approaches that use an algorithm to find optimal parameters—such as the probabilistic approach—having a larger ratio of matches to nonmatches will produce a better result.

3.5   Matching

The purpose of a record linkage algorithm is to examine pairs of records and make a prediction as to whether they correspond to the same underlying entity. (There are some sophisticated algorithms that examine sets of more than two records at a time [Steorts et al., 2014], but pairwise comparison remains the standard approach.) At the core of every record linkage algorithm is a function that compares two records and outputs a "score" that quantifies the similarity between those records. Mathematically, the match score is a function of the output from individual field comparisons: agreement in the first name field, agreement in the last name field, etc. Field comparisons may be binary—indicating agreement or disagreement—or they may output a range of values indicating different levels of agreement. There are a variety of methods in the statistical and computer science literature that can be used to generate a match score, including nearest-neighbor matching, regression-based matching, and propensity score matching. The probabilistic approach to record linkage defines the match score in terms of a likelihood ratio (Fellegi and Sunter, 1969).

Example: Matching in practice

Long strings, such as assignee and inventor names, are susceptible to typographical errors and name variations. For example, none of Sony Corporation, Sony Corporatoin, and Sony Corp. will match using simple exact matching. Similarly, David and Dave would not match (Ventura et al., 2015).

Comparing fields whose values are continuous is straightforward: often one can simply take the absolute difference as the comparison value. Comparing character fields in a rigorous way is more complicated. For this purpose, different mathematical definitions of the distance between two character fields have been defined. Edit distance, for example, is defined as the minimum number of edit operations—chosen from a set of allowed operations—needed to convert one string to another. When the set of allowed edit operations is single-character insertions, deletions, and substitutions, the corresponding edit distance is also known as the Levenshtein distance. When transposition of adjacent characters is allowed in addition to those operations, the corresponding edit distance is called the Levenshtein–Damerau distance.

Edit distance is appealing because of its intuitive definition, but it is not the most efficient string distance to compute. Another standard string distance, known as Jaro–Winkler distance, has been developed with record linkage applications in mind and is faster to compute. This is an important consideration because, in a typical record linkage application, most of the algorithm run time will be spent performing field comparisons. The definition of Jaro–Winkler distance is less intuitive than edit distance, but it works as expected: words with more characters in common will have a higher Jaro–Winkler value than those with fewer characters in common. The output value is normalized to fall between 0 and 1. Because of its history in record linkage applications, there are some standard variants of Jaro–Winkler distance that may be implemented in record linkage software. Some variants boost the weight given to agreement in the first few characters of the strings being compared. Others decrease the score penalty for letter substitutions that arise from common typos.

Once the field comparisons are computed, they must be combined to produce a final prediction of match status. In the following sections, we describe three types of record linkage algorithms: rule-based, probabilistic, and machine learning.

3.5.1   Rule-based approaches

A natural starting place is for a data expert to create a set of ad hoc rules that determine which pairs of records should be linked. In the classical record linkage setting where the two files have a number of identifying fields in common, this is not the optimal approach. However, if there are few fields in common but each file contains auxiliary fields that may inform a linkage decision, then an ad hoc approach may be appropriate.

Example: Linking in practice

Consider the problem of linking two lists of individuals where both lists contain first name, last name, and year of birth. Here is one possible linkage rule: link all pairs of records such that

- the Jaro–Winkler comparison of first names is greater than 0.9,
- the Jaro–Winkler comparison of last names is greater than 0.9,
- the first three digits of the year of birth are the same.

The result will depend on the rate of data errors in the year of birth field and typos in the name fields.

By *auxiliary field*, we mean data fields that do not appear on both data sets, but which may nonetheless provide information about whether records should be linked. Consider a situation in which the first list includes an occupation field and the second list includes educational history. In that case, one might create additional rules to eliminate matches where the education was deemed to be an unlikely fit for the occupation.

This method may be attractive if it produces a reasonable-looking set of links from intuitive rules, but there are several pitfalls. As the number of rules grows, it becomes more difficult to understand the ways that the different rules interact to produce the final set of links. There is no notion of a threshold that can be increased or decreased depending on the tolerance for false positive and false negative errors. The rules themselves are not chosen to satisfy any kind of optimality, unlike the probabilistic and machine learning methods. Instead, they reflect the practitioner's domain knowledge about the data sets.

3.5.2 Probabilistic record linkage

In this section, we describe the probabilistic approach to record linkage, also known as the Fellegi–Sunter algorithm (Fellegi and Sunter, 1969). This approach dominates in traditional record linkage applications and remains an effective and efficient way to solve the record linkage problem today.

In this section, we give a somewhat formal definition of the statistical model underlying the algorithm. By understanding this model, one is better equipped to define link keys and record comparisons in an optimal way.

Let A and B be two lists of individuals whom we wish to link. The product set $A \times B$ contains all possible pairs of records where

Example: Usefulness of probabilistic record linkage

In practice, it is typically the case that a researcher will want to combine two or more data sets containing records for the same individuals or units that possibly come from different sources. Unless the sources all contain the same unique identifiers, linkage will likely require matching on standardized text strings. Even standardized data are likely to contain small differences that preclude exact matching as in the matching example above. The Census Bureau's Longitudinal Business Database (LBD) links establishment records from administrative and survey sources. Exact numeric identifiers do most of the heavy lifting, but mergers, acquisitions, and other actions can break these linkages. Probabilistic record linkage on company names and/or addresses is used to fix these broken linkages that bias statistics on business dynamics (Jarmin and Miranda, 2002).

the first element of the pair comes from A and the second element of the pair comes from B. A fraction of these pairs will be matches, meaning that both records in the pair represent the same underlying individual, but the vast majority of them will be nonmatches. In other words, $A \times B$ is the disjoint union of the set of matches M and the set of nonmatches U, a fact that we denote formally by $A \times B = M \cup U$.

Let γ be a vector-valued function on $A \times B$ such that, for $a \in A$ and $b \in B$, $\gamma(a, b)$ represents the outcome of a set of field comparisons between a and b. For example, if both A and B contain data on individuals' first names, last names, and cities of residence, then γ could be a vector of three binary values representing agreement in first name, last name, and city. In that case, $\gamma(a, b) = (1, 1, 0)$ would mean that the records a and b agree on first name and last name, but disagree on city of residence.

For this model, the comparison outcomes in $\gamma(a, b)$ are not required to be binary, but they do have to be categorical: each component of $\gamma(a, b)$ should take only a finite number of values. This means that a continuous comparison outcome—such as output from the string comparator—has to be converted to an ordinal value representing levels of agreement. For example, one might create a three-level comparison, using one level for exact agreement, one level for approximate agreement defined as a Jaro–Winkler score greater than 0.85, and one level for nonagreement corresponding to a Jaro–Winkler score less than 0.85.

If a variable being used in the comparison has a significant number of missing values, it can help to create a comparison outcome level to indicate missingness. Consider two data sets that both have middle initial fields, and suppose that in one of the data sets the middle initial is included in only about half of the records. When comparing records, the case where both middle initial fields contain a letter but are not the same should be treated differently from the case where one of the middle initial fields is blank, because the first case provides more evidence that the records do not correspond to the same person. We handle this in the model by defining a three-level comparison for the middle initial, with levels to indicate "equal," "not equal," and "missing."

Probabilistic record linkage works by weighing the probability of seeing the result $\gamma(a, b)$ if (a, b) belongs to the set of matches M against the probability of seeing the result if (a, b) belongs to the set of nonmatches U. Conditional on M or U, the distribution of the individual comparisons defined by γ are assumed to be mutually independent. The parameters that define the marginal distributions of $\gamma | M$ are called *m-weights*, and similarly the marginal distributions of $\gamma | U$ are called *u-weights*.

In order to apply the Fellegi–Sunter method, it is necessary to choose values for these parameters, m-weights and u-weights. With labeled data—a pair of lists for which the match status is known—it is straightforward to solve for optimal values. Training data are not usually available, however, and the typical approach is to use expectation maximization to find optimal values.

We have noted that primary motivation for record linkage is to create a linked data set for analysis that will be richer than either of the input data sets alone. A natural application is to perform a linear regression using a combination of variables from both files as predictors. With all record linkage approaches, it is a challenge to understand how errors from the linkage process will manifest in the regression. Probabilistic record linkage has an advantage over rule-based and machine learning approaches in that there are theoretical results concerning coefficient bias and errors (Lahiri and Larsen, 2005; Scheuren and Winkler, 1993). More recently, Chipperfield and Chambers (2015) have developed an approach based on the bootstrap to account for record linkage errors when making inferences for cross-tabulated variables.

3.5.3 Machine learning approaches to record linkage

Computer scientists have contributed extensively in parallel literature focused on linking large data sets (Christen, 2012a). Their focus is on identifying potential links using approaches that are fast, adaptive, and scalable, and approaches are developed based on work in network analysis and machine learning.

While simple blocking, as described in Section 3.4, Indexing and blocking, is standard in Fellegi–Sunter applications, machine learning-based approaches are likely to use the more sophisticated clustering approach to indexing. Indexing also may use network information to include, for example, records for individuals that have a similar place in a social graph. When linking lists of researchers, one might specify that comparisons should be made between records that share the same address, have patents in the same patent class, or have overlapping sets of coinventors. These approaches are known as semantic blocking, and the computational requirements are similar to standard blocking (Christen, 2012a).

In recent years, machine learning approaches have been applied to record linkage. As Wick et al. (2013) note:

> ▶ This topic will be discussed in more detail in Chapter 7.

> Entity resolution, the task of automatically determining which mentions refer to the same real-world entity, is a crucial aspect of knowledge base construction and management. However, performing entity resolution at large scales is challenging because (1) the inference algorithms must cope with unavoidable system scalability issues and (2) the search space grows exponentially in the number of mentions. Current conventional wisdom declares that performing coreference at these scales requires decomposing the problem by first solving the simpler task of entity-linking (matching a set of mentions to a known set of KB entities), and then performing entity discovery as a post-processing step (to identify new entities not present in the KB). However, we argue that this traditional approach is harmful to both entity-linking and overall coreference accuracy. Therefore, we embrace the challenge of jointly modeling entity-linking and entity discovery as a single entity resolution problem.

Figure 3.2 provides a useful comparison between classical record linkage and learning-based approaches. In machine learning, there is a statistical model and an algorithm for "learning" the optimal set of parameters to use. The learning algorithm relies on a training data set. In record linkage, this would be a curated data set with true and false matches labeled as such. See Ventura et al. (2015) for an example and a discussion of how a training data set has been created for the problem of disambiguating inventors in the USPTO

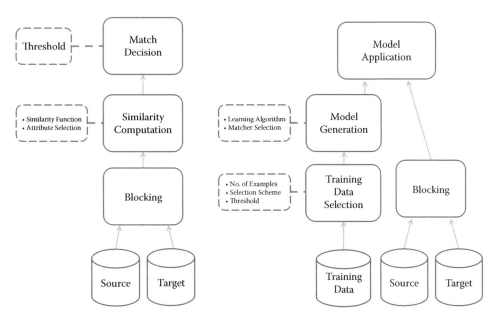

Figure 3.2. Probabilistic (left) vs. machine learning (right) approaches to linking. (From Köpcke et al. (2010). *Proceedings of the VLDB Endowment*, **3**(1–2):484–493.)

database. Once optimal parameters are computed from the training data, the predictive model can be applied to unlabeled data to find new links. The quality of the training data set is critical; the model is only as good as the data on which it is trained.

As shown in Figure 3.2, a major difference between probabilistic and machine learning approaches is the need for labeled training data to implement the latter approach. Usually, training data are created through a painstaking process of clerical review. After an initial round of record linkage, a sample of record pairs that are not clearly matches or nonmatches is given to a research assistant who makes the final determination. In some cases, it is possible to create training data by automated means, such as when there is a subset of the complete data that contains strongly identifying fields. Suppose that both of the candidate lists contain name and date of birth fields and that, in the first list, the date of birth data are complete, but in the second list only about 10% of records contain date of birth. For reasonably sized lists, name and date of birth together will be a nearly unique identifier. It is then possible to perform probabilistic record linkage on the subset of records with date of birth and be confident that the error rates would be small. If the subset of records with date of birth is representative of the complete data set, then the output from the probabilistic record linkage can be used as "truth" data.

Given a quality training data set, machine learning approaches may have advantages over probabilistic record linkage. There are many published studies on the effectiveness of random forests and other machine learning algorithms for record linkage. Christen (2012b) and Elmagarmid et al. (2007) provide some pointers.

3.5.4 Disambiguating networks

The problem of disambiguating entities in a network is closely related to record linkage: in both cases, the goal is to consolidate multiple records corresponding to the same entity. Rather than finding the same entity in two data sets, however, the goal in network disambiguation is to consolidate duplicate records in a network data set. By network, we mean that the data set contains not only typical record fields such as names and addresses but also information about how entities relate to one another: entities may be coauthors, coinventors, or simply friends in a social network.

The record linkage techniques that we have described in this chapter can be applied to disambiguate a network. To do so, one must convert the network to a form that can be used as input into a record linkage algorithm. For example, when disambiguating a social network, one might define a field comparison whose output produces the fraction of friends in common between two records. Ventura et al. (2015) demonstrate the relative effectiveness of the probabilistic method and machine learning approaches to disambiguating a database of inventors in the USPTO database. Another approach is to apply clustering algorithms from the computer science literature to identify groups of records that are likely to refer to the same entity. Huang et al. (2006) have developed a successful method based on an efficient computation of distance between individuals in the network. Then, these distances are fed into the DBSCAN clustering algorithm to identify unique entities.

3.6 Classification

Once the match score for a pair of records has been computed using the probabilistic or random forest method, a decision has to be made whether the pair should be linked. This requires classifying the pair as either a "true" or a "false" match. In most cases, a third classification is required—sending for manual review and classification.

3.6.1 Thresholds

In the probabilistic and random forest approaches, both of which output a "match score" value, a classification is made by establishing a threshold T such that all records with a match score greater than T are declared to be links. Because of the way these algorithms are defined, the match scores are not meaningful by themselves and the threshold used for one linkage application may not be appropriate for another application. Instead, the classification threshold must be established by reviewing the model output.

Typically, one creates an output file that includes pairs of records that have been compared along with the match score. The file is sorted by match score and the reviewer begins to scan the file from the highest match scores to the lowest. For the highest match scores, the record pairs will agree on all fields and there is usually no question about the records being linked. However, as the scores decrease, the reviewer will see more record pairs whose match status is unclear (or that are clearly nonmatches) mixed in with the clear matches. There are a number of ways to proceed, depending on the resources available and the goal of the project.

Rather than set a single threshold, the reviewer may set two thresholds, $T_1 > T_2$. Record pairs with a match score greater than T_1 are marked as matches and removed from further consideration. The set of record pairs with a match score between T_1 and T_2 are believed to contain significant numbers of matches and nonmatches. These are sent to clerical review, meaning that research assistants will make a final determination of match status. The final set of links will include clear matches with a score greater than T_1 as well as the record pairs that pass clerical review. If the resources are available for this approach and the initial threshold T_1 is set sufficiently high, then the resulting data set will contain a minimal number of false positive links. The collection of record pairs with match scores between T_1 and T_2 is sometimes referred to as the clerical review region.

The clerical review region generally contains many more pairs than the set of clear matches, and it can be expensive and time-consuming to review each pair. Therefore, a second approach is to establish tentative threshold T and send only a sample of record pairs with scores in a neighborhood of T to clerical review. This results in data on the relative numbers of true matches and true nonmatches at different score levels, as well as the characteristics of record pairs that appear at a given level. Based on the review and the relative tolerance for false positive errors and false negative errors, a final threshold T' is set such that pairs with a score greater than T' are considered to be matches.

After viewing the results of the clerical review, it may be determined that the parameters to the record linkage algorithm could be improved to create a clearer delineation between matches and non-matches. For example, a research assistant may determine that many potential false positives appear near the tentative threshold because the current set of record linkage parameters is giving too much weight to agreement in first name. In this case, the reviewer may decide to update the record linkage model to produce an improved set of match scores. The update may consist of an ad hoc adjustment of parameters, or the result of the clerical review may be used as training data and the parameter-fitting algorithm may be run again. This type of iterative approach is common when first linking two data sets because the clerical review process can improve one's understanding of the data sets involved.

Setting the threshold value higher will reduce the number of false positives (record pairs for which a link is incorrectly predicted) while increasing the number of false negatives (record pairs that should be linked but for which a link is not predicted). The proper tradeoff between false positive and false negative error rates will depend on the particular application and the associated loss function, but there are some general concerns to keep in mind. Both types of errors create bias, which can impact the generalizability of analyses conducted on the linked data set. Consider a simple regression on the linked data that includes fields from both data sets. If the threshold is too high, then the linked data will be biased toward records with no data entry errors or missing values, and whose fields did not change over time. This set of records may not be representative of the population as a whole. If a low threshold is used, then the set of linked records will contain more pairs that are not true links and the variables measured in those records are independent of each other. Including these records in a regression amounts to adding statistical noise to the data.

3.6.2   One-to-one links

In the probabilistic and machine learning approaches to record linkage that we have described, each record pair is compared and a link is predicted independently of all other record pairs. Because of the independence of comparisons, one record in the first file may be predicted to link to multiple records in the second file. Under the assumption that each input file has been deduplicated, at most one of these predictions can correspond to a true link. For many applications, it is preferable to extract a set of "best" links with the property that each record in one file links to at most one record in the second file. A set of links with this property is said to be one-to-one.

One possible definition of "best" is a set of one-to-one links such that the sum of the match scores of all included links is maximal. This is an example of what is known as the *assignment problem* in combinatorial optimization. In the linear case above, where we care about the sum of match scores, the problem can be solved exactly using the Hungarian algorithm (Kuhn, 2005).

3.7 Record linkage and data protection

In many social science applications data sets, there is no need for data to include identifying fields such as names and addresses. These fields may be omitted intentionally out of concern for privacy, or they may simply be irrelevant to the research question. For record linkage, however, names and addresses are among the best possible identifiers. We describe two approaches to the problem of balancing needs for both effective record linkage and privacy.

▶ See Chapter 12.

The first approach is to establish a trusted third party or safe center. The concept of trusted third parties (TTPs) is well known in cryptography. In the case of record linkage, a third party takes a place between the data owners and the data users, and it is this third party that actually performs the linkage work. Both the data owners and data users trust the third party in the sense that it assumes responsibility for data protection (data owners) and data competence (data users) at the same time. No party other than the TTP learns about the private data of the other parties. After record linkage, only the linked records are revealed, with no identifiers attached. The TTP ensures that the released linked data set cannot be relinked to any of the source data sets. Possible third parties are safe centers, which are operated by lawyers, or official trusted institutions, such as the US Census Bureau.

The second approach is known as privacy-preserving record linkage. The goal of this approach is to find the same individual in separate data files without revealing the identity of the individual (Clifton et al., 2006). In privacy-preserving record linkage, cryptographic procedures are used to encrypt or hash identifiers before they are shared for record linkage. Many of these procedures require exact matching of the identifiers, however, and do not tolerate any errors in the original identifiers. This leads to information loss because it is not possible to account for typos or other small variations in hashed fields. To account for this, Schnell has developed a method to calculate string similarity of encrypted fields using bloom filters (Schnell, 2014; Schnell et al., 2009).

In many countries, these approaches are combined. For example, when the UK established the ADRN, the latter established the concept of trusted third parties. That third party is provided with data in which identifying fields have been hashed. This solves the challenge of trust between the different parties. Some authors argue that transparency of data use and informed consent will help to build trust. In the context of big data, this is more challenging.

▶ This topic will be discussed in more detail in Chapter 12.

3.8 Summary

Accurate record linkage is critical to creating high-quality data sets for analysis. However, outside of a few small centers for record linkage research, linking data sets historically has relied on artisan approaches, particularly for parsing and cleaning data sets. As the creation and use of big data increases, so does the need for systematic record linkage. The history of record linkage is long by computer science standards, but new data challenges encourage the development of new approaches including machine learning methods, clustering algorithms, and privacy-preserving record linkage.

Record linkage stands on the boundary between statistics, information technology, and privacy. We are confident that there will continue to be exciting developments in this field in the years to come.

3.9 Resources

Out of many excellent resources on the subject, we note the following.

- We strongly recommend Christen's book (Christen, 2012a).

- There is a wealth of information available on the ADRN website (Economic and Social Research Council, 2016).

- Winkler has a series of high-quality survey articles (Winkler, 2014).

- The German Record Linkage Center is a resource for research, software, and ongoing conference activities (Schnell, 2016).

- The *Record Linkage* workbook in Chapter 13 provides an intro-duction to probabilistic record linkage with Python.

▶ See https://workbooks.coleridgeinitiative.org.

Chapter 4

Databases

Ian Foster and Pascal Heus

After data have been collected and linked, it is necessary to store and organize them. Many social scientists are accustomed to working with one analytical file, often in SAS, Stata, SPSS, or R. But most organizations store (or should store) their data in databases, which makes it critical for social scientists to learn how to create, manage, and use databases for data storage and analysis. This chapter describes the concept of databases and introduces different types of databases and analysis languages (in particular, relational databases and SQL, respectively) that allow storing and organizing of data for rapid and efficient data exploration and analysis.

4.1   Introduction

We turn now to the question of how to store, organize, and manage the data used in data-intensive social science. As the data with which you work grow in volume and diversity, effective data management becomes increasingly important to avoid scale and complexity from overwhelming your research processes. In particular, when you handle data that are frequently updated, with changes made by different people, you will want to use database management systems (DBMSs) instead of maintaining data in text files or within siloed statistical packages such as SAS, SPSS, Stata, and R. Indeed, we go so far as to say: if you take away *just one thing* from this book (or at least from this chapter), it should be this: *Use a database!*

As we explain in this chapter, DBMSs greatly simplify data management. They require a little effort to set up, but are worth it. They permit: large amounts of data to be organized in multiple ways that allow for efficient and rapid exploration via query languages; durable and reliable storage that maintains data consistency; scaling to large data sizes; and intuitive analysis, both within the DBMS itself and via connectors to other data analysis packages and tools. The DBMSs have become a critical component of most real-world systems, from handling transactions in financial systems to delivering data to power websites, dashboards, and software that we use every day. If you are using a production-level enterprise system, chances are there is a database in the back-end. They are multipurpose and well suited for organizing social science data and for supporting data exploration and analysis.

The DBMSs make many easy things trivial, and many difficult things easy. They are easy to use but can appear daunting at first. A basic understanding of databases and of when and how to use DBMSs is an important element of the social data scientist's knowledge base. Therefore, we provide in this chapter an introduction to databases and how to use them. We describe different types of databases and their various features, and how they can be used in different contexts. We describe basic features and "how to" steps, including getting started, setting up a database schema, ingesting data, querying and analyzing data within a database, and outputting results. We also discuss how to link from databases to other tools, such as Python, R, and (if you must) Stata.

4.2 The DBMS: When and why

Consider the following three data sets:

1. 10,000 records describing research grants, each specifying the principal investigator, institution, research area, proposal title, award date, and funding amount in a comma-separated-value (CSV) format;

2. 10 million records in a variety of formats from funding agencies, web APIs, and institutional sources describing people, grants, funding agencies, and patents;

3. 10 billion Twitter messages and associated metadata—around 10 terabytes (10^{13} bytes) in total, and increasing at a terabyte a month.

Which tools should you use to manage and analyze these data sets? The answer depends on the specifics of the data, the analyses that

Table 4.1. When to use different data management and analysis technologies

Text files, spreadsheets, and scripting language
• Your data are small • Your analysis is simple • You do not expect to repeat analyses over time
Statistical packages
• Your data are modest in size • Your analysis maps well to your chosen statistical package
Relational database
• Your data are structured • Your data are large • You will be analyzing changed versions of your data over time • You want to share your data and analyses with others
NoSQL database
• Your data are unstructured • Your data are extremely large • Your analysis primarily will occur outside the database in a programming language

you want to perform, and the life cycle within which data and analyses are embedded. Table 4.1 summarizes relevant factors, which we now discuss.

In the case of data set 1 (10,000 records describing research grants), it may be feasible to leave the data in their original file, use spreadsheets, pivot tables, or write programs in scripting languages* such as Python or R to analyze the data in those files. For example, someone familiar with such languages can quickly create a script to extract from data set 1 all grants awarded to one investigator, compute average grant size, and count grants made each year in different areas.

★ A scripting language is a programming language used to automate tasks that otherwise the user could perform one by one.

However, this approach also has disadvantages. Scripts do not provide inherent control over the file structure. This means that, if you obtain new data in a different format, your scripts need to be updated. You cannot simply run them over the newly acquired file. Scripts also can easily become unreasonably slow as data volumes grow. A Python or R script will not take long to search a list of 1,000 grants to find those that pertain to a particular institution. But, what if you have information about 1 million grants, and for each grant you want to search a list of 100,000 investigators, and for each investigator you want to search a list of 10 million papers to see whether that investigator is listed as an author of each paper? You now have $1,000,000 \times 100,000 \times 10,000,000 = 10^{18}$ comparisons to perform. Your simple script may now run for hours or even days.

You can speed up the search process by constructing indices, so that, for example, when given a grant, you can find the associated investigators in constant time rather than in time proportional to the number of investigators. However, the construction of such indices is itself a time-consuming and error-prone process.

For these reasons, the use of scripting languages alone for data analysis is rarely to be recommended. This is not to say that all analysis computations can be performed in database systems. Often, a programming language also will be needed. But, many data access and manipulation computations are best handled in a database.

Researchers in the social sciences frequently use statistical packages such as R, SAS, SPSS, and Stata for data analysis. Because these systems integrate some data management, statistical analysis, and graphics capabilities in a single package, a researcher often can conduct a data analysis project of modest size within the same environment. However, each of these systems has limitations that hinder its use for modern social science research, especially as data grow in size and complexity.

Consider Stata, for example. Stata loads the entire data set into the computer's working memory, and thus you would have no problems loading data set 1. However, depending on your computer's memory, it could have problems handling data set 2 and most likely would not be able to handle data set 3. In addition, you would need to perform this data loading step each time you start working on the project, and your analyses would be limited to what Stata can do. In comparison, SAS can manage larger data sets, but it is renowned for being difficult to learn and use. Of course, there are workarounds in statistical packages. For example, in Stata you can handle larger file sizes by choosing to only load the variables or cases that you need for the analysis (Kohler and Kreuter, 2012). Likewise, you can handle more complex data by creating a system of files that each can be linked as needed for a particular analysis through a common identifier variable.

▶ For example, the Panel Study of Income Dynamics (http://psidonline.isr.umich.edu) has a series of files that are related and can be combined through common identifier variables (Institute for Social Research, 2013).

The ad hoc approaches to problems of scale mentioned in the preceding paragraph are provided as core functions in most DBMSs, and thus, rather than implement such inevitably limited workarounds, usually you will be well advised to set up a database. A database is especially valuable if you find yourself in a situation where the data set is constantly updated by different users, if groups of users have different rights to use your data or should only have access to subsets of the data, or if the analysis takes place on a server that sends results to a client (browser). Some statistics packages also have difficulty working with more than one data source at a time—something that DBMSs are designed to do well.

Alas, databases are not perfectly suited for every need. For example, in social science research, the reproducibility of our analysis is critical and hence versioning of the data used for analysis is critical. Most databases do not provide versioning "out of the box." Typically, they do keep a log of all operations performed (inserting, updating, and updating data, for example), which can facilitate versioning and rollbacks, but often we need to configure the database to allow versioning and support reproducibility.

These considerations bring us to the topic of this chapter, database management systems. A DBMS[*] handles all of the issues listed above, and more. As we will see when we consider specific examples, a DBMS allows us to define a logical design that fits the structure of our data. The DBMS then creates a *data model* (which will be discussed next) that allows these data to be stored, queried, and updated efficiently and reliably on disk, thus providing independence from underlying physical storage. It supports efficient access to data through *query languages* and (somewhat) automatic optimization of those queries to permit fast analysis. Importantly, it also supports concurrent access by multiple users, which is not an option for file-based data storage. It supports *transactions*, meaning that any update to a database is performed in its entirety or not at all, even in situations experiencing computer failures or multiple concurrent updates. It also reduces the time spent both by analysts, by making it easy to express complex analytical queries concisely, and on data administration, by providing simple and uniform data administration interfaces.

A *database* is a structured collection of data about entities and their relationships. It models real-world objects—both entities (e.g., grants, investigators, universities) and relationships (e.g., "Steven Weinberg" works at "University of Texas at Austin")—and captures structure in ways that allow these entities and relationships to be queried for analysis. A *database management system* is a software suite designed to safely store and efficiently manage databases, and to assist with the maintenance and discovery of the relationships that database represents. In general, a DBMS encompasses three key components: its *data model* (which defines how data are represented; see Box 4.1), its *query language* (which defines how the user interacts with the data), and support for *transactions and crash recovery* (to ensure reliable execution despite system failures).[*]

★ A DBMS is a system that interacts with users, other applications, and the database itself to capture and analyze data.

★ Some key DBMS features often are lacking in standard statistical packages: a standard query language (with commands that allow analyses or data manipulation on a subgroup of cases defined during the analysis, for example, "group by ...," "order by ..."), keys (for speed improvement), and an explicit model of a relational data structure.

Box 4.1: Data model

A *data model* specifies the data elements associated with the domain we are analyzing, the properties of those data elements, and how those data elements relate to one another. In developing a data model, we commonly first identify the entities that are to be modeled and then define their properties and relationships. For example, when working on the science of science policy (see Figure 1.3), the entities include people, products, institutions, and funding, each of which has various properties (e.g., for a person, their name, address, employer); relationships include "is employed by" and "is funded by." Then, this conceptual data model can be translated into relational tables or some other database representation, as we describe next.

Hundreds of different open source, commercial, and cloud-hosted versions of DBMSs are available and new ones pop up every day. However, you only need to understand a relatively small number of concepts and major database types to make sense of this diversity. Table 4.2 defines the major classes of DBMSs that we will consider in this chapter. We consider only a few of these in any detail.

Table 4.2. Types of databases: relational (first row) and various types of NoSQL (other rows)

Type	Examples	Advantages	Disadvantages	Uses
Relational database	MySQL, PostgreSQL, Oracle, SQL Server, Teradata	Consistency (ACID)	Fixed schema; typically more difficult to scale	Transactional systems: order processing, retail, hospitals, etc.
Key–value store	Dynamo, Redis	Dynamic schema; easy scaling; high throughput	Not immediately consistent; no higher-level queries	Web applications
Column store	Cassandra, HBase	Same as key–value; distributed; better compression at column level	Not immediately consistent; using all columns is inefficient	Large-scale analysis
Document store	CouchDB, MongoDB	Index entire document (JSON)	Not immediately consistent; no higher-level queries	Web applications
Graph database	Neo4j, InfiniteGraph	Graph queries are fast	May not be efficient to perform non-graph analysis	Recommendation systems, networks, routing

Relational DBMSs are the most widely used systems, and they will be the optimal solution for many social science data analysis purposes. We will describe relational DBMSs in detail, but in brief, they allow for the efficient storage, organization, and analysis of large quantities of *tabular* data: data organized as tables, in which rows represent entities (e.g., research grants) and columns represent attributes of those entities (e.g., principal investigator, institution, funding level). Then, the associated Structured Query Language (SQL) can be used to perform a wide range of tasks, which are executed with high efficiency due to sophisticated indexing and query planning techniques.

▶ As discussed in Chapter 3, sometimes the links are one to one and sometimes one to many.

While relational DBMSs have dominated the database world for decades, other database technologies have become popular for various classes of applications in recent years. As we will show, the design of these alternative *NoSQL DBMSs* typically is motivated by a desire to scale the quantities of data and/or number of users that can be supported and/or to manage unstructured data that are not easily represented in tabular form. For example, a key–value store can organize large numbers of records, each of which associates an arbitrary key with an arbitrary value. These stores, and in particular variants called *document stores* that permit text search on the stored values, are widely used to organize and process the billions of records that can be obtained from web crawlers. We will review some of these alternatives and the factors that may motivate their use.

Relational and NoSQL databases (and indeed other solutions, such as statistical packages) also can be used together. Consider, for example, Figure 4.1, which depicts data flows commonly encountered in large research projects. Diverse data are being collected from different sources: JSON documents from web APIs, web pages from web scraping, tabular data from various administrative databases, Twitter data, and newspaper articles. There may be hundreds or even thousands of data sets in total, some of which may be extremely large. We initially have no idea of what schema[*] to use for the different data sets, and indeed it may not be feasible to define a unified set of schema, as the data may be diverse and new data sets may be acquired continuously. Furthermore, the way we organize the data may vary according to our intended purpose. Are we interested in geographic, temporal, or thematic relationships among different entities? Each type of analysis may require a different way of organizing.

★ A schema defines the structure of a database in a formal language defined by the DBMS; see Section 4.3.3.

For these reasons, a common storage solution may be to first load all data into a large NoSQL database. This approach makes all

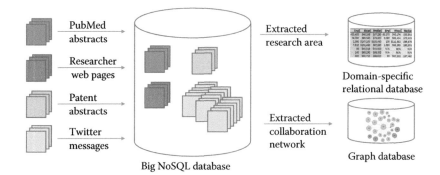

Figure 4.1. A research project may use a NoSQL database to accumulate large amounts of data from many different sources, and then extract selected subsets to a relational or other database for more structured processing

data available via a common (albeit limited) query interface. Then, researchers can extract from this database the specific elements that are of interest for their work, loading those elements into a relational DBMS, another specialized DBMS (e.g., a graph database), or a package for more detailed analysis. As part of the process of loading data from the NoSQL database into a relational database, the researcher will necessarily define schemas, relationships between entities, and so forth. Analysis results can be stored in a relational database or back into the NoSQL store.

4.3   Relational DBMSs

We now provide a more detailed description of relational DBMSs. Relational DBMSs implement the relational data model, in which data are represented as sets of records organized in tables. This model is particularly well suited for the structured data with which we frequently deal in the social sciences; we discuss in Section 4.5 alternative data models, such as those used in NoSQL databases.

We use the data shown in Figure 4.2 to introduce key concepts. These two CSV format files describe grants made by the US National Science Foundation (NSF). One file contains information about grants, the other information about investigators. How should you proceed to manipulate and analyze these data?

The main concept underlying the relational data model is a *table* (also referred to as a *relation*): a set of rows (also referred to as tuples, records, or observations), each with the same columns (also referred to as fields, attributes, or variables). A database consists of

The file **grants.csv**

```
# Identifier,Person,Funding,Program
1316033,Steven Weinberg,666000,Elem. Particle Physics/Theory
1336199,Howard Weinberg,323194,ENVIRONMENTAL ENGINEERING
1500194,Irving Weinberg,200000,Accelerating Innovation Rsrch
1211853,Irving Weinberg,261437,GALACTIC ASTRONOMY PROGRAM
```

The file **investigators.csv**

```
# Name,Institution,Email
Steven Weinberg,University of Texas at Austin,weinberg@utexas.edu
Howard Weinberg,University of North Carolina Chapel Hill,
Irving Weinberg,University of Maryland College Park,irving@ucmc.edu
```

Figure 4.2. The CSV files representing grants and investigators. Each line in the first table specifies a grant number, investigator name, total funding amount, and NSF program name; each line in the second gives an investigator name, institution name, and investigator email address

Number	Person	Funding	Program
1316033	1	660,000	Elem. Particle Physics/Theory
1336199	2	323,194	ENVIRONMENTAL ENGINEERING
1500194	3	200,000	Accelerating Innovation Rsrch
1211853	3	261,437	GALACTIC ASTRONOMY PROGRAM

ID	Name	Institution	Email
1	Steven Weinberg	University of Texas at Austin	weinberg@utexas.edu
2	Howard Weinberg	University of North Carolina Chapel Hill	
3	Irving Weinberg	University of Maryland College Park	irving@ucmc.edu

Figure 4.3. Relational tables "Grants" and "Investigators" corresponding to the grants.csv and investigators.csv data in Figure 4.2, respectively. The only differences are the representation in a tabular form, the introduction of a unique numerical investigator identifier ("ID") in the "Investigators" table, and the substitution of that identifier for the investigator name in the "Grants" table

multiple tables. For example, we show in Figure 4.3 how the data contained in the two CSV files of Figure 4.2 may be represented as two tables. The Grants table contains one tuple for each row in grants.csv, with columns GrantID, Person, Funding, and Program. The table contains one tuple for each row in investigators.csv, with columns ID, Name, Institution, and Email. The CSV files and tables contain essentially the same information, albeit with important differences (the addition of an ID field in the Investigators table, the substitution of an ID column for the Person column in the Grants table) that we will explain next.

The use of the relational data model provides for physical independence: a given table can be stored in many different ways. The SQL queries are sets of instructions to execute commands written in terms of the logical representation of tables (i.e., their schema definition). Consequently, even if the physical organization of the data changes (e.g., a different layout is used to store the data on disk, or a new index is created to speed up access for some queries), the queries need not change. Another advantage of the relational data model is that, because a table is a *set*, in a mathematical sense, simple and intuitive set operations (e.g., union, intersection) can be used to manipulate the data, as we will discuss. We can easily, for example, determine the intersection of two relations (e.g., grants that are awarded to a specific institution), as we describe in the following. The database further ensures that the data comply with the model (e.g., data types, key uniqueness, entity relationships), essentially providing core quality assurance.

4.3.1   Structured Query Language

We use query languages to manipulate data in a database (e.g., to add, update, or delete data elements) and to retrieve (raw and aggregated) data from a database (e.g., data elements that represent certain properties). Most relational DBMSs support SQL, a simple, powerful query language with a strong formal foundation based on logic, a foundation that allows relational DBMSs to perform a wide variety of sophisticated optimizations. Primarily, SQL is used for three purposes:

- Data definition: for example, creation of new tables,

- Data manipulation: queries and updates,

- Control: creation of assertions to protect data integrity.

We introduce each of these features in the following, although not in that order, and certainly not completely. Our goal here is to share enough information to provide the reader with insights into how relational databases work and what they do well; an in-depth SQL tutorial is beyond the scope of this book, but we highly recommend checking the references at the end of this chapter.

4.3.2   Manipulating and querying data

The SQL and other query languages used in DBMSs support the concise, declarative specification of complex queries. Because we are eager to show you something immediately useful, we cover these features first, before discussing how to define data models.

Example: Identifying grants of more than $200,000

Here is an SQL query to identify all grants with total funding of at most $200,000:

```
select * from Grants
where Funding <= 200000;
```

Notice SQL's declarative nature: this query can be read almost as the English language statement, "select all rows from the Grants table for which the Funding column has a value less than or equal to 200,000." This query is evaluated as follows.

1. The input table specified by the from clause, Grants, is selected.

2. The condition in the where clause, Funding <= 200000, is checked against all rows in the input table to identify those rows that match.

3. The select clause specifies which columns to keep from the matching rows, that is, which columns constitute the schema of the output table. (The "*" indicates that all columns should be kept.)

The answer, given the data in Figure 4.3, is the following single-row table. (The fact that an SQL query returns a table is important when it comes to creating more complex queries: the result of a query can be stored in the database as a new table, or passed to another query as input.)

Number	Person	Funding	Program
1500194	3	200000	Accelerating Innovation Rsrch

The DBMSs automatically optimize declarative queries such as the example we have presented, translating them into a set of low-level data manipulations (an imperative *query plan*) that can be evaluated efficiently. This feature allows users to write queries without having to worry too much about performance issues—the database does the worrying for you. For example, a DBMS need not consider every row in the Grants table in order to identify those with funding less than $200,000, a strategy that would be slow if the Grants table were large; instead, it can use an index to retrieve the relevant records much more quickly. We will discuss indices in more detail in Section 4.3.6.

The querying component of SQL supports a wide variety of manipulations on tables, whether referred to explicitly by a table name (as in the example just shown) or constructed by another query. We have demonstrated how to use the `select` operator to pick both certain rows (what is termed *selection*) and certain columns (what is called *projection*) from a table.

Example: Finding grants awarded to an investigator

We want to find all grants awarded to the investigator with the name "Irving Weinberg." The information required to answer this question is distributed over two tables, `Grants` and `Investigators`, and so we *join* the two tables to combine tuples from both.

```
select Number, Name, Funding, Program
from Grants, Investigators
where Grants.Person = Investigators.ID
and Name = "Irving Weinberg";
```

This query combines tuples from the `Grants` and `Investigators` tables for which the `Person` and `ID` fields match. It is evaluated in a manner similar to the query presented above, except for the `from` clause: when multiple tables are listed, as here, the conditions in the `where` clause are checked for all different combinations of tuples from the tables defined in the `from` clause (i.e., the cartesian product of these tables)—in this case, a total of $3 \times 4 = 12$ combinations. We thus determine that Irving Weinberg has two grants. The query further selects the `Number`, `Name`, `Funding`, and `Program` fields from the result, yielding the following.

Number	Person	Funding	Program
1500194	Irving Weinberg	200000	Accelerating Innovation Rsrch
1211853	Irving Weinberg	261437	GALACTIC ASTRONOMY PROGRAM

This ability to join two tables in a query is one example of how SQL permits concise specifications of complex computations. This joining of tables via a cartesian product operation is formally called a *cross join*. Other types of join also are supported. We will describe one such type, the *inner join*, in Section 4.6.

The SQL aggregate functions allow for the computation of aggregate statistics over tables. For example, we can use the following query to determine the total number of grants and their total and average funding levels.

```
select count(*) as 'Number', sum(Funding) as 'Total',
       avg(Funding) as 'Average'
from Grants;
```

This yields the following.

Number	Total	Average
4	1444631	361158

The `group by` operator can be used in conjunction with the aggregate functions to group the result set by one or more columns. For example, we can use the following query to create a table with three columns: investigator name, the number of grants associated with the investigator, and the aggregate funding.

```
select Name, count(*) as 'Number',
       avg(Funding) as 'Average funding'
from Grants, Investigators
where Grants.Person = Investigators.ID
group by Name;
```

We obtain the following.

Name	Number	Average funding
Steven Weinberg	1	666000
Howard Weinberg	1	323194
Irving Weinberg	2	230719

4.3.3 Schema design and definition

When using a pre-existing database, you will be given the database design that includes tables, rows, and columns. But, when you are starting with your own data and need to create a database, the first step is to determine the design of the database.

We have shown that a relational database comprises a set of tables. The task of specifying the structure of the data to be stored in a database is called *logical design*. This task may be performed by a database administrator, in the case of a database to be shared by many people, or directly by users, if they are creating databases themselves. More specifically, the logical design process involves defining a *schema*. A schema comprises a set of tables (including, for each table, its columns and their types), their relationships, and integrity constraints. See, for example, Figure 4.4.

The first step in the logical design process is to identify the entities that need to be modeled. In our example, we identify two important classes of entity: "grants" and "investigators." We thus define a table for each; each row in these two tables will correspond to a unique grant or investigator, respectively. (In a more complete and realistic design, we would likely also identify other entities, such as institutions and research products.) During this step, often we will break up information into multiple tables, so as to avoid duplicating information.

For example, imagine that we are provided grant information in the form of one CSV file rather than two, with each line providing

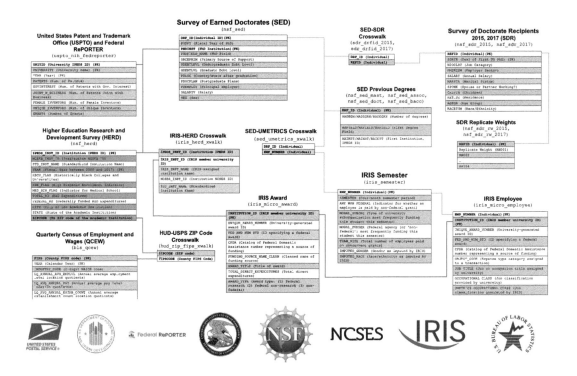

Figure 4.4. A database schema can show the ways in which many tables are linked. Here, there are individual-links (shown in green) as well as institution-level links (shown in red) and location-level links (shown in blue)

a grant number, investigator, funding, program, institution, and email. In this file, the name, institution, and email address for Irving Weinberg would appear twice, because he has two grants, which can lead to errors when updating values and make it difficult to represent certain information. If, however, we want to add more information, e.g., an investigator who does not yet have a grant, we will need to create a tuple (row) with empty slots for all columns (variables) associated with grants. Thus, we would want to break up the single big table into the two tables that we have defined here. This breaking up of information across different tables to avoid repetition of information is referred to as normalization.[*]

The second step in the design process is to define the columns that are to be associated with each entity. For each table, we define a set of columns. For example, given the data in Figure 4.2, the grant table should include columns for award identifier, title, investigator, and award amount; for an investigator, the columns will

> ★ Normalization involves organizing columns and tables of a relational database to minimize data redundancy.

be name, university, and email address. In general, we will want to ensure that each row in our table has a key: a set of columns that uniquely identifies that row. In our example tables, grants are uniquely identified by `Number` and investigators by `ID`.

The third step in the design process is to capture relationships between entities. In our example, we are concerned with only one relationship, i.e., between grants and investigators: each grant has an investigator. We represent this relationship between tables by introducing a `Person` column in the `Grants` table, as shown in Figure 4.3. Note that we do not simply duplicate the investigator names in the two tables, as was the case in the two CSV files shown in Figure 4.2: these names might not be unique, and the duplication of data across tables can lead to later inconsistencies if a name is updated in one table but not the other.

The final step in the design process is to represent integrity constraints (or rules) that must hold for the data. In our example, we may want to specify: that each grant must be awarded to an investigator; that each value of the grant identifier column must be unique (i.e., there cannot be two grants with the same number); and that total funding can never be negative. Such restrictions can be achieved by specifying appropriate constraints at the time of schema creation, as we show in Listing 4.1, which contains the code used to create the two tables that compose our schema.

Listing 4.1 contains four SQL statements. The first two statements, lines 1 and 2, simply set up our new database. The `create table` statement in lines 1 and 2 creates our first table. It specifies the table name (`Investigators`) and, for each of the four columns, the column name and its type.[*] Relational DBMSs offer a rich set of types from which to choose when designing a table: for example, `int` or `integer` (synonyms); `real` or `float` (synonyms); `char(n)`, a fixed-length string of n characters; and `varchar(n)`, a variable-length string of up to n characters. Types are important for several reasons. First, they allow for more efficient encoding of data. For example, the `Funding` field in the grants.csv file of Figure 4.2 could be represented as a string in the `Grants` table, `char(15)`, e.g., to allow for large grants. By representing it as a floating point number instead (line 15 in Listing 4.1), we reduce the space requirement per grant to only four bytes. Second, types allow for integrity checks on data as they are added to the database: for example, that same type declaration for `Funding` ensures that only valid numbers will be entered into the database. Third, types allow for type-specific operations on data, such as arithmetic operations on numbers (e.g., min, max, sum).

► Normalization can be done in statistical packages as well. For example, as noted above, PSID splits its data into different files linked through ID variables. The difference here is that the DBMS makes creating, navigating, and querying the resulting data particularly easy.

★ These storage types will be familiar to many of you from statistical software packages.

```
   create database grantdata;
 2 use grantdata;

 4 create table Investigators (
       ID int auto_increment,
 6     Name varchar(100) not null,
       Institution varchar(256) not null,
 8     Email varchar(100),
       primary key(ID)
10 );

12 create table Grants (
       Number int not null,
14     Person int not null,
       Funding float unsigned not null,
16     Program varchar(100),
       primary key(Number)
18 );
```

Listing 4.1. Code to create the `grantdata` database and its `Investigators` and `Grants` tables

Other SQL features allow for the specification of additional constraints on the values that can be placed in the corresponding column. For example, the `not null` constraints for `Name` and `Institution` (lines 6, 7) indicate that each investigator must have a name and an institution, respectively. (The lack of such a constraint on the `Email` column shows that an investigator need not have an email address.)

4.3.4   Loading data

To this point, we have created a database and two empty tables. The next step is to add data to the tables. We can, of course, do that manually, row by row, but in most cases we will import data from another source, such as a CSV file. Listing 4.2 shows the two statements that load the data of Figure 4.2 into our two tables. (Here and elsewhere in this chapter, we use the MySQL DBMS. The SQL syntax used by different DBMSs differs in various, mostly minor ways.) Each statement specifies the name of the file from which data are to be read and the table into which they are to be loaded. The `fields terminated by ","` statement tells SQL that values are separated by columns, and `ignore 1 lines` tells SQL to skip the header. The list of column names is used to specify how values from the file are to be assigned to columns in the table.

For the `Investigators` table, the three values in each row of the investigators.csv file are assigned to the `Name`, `Institution`, and

```
   load data local infile "investigators.csv"
2      into table Investigators
       fields terminated by ","
4      ignore 1 lines
       (Name, Institution, Email);
6
   load data local infile "grants.csv" into table Grants
8      fields terminated by ","
       ignore 1 lines
10     (Number, @var, Funding, Program)
   set Person = (select ID from Investigators
12               where Investigators.Name=@var);
```

Listing 4.2. Code to load data into the `Investigators` and `Grants` tables

`Email` columns of the corresponding database row. Importantly, the `auto_increment` declaration on the `ID` column (line 5 in Listing 4.1) causes values for this column to be assigned automatically by the DBMS, as rows are created, starting at `1`. This feature allows us to assign a unique integer identifier to each investigator as its data are loaded.

For the `Grants` table, the `load data` call (lines 7–12) is somewhat more complex. Rather than loading the investigator name (the second column of each line in our data file, represented here by the variable `@var`) directly into the database, we use an SQL query (the `select` statement in lines 11–12) to retrieve from the `Investigators` table the `ID` corresponding to that name. By thus replacing the investigator name with the unique investigator identifier, we avoid replicating the name across the two tables.

4.3.5 Transactions and crash recovery

A DBMS protects the data that it stores from computer crashes: if your computer stops running suddenly (e.g., your operating system crashes or you unplug the power), the contents of your database are not corrupted. It does so by supporting *transactions*. A transaction is an atomic sequence of database actions. In general, every SQL statement is executed as a transaction. You also can specify sets of statements to be combined into a single transaction, but we do not discuss that capability here. The DBMS ensures that each transaction is executed completely even in the case of failure or error: if the transaction succeeds, the results of all operations are recorded permanently ("persisted") in the database; if it fails, all operations are "rolled back" and no changes are committed. For example, suppose we run the following SQL statement to convert the funding amounts in the table from dollars to euros, by scaling each number by 0.9. The `update` statement specifies the table to be updated and the operation to be performed, which in this case is to update the `Funding` column of each row. The DBMS will ensure that either no rows are altered or all are altered.

```
update Grants set Grants.Funding = Grants.Funding*0.9;
```

Transactions are also key to supporting multi-user access. The *concurrency control* mechanisms in a DBMS allow multiple users to operate on a database concurrently, as if each were the only user of the system: transactions from multiple users can be interleaved to ensure fast response times, while the DBMS ensures that the database remains consistent. While entire books could be (and have been) written on concurrency in databases, the key point is that read operations can proceed concurrently, while update operations are typically serialized.

4.3.6   Database optimizations

A relational DBMS applies query planning and optimization methods with the goal of evaluating queries as efficiently as possible. For example, if a query asks for rows that fit two conditions, one cheap to evaluate and one expensive, a relational DBMS may filter first on the basis of the first condition, and then apply the second conditions only to the rows identified by that first filter. These types of optimization are what distinguish SQL from other programming languages, as they allow the user to write queries declaratively and rely on the DBMS to determine an efficient execution strategy.

Nevertheless, the user can help the DBMS to improve performance. The single most powerful performance improvement tool is the index, an internal data structure that the DBMS maintains to speed up queries. While various types of indices can be created, with different characteristics, the basic idea is simple. Consider the column in our table. Assume that there are N rows in the table. In the absence of an index, a query that refers to a column value (e.g., `where ID=3`) would require a linear scan of the table, taking on average $N/2$ comparisons and in the worst case N comparisons. A binary tree index allows the desired value to be found with simply $\log_2 N$ comparisons.

Example: Using indices to improve database performance

Consider the following query.

```
select ID, Name, sum(Funding) as TotalFunding
  from Grants, Investigators
    where Investigators.ID=Grants.Person
  group by ID;
```

This query joins our two tables to link investigators with the grants that they hold, groups grants by investigator (using `group by`), and finally sums the funding associated with the grants held by each investigator. The result is the following.

ID	Name	Total funding
1	Steven Weinberg	666000
2	Howard Weinberg	323194
3	Irving Weinberg	230719

In the absence of indices, the DBMS must compare each row in `Investigators` with each row in `Grants`, checking for each pair whether `Investigators.ID = Grants.Person` holds. As the two tables in our sample database have only three and four rows, respectively, the total number of comparisons is only $3 \times 4 = 12$. But if we had 1 million investigators and 1 million grants, then the DBMS would have to perform 1 trillion comparisons, which would take a long time. (More importantly, it would have to perform a large number of disk I/O operations, if the tables did not fit in memory.) An index on the `ID` column of the `Investigators` table reduces the number of operations dramatically, as the DBMS can then take each of the 1 million rows in the `Grants` table and, for each row, identify the matching row(s) in `Investigators` via an index lookup rather than a linear scan.

In our example table, the `ID` column has been specified to be a `primary key`, and thus an index is created for it automatically. If it were not, we could easily create the desired index as follows.

```
alter table Investigators add index(ID);
```

It can be difficult for the user to determine when an index is required. A good general rule is to create an index for any column that is queried often, that is, appears on the right side of a `where` statement that is to be evaluated frequently. However, the presence of indices makes updates more expensive, as every change to a column value requires that the index be rebuilt to reflect the change. Thus, if your data are highly dynamic, you should carefully select which indices to create. (For bulk load operations, a common practice is to drop indices prior to the data import, and recreate them after the load is completed.) Also, indices require disk space, so you need to consider the tradeoff between query efficiency and resources.

The `explain` command can be useful for determining when indices are required. For example, we show in the following some of the output produced when we apply `explain` to our query. (For this example, we have expanded the two tables to 1,000 rows each, as our original tables are too small for MySQL to consider the use of indices.) The output provides useful information such as: the key(s) that could be used, if indices exist (`Person` in the `Grants` table, and the primary key, `ID`, for the `Investigators` table); the key(s) that are actually used (the primary key, `ID`, in the `Investigators` table); the column(s) that are compared to the index (`Investigators.ID` is compared with `Grants.Person`); and the number of rows that must be considered (each of the 1,000 rows in `Grants` is compared with one row in `Investigators`, for a total of 1,000 comparisons).

```
mysql> explain select ID, Name, sum(Funding) as TotalFunding
       from Grants, Investigators
       where Investigators.ID=Grants.Person group by ID;
+---------------+---------------+---------+---------------+------+
| table         | possible_keys | key     | ref           | rows |
+---------------+---------------+---------+---------------+------+
| Grants        | Person        | NULL    | NULL          | 1000 |
| Investigators | PRIMARY       | PRIMARY | Grants.Person |    1 |
+---------------+---------------+---------+---------------+------+
```

Contrast this output with the output obtained for equivalent tables in which `ID` is not a primary key. In this case, no keys are used and thus 1,000 × 1,000 = 1,000,000 comparisons, and the associated disk reads, must be performed.

```
+---------------+---------------+------+------+------+
| table         | possible_keys | key  | ref  | rows |
+---------------+---------------+------+------+------+
| Grants        | Person        | NULL | NULL | 1000 |
| Investigators | ID            | NULL | NULL | 1000 |
+---------------+---------------+------+------+------+
```

A second way in which the user can contribute to performance improvement is by using appropriate table definitions and data types. Most DBMSs store data on disk. Data must be read from disk into memory before they can be manipulated. Memory accesses are fast, but loading data into memory is expensive: accesses to main memory can be a million times faster than accesses to disk. Therefore, to ensure queries are efficient, it is important to minimize the number of disk accesses. A relational DBMS automatically optimizes queries: based on how the data are stored, it transforms a SQL query into a query plan that can be executed efficiently, and chooses an execution strategy that minimizes disk accesses. Even so, users can contribute to making queries efficient. As discussed, the choice of types made when defining schemas can make a big difference. As a general rule, only use as much space as needed for your data: the smaller your records, the more records can be transferred to main memory using a single disk access. The design of relational tables is also important. If you put all columns in a single table (i.e., you do not normalize), more data may come into memory than is required.

4.3.7   Caveats and challenges

It is important to keep the following caveats and challenges in mind when using SQL technology with social science data.

4.3.7.1   Data cleaning

Data created outside an SQL database, such as data in files, are not always subject to strict constraints: data types may not be correct or consistent (e.g., numeric data stored as text) and consistency or integrity may not be enforced (e.g., absence of primary keys, missing foreign keys). Indeed, as the reader probably knows well from experience, data are rarely perfect. As a result, the data may fail to comply with strict SQL schema requirements and fail to load, in which case either data must be cleaned before or during loading, or the SQL schema must be relaxed.

4.3.7.2   Missing values

Care must be taken when loading data in which some values may be missing or blank. The SQL engines represent and refer to a missing or blank value as the built-in constant `null`. Counterintuitively, when loading data from text files (e.g., CSV), many SQL engines require that missing values be represented explicitly by the term `null`; if a data value is simply omitted, it may fail to load or be incorrectly represented, e.g., as zero or the empty string (`""`) instead of `null`. Thus, for example, the second row in the investigators.csv file of Figure 4.2:

```
Howard Weinberg,University of North Carolina Chapel Hill,
```

may need to be rewritten as:

```
Howard Weinberg,University of North Carolina Chapel Hill,null
```

4.3.7.3   Metadata for categorical variables

The SQL engines are metadata poor: they do not allow extra information to be stored about a variable (field) beyond its base name and type (int, char, etc., as introduced in Section 4.3.3). They cannot, for example, record directly the fact that the column class can only take one of three values, animal, vegetable, or mineral, or what these values mean. Thus, common practice is to store information about possible values in another table (commonly referred to as a *dimension table*) that can be used as a lookup and constraint, as in the following.

Table **class_values**

Value	Description
animal	Is alive
vegetable	Grows
mineral	Isn't alive and doesn't grow

A related concept is that a column or list of columns may be declared primary key or unique. Such a statement specifies that no two tuples of the table may agree in the specified column—or, if a list of columns is provided, in all of those columns. There can be only one primary key for a table, but several unique columns. No column of a primary key can ever be null in any tuple. But, columns declared unique may have nulls, and there may be several tuples with null.

4.4   Linking DBMSs and other tools

Query languages such as SQL are not general-purpose programming languages; they support easy efficient access to large data sets, are extremely efficient for specific types of analysis, but may not be the right choice for all analysis. When complex computations are required, one can embed query language statements into a programming language or statistical package. For example, we might want to calculate the interquartile range of funding for all grants. While this calculation can be accomplished in SQL, the resulting SQL code will be complicated (depending on which type of SQL your database supports). Languages such as Python make these statistical calculations straightforward, so it is natural to write a Python (or R, SAS, Stata, etc.) program that connects to the DBMS that contains our data, fetches the required data from the DBMS, and then calculates the interquartile range of those data. Then, the program can, if desired, store the result of this calculation back into the database.

Many relational DBMSs also have built-in analytical functions or often now support different programming languages, providing

```
   from mysql.connector import MySQLConnection, Error
 2 from python_mysql_dbconfig import read_db_config

 4 def retrieve_and_analyze_data():
       try:
 6         # Open connection to the MySQL database
           dbconfig = read_db_config()
 8         conn = MySQLConnection(**dbconfig)
           cursor = conn.cursor()
10         # Transmit the SQL query to the database
           cursor.execute('select Funding from Grants;')
12         # Fetch all rows of the query response
           rows = [row for row in cur.fetchall()]
14         calculate_inter_quartile_range(rows)
       except Error as e:
16         print(e)
       finally:
18         cursor.close()
           conn.close()
20
   if __name__ == '__main__':
22     retrieve_and_analyze_data()
```

Listing 4.3. Embedding SQL in Python

significant in-database statistical and analytical capabilities and alleviating the need for external processing.

Example: Embedding database queries in Python

The Python script in Listing 4.3 shows how this embedding of database queries in Python is accomplished. This script establishes a connection to the database, transmits the desired SQL query to the database (line 7–9), retrieves the query results into a Python array (line 11), and calls a Python procedure (not given) to perform the desired computation (line 14). A similar program could be used to load the results of a Python (or R, SAS, Stata, etc.) computation into a database.

Example: Loading other structured data

We have shown in Listing 4.2 how to load data from CSV files into SQL tables. Data in other formats, such as the commonly used JSON, also can be loaded into a relational DBMS. Consider, for example, the following JSON format data, a simplified version of data shown in Chapter 2.

```
1   [
2     {
3       institute : Janelia Campus,
4       name : Laurence Abbott,
5       role : Senior Fellow,
6       state : VA,
7       town : Ashburn
8     },
9     {
10      institute : Jackson Lab,
11      name : Susan Ackerman,
12      role : Investigator,
13      state : ME,
14      town : Bar Harbor
15    }
16  ]
```

Although some relational DBMSs (such as PostgreSQL) provide built-in support for JSON objects, we assume here that we want to convert these data into normal SQL tables. Using one of the many utilities for converting JSON into CSV, we can construct the following CSV file, which we can load into an SQL table using the method shown previously.

```
institute,name,role,state,town
Janelia Campus,Laurence Abbott,Senior Fellow,VA,Ashburn
Jackson Lab,Susan Ackerman,Investigator,ME,Bar Harbor
```

But into what table? The two records each combine information about a person with information about an institute. Following the schema design rules given in Section 4.3.3, we should *normalize* the data by reorganizing them into two tables, one describing people and one describing institutes. Similar problems arise when JSON documents contain nested structures. For example, consider the following alternative JSON representation of this data. Here, the need for normalization is even more apparent.

```
1   [
2     {
3       name : Laurence Abbott,
4       role : Senior Fellow,
5       employer : { institute : Janelia Campus,
6                    state : VA,
7                    town : Ashburn}
8     },
9     {
10      name : Susan Ackerman,
11      role : Investigator,
12      employer: { institute : Jackson Lab,
13                   state : ME,
14                   town : Bar Harbor}
15    }
16  ]
```

Thus, the loading of JSON data into a relational database usually requires work on both schema design (Section 4.3.3) and data preparation.

4.5   NoSQL databases

While relational DBMSs have dominated the database world for several decades, other database technologies exist and indeed have become popular for various classes of applications in recent years. As we will show, these alternative technologies typically have been motivated by a desire to scale the quantities of data and/or number of users that can be supported, and/or to support specialized data types (e.g., unstructured data, graphs). Here, we review some of these alternatives and the factors that may motivate their use.

4.5.1   Challenges of scale: The CAP theorem

For many years, the big relational database vendors (Oracle, IBM, Sybase, Microsoft) were the mainstay of how data were stored. During the Internet boom, startups looking for low-cost alternatives to commercial relational DBMSs turned to MySQL and PostgreSQL. However, these systems proved inadequate for big websites as they could not cope well with large traffic spikes, e.g., when many customers all suddenly wanted to order the same item. In other words, they did not *scale*.

An obvious solution to scaling databases is to partition data across multiple computers, such as by distributing different tables, or different rows from the same table, over multiple computers. We also may want to replicate popular data, by placing copies on more than one computer. However, partitioning and replication also introduce challenges, as we now explain. Let us first define some terms. In a system comprised of multiple computers:

- Consistency indicates that all computers see the same data at the same time,

- Availability indicates that every request receives a response about whether it succeeded or failed,

- Partition tolerance indicates that the system continues to operate even if a network failure prevents computers from communicating.

An important result in distributed systems (the so-called "CAP theorem," Brewer, 2012) observes that it is not possible to create a distributed system with all three properties. This situation creates a challenge with large transactional data sets. Partitioning and replication are needed in order to achieve high performance, but, as the number of computers grows, so too does the likelihood of network disruption among pair(s) of computers. A network disruption can prevent some replicas of a data item from being updated, compromising consistency. Because strict consistency cannot be achieved at the same time as availability and partition tolerance, the DBMS designer must choose between high consistency and high availability for a particular system.

The right combination of availability and consistency will depend on the needs of the service. For example, in an e-commerce setting, it makes sense to choose high availability for a checkout process, in order to ensure that requests to add items to a shopping cart (a revenue-producing process) can be honored. Errors can be hidden from the customer and addressed later. However, for order submission—when a customer submits an order—it makes sense to favor consistency because several services (credit card processing, shipping and handling, reporting) need to access the data simultaneously. However, in almost all cases, availability is chosen over consistency.

4.5.2 NoSQL and key–value stores

Relational DBMSs were traditionally motivated by the need for transaction processing and analysis, which led them to put a premium on consistency and availability. This led the designers of these systems to provide a set of properties summarized by the acronym ACID (Gray, 1981; Silberschatz et al., 2010):

- Atomic: All work in a transaction completes (i.e., is committed to stable storage) or none of it completes;

- Consistent: A transaction transforms the database from one consistent state to another consistent state;

- Isolated: The results of any changes made during a transaction are not visible until the transaction has committed;

- Durable: The results of a committed transaction survive failures.

The need to support extremely large quantities of data and numbers of concurrent clients has led to the development of a range of alternative database technologies that relax consistency and thus these ACID properties in order to increase scalability and/or availability. These systems are commonly referred to as NoSQL (for "not SQL"—or, more recently, "not only SQL," to communicate that they may support SQL-like query languages) because they usually do not require a fixed table schema nor support joins and other SQL features. Such systems are sometimes referred to as BASE (Fox et al., 1997): Basically Available (the system seems to work all of the time), Soft state (it does not have to be consistent all of the time), and Eventually consistent (it becomes consistent at some later time). The data systems used in essentially all large Internet companies (Google, Yahoo!, Facebook, Amazon, eBay) are BASE.

Dozens of different NoSQL DBMSs exist, with widely varying characteristics as summarized in Table 4.2. The simplest are *key-value stores*, such as Redis, Amazon Dynamo, Apache Cassandra, and Project Voldemort. We can think of a key–value store as a relational database with a single table that has only two columns, key and value, and that supports only two operations: store (or update) a key–value pair and retrieve the value for a given key.

Example: Representing investigator data in a NoSQL database

We might represent the contents of the investigators.csv file of Figure 4.2 (in a NoSQL database) as follows.

Key	Value
Investigator_StevenWeinberg_Institution	University of Texas at Austin
Investigator_StevenWeinberg_Email	weinberg@utexas.edu
Investigator_HowardWeinberg_Institution	University of North Carolina Chapel Hill
Investigator_IrvingWeinberg_Institution	University of Maryland College Park
Investigator_IrvingWeinberg_Email	irving@ucmc.edu

A client then can read and write the value associated with a given *key* by using operations such as the following.

- **Get**(*key*) returns the value associated with *key*.

- **Put**(*key*, *value*) associates the supplied *value* with *key*.

- **Delete**(*key*) removes the entry for *key* from the data store.

Thus, key–value stores are particularly easy to use. Furthermore, because there is no schema, there are no constraints on what values can be associated with a key. This lack of constraints can be useful if we want to store arbitrary data. For example, it is trivial to add the following records to a key–value store; adding this information to a relational table would require schema modifications.

Key	Value
Investigator_StevenWeinberg_FavoriteColor	Blue
Investigator_StevenWeinberg_Awards	Nobel

Another advantage is that, if a given key would have no value (e.g., Investigator_HowardWeinberg_Email), we need not create a record. Thus, a key–value store can achieve a more compact representation of sparse data, which would have many empty fields if expressed in relational form.

A third advantage of the key–value approach is that a key–value store is easily partitioned and thus can scale to extremely large sizes. A key–value DBMS can partition the space of keys (e.g., via a hash on the key) across different computers for scalability. It also can replicate key–value pairs across multiple computers for availability. Adding, updating, or querying a key–value pair requires simply sending an appropriate message to the computer(s) that hold that pair.

The key–value approach also has disadvantages. As we can see from the example, users must be careful in their choice of keys if they are to avoid name collisions. The lack of schema and constraints can make it difficult to detect erroneous keys and values. Key–value stores typically do not support join operations (e.g., "which investigators have the Nobel and live in Texas?"). Many key–value stores also relax consistency constraints and do not provide transactional semantics.

4.5.3   Other NoSQL databases

The simple structure of key–value stores allows for extremely fast and scalable implementations. However, as we have shown, many interesting data cannot be easily modeled as key–value pairs. Such concerns have motivated the development of a variety of other NoSQL systems that offer, for example, richer data models: document-based (CouchDB and MongoDB), graph-based (Neo4 J) and column-based (Cassandra, HBase) databases, and graph databases.

In document-based databases, the value associated with a key can be a structured document, e.g., a JSON document, permitting the following representation of our investigators.csv file plus the additional information that we just introduced.

Key	Value
Investigator_StevenWeinberg	{ institution : University of Texas at Austin, email : weinberg@utexas.edu, favcolor : Blue, award : Nobel }
Investigator_HowardWeinberg	{ institution : University of North Carolina Chapel Hill }
Investigator_IrvingWeinberg	{ institution : University of Maryland College Park, email : irving@ucmc.edu }

Associated query languages may permit queries within the document, such as regular expression searches, and retrieval of selected fields, providing a form of a relational DBMS's selection and projection capabilities (Section 4.3.2). For example, MongoDB allows us to ask for documents in a collection called that have "University of Texas at Austin" as their institution and the Nobel as an award.

```
db.investigators.find(
    { institution:  'University of Texas at Austin',
      award:  'Nobel' }
)
```

A column-oriented DBMS stores data tables by columns rather than by rows, as is common practice in relational DBMSs. This approach has advantages in settings where aggregates must frequently be computed over many similar data items, e.g., in clinical data analysis. Google Cloud BigTable and Amazon RedShift are two cloud-hosted column-oriented NoSQL databases. HBase and Cassandra are two open source systems with similar characteristics. (Confusingly, the term *column oriented* also often is used to refer to SQL database engines that store data in columns instead of rows, e.g., Google BigQuery, HP Vertica, Terradata, and the open source MonetDB. Such systems are not to be confused with column-based NoSQL databases.)

Graph databases store information about graph structures in terms of nodes, edges that connect nodes, and attributes of nodes and edges. Proponents argue that they permit particularly straightforward navigation of such graphs, as when answering queries such as "find all the friends of the friends of my friends"—a task that would require multiple joins in a relational database.

4.6 Spatial databases

Social science research commonly involves spatial data. Socioeconomic data may be associated with census tracts, data about the distribution of research funding and associated jobs with cities and states, and crime reports with specific geographic locations. Furthermore, the quantity and diversity of such spatially resolved data are growing rapidly, as are the scale and sophistication of the systems that provide access to these data. For example, just one urban data store, Plenario, contains many hundreds of data sets about the City of Chicago (Catlett et al., 2014).

Researchers who work with spatial data need methods for representing those data and then for performing various queries against them. Does crime correlate with weather? Does federal spending on research spur innovation within the locales where research occurs? These and many other questions require the ability to quickly determine such things as which points exist within which regions, the areas of regions, and the distance between two points. Spatial databases address these and many other related requirements.

Example: Spatial extensions to relational databases

Spatial extensions have been developed for many relational databases, e.g., Oracle Spatial, DB2 Spatial, and SQL Server Spatial. We use the PostGIS extensions to the PostgreSQL relational database here. These extensions implement support for spatial data types such as `point`, `line`, and `polygon`, and operations such as `st_within` (returns `true` if one object is contained within another), `st_dwithin` (returns `true` if two objects are within a specified distance of each other), and `st_distance` (returns the distance between two objects). Thus, for example, given two tables with rows for schools and hospitals in Illinois (`illinois_schools` and `illinois_hospitals`, respectively; in each case, the column `the_geom` is a polygon for the object in question) and a third table with a single row representing the City of Chicago (`chicago_citylimits`), we can easily find the names of all schools within the Chicago city limits.

```
select illinois_schools.name
  from illinois_schools, chicago_citylimits
  where st_within(illinois_schools.the_geom,
                  chicago_citylimits.the_geom);
```

We join the two tables with the constraint constraining the selected rows to those representing schools within the city limits. Here, we use the inner join introduced in Section 4.3.2. This query also could be written as follows.

```
select illinois_schools.name
  from illinois_schools left join chicago_citylimits
  on st_within(illinois_schools.the_geom,
               chicago_citylimits.the_geom);
```

In addition, we can determine the names of all schools that do *not* have a hospital within 3,000 m:

```
select s.name as 'School Name'
    from illinois_schools as s
       left join illinois_hospitals as h
         on st_dwithin(s.the_geom, h.the_geom, 3000)
    where h.gid is null;
```

Here, we use an alternative form of the join operator, the *left join*—or, more precisely, the *left excluding join*. The expression

```
table1 left join table2 on constraint
```

Inner join	Left join	Left excluding join
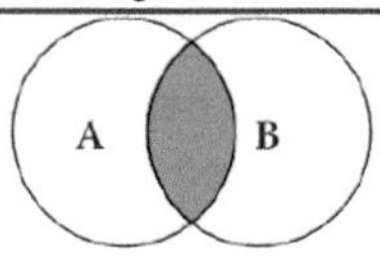	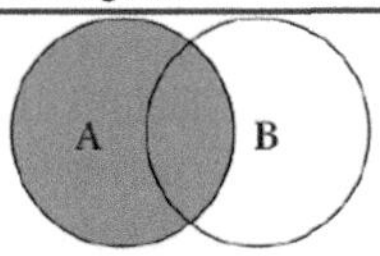	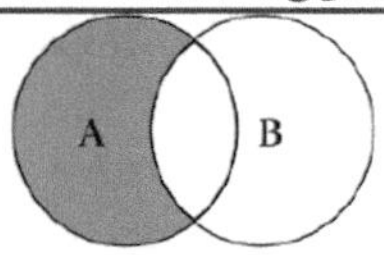

`select columns`
`from Table_A A`
`inner join Table_B B`
`on A.Key = B.Key`
	`select columns`
	`from Table_A A`
	`left join Table_B B`
	`on A.Key = B.Key`
		`select columns`
		`from Table_A A`
		`left join Table_B B`
		`on A.Key = B.Key`
		`where B.Key is null`

Figure 4.5. Three types of *join* illustrated: the inner join, the left join, and the left excluding join

returns all rows from the left table (`table1`) with the matching rows in the right table (`table2`), with the result being `null` in the right side when there is no match. This selection is illustrated in the middle column of Figure 4.5. The addition of the `where h.gid is null` then selects only those rows in the left table with no right-side match, as illustrated in the right-side column of Figure 4.5. Note also the use of the `as` operator to rename the columns `illinois_schools` and `illinois_hospitals`. In this case, we rename them simply to make our query more compact.

4.7   Which database to use?

The question of which DBMS to use for a social science project depends on many factors. We have introduced some relevant rules in Table 4.1, and we will expand on those considerations here.

4.7.1   Relational DBMSs

If your data are structured into rows and columns, then a relational DBMS is almost certainly the correct technology to use. Many open source, commercial, and cloud-hosted relational DBMSs exist. Among the open source DBMSs, MySQL and PostgreSQL (often simply called Postgres) are particularly widely used. MySQL is the most popular. It is particularly easy to install and use, but does not support all features of the SQL standard. PostgreSQL is fully standard compliant and supports useful features such as full text search and the PostGIS extensions mentioned in the previous section. We recommend you start with Postgres.

Popular commercial relational DBMSs include Microsoft SQL Server, Oracle, IBM DB2, Teradata, and Sybase. These systems are heavily used in commercial settings. There are free community editions, and some large science projects use enterprise features via licensing, e.g., the Sloan Digital Sky Survey uses Microsoft SQL Server (Szalay et al., 2002) and the CERN high-energy physics lab uses Oracle (Girone, 2008).

We also see increasing use being made of cloud-hosted relational DBMSs, such as Amazon Relational Database Service (RDS; this supports MySQL, PostgreSQL, and various commercial DBMSs), Microsoft Azure, and Google Cloud SQL. These systems obviate the need to install local software, administer a DBMS, or acquire hardware to run and scale your database. Particularly if your database is bigger than can fit on your workstation, a cloud-hosted solution can be a good choice, both for scalability but also for ease of configuration and management.

4.7.2 NoSQL DBMSs

Some social science problems have the scale that might motivate the use of a NoSQL DBMS. Furthermore, while defining and enforcing a schema can involve some effort, the benefits of doing so are considerable. Thus, the use of a relational DBMS usually is recommended.

Nevertheless, as noted in Section 4.2, there are occasions when a NoSQL DBMS can be a highly effective choice, such as when working with large quantities of unstructured data. For example, researchers analyzing large collections of Twitter messages frequently store the messages in a NoSQL document-oriented database such as MongoDB. NoSQL databases also often are used to organize large numbers of records from many different sources, as illustrated in Figure 4.1.

4.8 Summary

A key message of this book is that you should, whenever possible, use a database. Database management systems are one of the great achievements of information technology, permitting large amounts of data to be stored and organized so as to allow rapid and reliable exploration and analysis. They have become a central component of a great variety of applications, from handling transactions in financial systems to serving data published in websites.

They are particularly well suited for organizing social science data and for supporting analytics for data exploration.

The DBMSs provide an environment that greatly simplifies data management and manipulation. They make many easy things trivial, and many difficult things easy. They automate many other error-prone, manual tasks associated with query optimization. Although they can be daunting to those unfamiliar with their concepts and workings, they are, in fact, easy to use. A basic understanding of databases, and of when and how to use DBMSs, is an important element of the social data scientist's knowledge base.

4.9 Resources

The enormous popularity of DBMSs means that there are many good books to be found. Classic textbooks, such as those by Silberschatz et al. (2010) and Ramakrishnan and Gehrke (2002), provide a great deal of technical detail. The DB Engines website collects information on DBMSs. There are also many useful online tutorials, and of course StackExchange and other online forums often have answers to your technical questions.

▶ http://db-engines.com/en/

Turning to specific technologies, the *SQL Cookbook* (Molinaro, 2005) provides a wonderful introduction to SQL. We also recommend the SQL Cheatsheet and a useful visual depiction of different SQL join operators (Moffatt, 1999). Two good books on the PostGIS geospatial extensions to the PostgreSQL database are the *PostGIS Cookbook* (Corti et al., 2014) and *PostGIS in Action* (Obe and Hsu, 2015). The online documentation is also excellent. The monograph *NoSQL Databases* (Strauch, 2009) provides much useful technical detail.

▶ http://www.sql-tutorial
.net/SQL-Cheat-Sheet.pdf

▶ http://postgis.net/doc

In this chapter, we have not considered the native extensible Markup Language (XML) and Resource Description Framework (RDF) triple stores, because these are not typically used for data management. However, they do play a fundamental role in metadata and knowledge management; see, for example, Sesame (Broekstra et al., 2002).

▶ http://rdf4j.org

If you are interested in the state of database and data management research, recent Seattle Report (Abadi et al., 2020) provides a useful perspective.

The *Databases* notebook of Chapter 13 provides an introduction to working with SQL.

▶ See https://workbooks.
coleridgeinitiative.org.

Chapter 5

Scaling up through Parallel and Distributed Computing

Huy Vo and Claudio Silva

This chapter provides an overview of techniques that allow us to analyze large amounts of data using distributed computing (multiple computers concurrently). While the focus is on a widely used framework, called MapReduce, and popular implementations, such as Apache Hadoop and Spark, the goal of the chapter is to provide a conceptual and practical framework to manage large amounts of data that may not fit in memory or may require too much time to analyze on a single computer. It is important to note that these frameworks do not result in analysis that is better—they are useful because they allow us to process large amounts of data faster and/or without obtaining access to a single massive computer with extensive memory (RAM) and processing power (CPU).

5.1   Introduction

As the amount of data available for social science research increases, we have to determine how to perform our analyses quickly and efficiently. One way to handle large amounts of data that may not fit in memory or may require too much time to analyze on a single computer is to subsample the data or to simplify the analysis. Another approach is to use all of the data by making use of multiple computers concurrently to perform the analysis. The use of parallel computing to manage large amounts of data has been a common approach in physical sciences. Data analysts have routinely been working on data sets much larger than a single machine can handle for several decades, especially at the DOE National Laboratories (Sethian et al., 1991; Crossno et al., 1993) where high-performance computing has been a significant technology trend. This trend also is demonstrated by the history of research in distributed computing and data management since the 1980s.

There are many ways to perform distributed and parallel computing, ranging from completely flexible (but more complex to use) approaches, such as Message Passing Interface (MPI) (Gropp et al., 2014), to more restrictive (but much easier to use) approaches, such as MapReduce. The MPI allows you to do anything with as much efficiency as your MPI skills allow you to code, while MapReduce allows a more restrictive set of analysis to be done (possibly less efficiently) but is much easier to learn and implement.

This chapter focuses on one such framework, called MapReduce, to perform large-scale data analysis distributed across multiple computers. We describe the MapReduce framework, work through an example using it, and highlight one implementation of the framework, called Hadoop, in detail.

▶ If you have examples from your own research using the methods we describe in this chapter, please submit a link to the paper (and/or code) here: https://textbook.coleridge initiative.org/submitexamp les.

Box 5.1: Parallel computing examples

Al Aghbari et al. (2019) introduce GeoSim, an algorithm used for clustering users in any social network site into communities based on the semantic meaning of the nodes of interest as well as their relationships with each other. The parallelized version of GeoSim utilizes the MapReduce model to run on multiple machines simultaneously and produce results faster.

Kolb et al. (2012) develop a tool, DeDoop, that uses Hadoop to perform efficient record linkage (remember Chapter 3?) and scale to large data sets. Tasks, such as record linkage where we can easily break down the larger task into smaller chunks (such as comparing two records to see if they belong to the same entity), that can be done in parallel are ideally suited for MapReduce frameworks.

Ching et al. (2012) describe the data infrastructure at Facebook with MapReduce at the core of Facebook's data analytics engine. More than half a petabyte of new data arrives in the warehouse every 24 hours, and ad-hoc queries, data pipelines, and custom MapReduce jobs process this raw data around the clock to generate more meaningful features and aggregations.

5.2 MapReduce

The MapReduce framework has been proposed by Jeffrey Dean and Sanjay Ghemawat of Google (Dean and Ghemawat, 2004), and its origins are derived from conceptually similar approaches first described in the early 1980s. Using the MapReduce framework requires turning the analysis problem we have into operations that the framework supports—these are map and reduce. The "map" operation takes the input and splits up the task into multiple (parallel) components, and the "reduce" operation consolidates the results of the parallel "mapped" tasks and produces the final output. In order to use the MapReduce framework, we need to break up our tasks into map and reduce operations and implement these two operations.

Example: Counting NSF awards

To gain a better understanding of these MapReduce operations, let's take a trivial task that may need to be done on billions of records, causing scalability challenges. Imagine that we have a list of NSF principal investigators, along with their email information and award IDs as shown. Our task is to count the number of awards for each institution. For example, given these four records, we will discover that the Berkeley Geochronology Center has two awards, while New York University and the University of Utah each have one.

```
AwardId,FirstName,LastName,EmailAddress
0958723,Roland,Mundil,rmundil@bgc.org
0958915,Randall,Irmis,irmis@umnh.utah.edu
1301647,Zaher,Hani,zh8@nyu.edu
1316375,David,Shuster,dshuster@bgc.org
```

We observe that institutions can be distinguished by their email address domain name. Thus, we adopt a strategy of first grouping all award IDs by domain names, and then counting the number of distinct awards within each group. In order to do this, we first set the function to scan input lines and extract institution information and award IDs. Then, in the function, we simply count unique IDs on the data, because everything is already grouped by institution. Python pseudo-code is provided in Listing 5.1.

In the map phase, the input will be transformed into tuples of institutions and award IDs.

```
"0958723,Roland,Mundil,rmundil@bgc.org"        →    ("bgc.org", 0958723)
"0958915,Randall,Irmis,irmis@umnh.utah.edu"    →    ("utah.edu", 0958915)
```

```
 1   # Input   : a list of text lines
     # Output  : a list of domain name and award ids
 3   def MAP(lines):
         for line in lines:
 5           fields     = line.strip('\n').split(',')
             awardId    = fields[0]
 7           domainName = fields[3].split('@')[-1].split('.')[-2:]
             yield (domainName, awardId)
 9
     # Input   : a list of domain name and award ids
11   # Output  : a list of domain name and award count
     def REDUCE(pairs):
13       for (domainName, awardIds) in pairs:
             count = len(set(awardIds))
15           yield (domainName, count)
```

Listing 5.1. Python pseudo-code for the map and reduce functions to count the number of awards per institution

```
"1301647,Zaher,Hani,zh8@nyu.edu"          →   ("nyu.edu", 1301647)
"1316375,David,Shuster,dshuster@bgc.org"  →   ("bgc.org", 1316375)
```

Then, the tuples will be grouped by institutions and counted by the function.

```
("bgc.org", [0958723,1316375])   →   ("bgc.org", 2)
("utah.edu", [0958915])          →   ("utah.edu", 1)
("nyu.edu", [1301647])           →   ("nyu.edu", 1)
```

As we have demonstrated, the MapReduce programming model is quite simple and straightforward, yet it supports a simple parallelization model. In fact, it has been said to be *too* simple and criticized as "a major step backwards" (DeWitt and Stonebraker, 2008) for large-scale data-intensive applications. It is difficult to argue that MapReduce is offering something truly innovative when MPI has been offering similar scatter and reduce operations since 1995, and Python has had high-order functions (map, reduce, filter, and lambda) since its 2.2 release in 1994. However, the biggest strength of MapReduce is its simplicity. Its simple programming model has brought many non-expert users to analyzing large amounts of data. Its simple architecture also has inspired many developers to develop advanced capabilities, such as support for distributed computing, data partitioning, and streaming processing. A downside of this diversity of interest is that available features and capabilities can vary considerably, depending on the specific implementation of MapReduce that is being used.

As mentioned above, MapReduce is a programming model. In order to implement an analysis in MapReduce, we need to select an implementation of MapReduce. The two most commonly used implementations of the MapReduce model are Hadoop and Spark, which we will describe in more detail next.

5.3 Apache Hadoop MapReduce

Apache Hadoop (or Hadoop)[*] was originally designed to run in environments with thousands of machines. Supporting such a large computing environment puts several constraints on the system; for instance, with so many machines, the system had to assume computing nodes would fail. Hadoop became an enhanced MapReduce implementation with the support for fault tolerance, distributed storage, and data parallelism through two added key design features: (1) a distributed file system, called the Hadoop Distributed File System (HDFS), and (2) a data distribution strategy that allows computation to be moved to the data during execution.

> ★ The term *Hadoop* refers to the creator's son's toy elephant.

5.3.1 The Hadoop Distributed File System

The Hadoop Distributed File System (Apache Hadoop, nd) is a distributed file system that stores data across all the nodes (machines) of a Hadoop cluster. The HDFS splits large data files into smaller blocks (chunks of data) which are managed by different nodes in a cluster. Each block is also replicated across several nodes as an attempt to ensure that a full copy of the data are still available even in the case of computing node failures. The block size as well as the number of replications per block are fully customized by users when they create files on HDFS. By default, the block size is set to 64 MB with a replication factor of 3, meaning that the system may encounter at least two concurrent node failures without losing any data. The HDFS also actively monitors failures and re-replicates blocks on failed nodes to ensure that the number of replications for each block always stays at the user-defined settings. Thus, if a node fails, and only two copies of some data exist, the system will quickly copy those data to a working node, thereby raising the number of copies to three again. This dynamic replication is the primary mechanism for fault tolerance in Hadoop.

> ▶ https://hadoop.apache.org/docs/stable/

Note that data blocks are replicated and distributed across several machines. This could create a problem for users, because, if they have to manage the data manually, they might, for example, have to access more than one machine to fetch a large data file. Fortunately, Hadoop provides infrastructure for managing this complexity seamlessly, including command line programs as well as an API that users can employ to interact with HDFS as if it were a local file system. For example, one can run simple Linux commands, such as ls and mkdir, to list and create a directory on HDFS, or even use to inspect file contents the same way as one would do in a Linux file system. The following code shows some examples of interacting with HDFS.

```
# Creating a folder
hadoop dfs -mkdir /hadoopiseasy

# Upload a CSV file from our local machine to HDFS
hadoop dfs -put importantdata.csv /hadoopiseasy

# Listing all files under hadoopiseasy folder
hadoop dfs -ls /hadoopiseasy

# Download a file to our local machine
hadoop dfs -get /hadoopiseasy/importantdata.csv
```

5.3.2   Hadoop setup: Bringing compute to the data

There are two parts of the computing environment when using Hadoop: (1) a *compute cluster* with substantial computing power (e.g., thousands of computing cores) and (2) a *storage cluster* with lots of disk space, capable of storing and serving data quickly to the compute cluster.

These two clusters have quite different hardware specifications: the first is optimized for CPU performance and the second for storage. Typically, the two systems are configured as separate physical hardware.

Running compute jobs on such hardware often proceeds as follows. When a user requests to run an intensive task on a particular data set, the system will first reserve a set of computing nodes. Then, the data are partitioned and copied from the storage server into these computing nodes before the task is executed. This process is illustrated in Figure 5.1 (top). This computing model will be referred to as *bringing data to computation*. In this model, if a data set is being analyzed in multiple iterations, it is very likely that the data will be copied multiple times from the storage cluster to the compute nodes without reusability. This is because the compute node scheduler normally does not have or keep knowledge of where

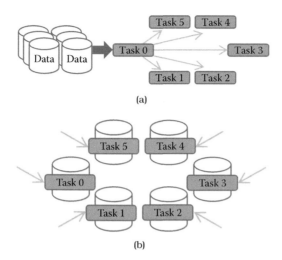

Figure 5.1. (a) The traditional parallel computing model where data are brought to the computing nodes. (b) Hadoop's parallel computing model: bringing compute to the data (Lockwood, 2015)

data have been held previously. The need to copy data multiple times tends to make such a computation model inefficient, and I/O becomes the bottleneck when all tasks constantly pull data from the storage cluster (the red arrow). This in turn leads to poor scalability; adding more nodes to the computing cluster would not increase its performance.

To solve this problem, Hadoop implements a *bring compute to the data* strategy that combines both computing and storage at each node of the cluster. In this setup, each node offers both computing power and storage capacity. As shown in Figure 5.1 (bottom), when users submit a task to be run on a data set, the scheduler will first look for nodes that contain the data, and, if the nodes are available, it will schedule the task to run directly on those nodes. If a node is busy with another task, data will still be copied to available nodes, but the scheduler will maintain records of the copy for subsequent use of the data. In addition, data copying can be minimized by increasing the data duplication in the cluster, which also increases the potential for parallelism, because the scheduler has more choices to allocate computing without copying. Because both the compute and data storage are closely coupled for this model, it is best suited for data-intensive applications.

Given that Hadoop has been designed for batch data processing at scale, this model fits the system nicely, especially with the
support of HDFS. However, in an environment where tasks are more
compute-intensive, a traditional high-performance computing environment is probably best, because it would tend to spend more
resources on CPU cores. It should be clear now that the Hadoop
model has hardware implications, and computer architects have
optimized systems for data-intensive computing.

5.3.3   Hardware provisioning

Hadoop requires a distributed cluster of machines to operate efficiently. (It can be set up to run entirely on a single computer,
but this should only be done for technology demonstration purposes.) This is mostly because the MapReduce performance heavily
depends on the total I/O throughput (i.e., disk read and write) of
the entire system. Having a distributed cluster, where each machine
has its own set of hard drives, is one of the most efficient ways to
maximize this throughput.

A typical Hadoop cluster consists of two types of machine: mas-
ters and workers. Master machines are those exclusively reserved
for running services that are critical to the framework operations.
Some examples are the NameNode and the JobTracker services,
which are tasked to manage how data and tasks are distributed
among the machines, respectively. The worker machines are
reserved for data storage and for running actual computation tasks
(i.e., map and reduce). It is normal to have worker machines that
can be included or removed from an operational cluster on demand.
This ability to vary the number of worker nodes makes the overall
system more tolerant of failure. However, master machines are
usually required to be running uninterrupted.

Provisioning and configuring the hardware for Hadoop, like any
other parallel computing, are some of the most important and complex tasks in setting up a cluster, and they often require a lot of
experience and careful consideration. Major big data vendors provide guidelines and tools to facilitate the process (Apache Software
Foundation, nd; Cloudera, nd). Most decisions will be based on the
types of analysis to be run on the cluster, for which only you, as the
user, can provide the best input.

5.3.4 Programming in Hadoop

Now that we are equipped with the knowledge that Hadoop is a MapReduce implementation that runs on HDFS and a bring-compute-to-the-data model, we review the design of a Hadoop MapReduce job. A MapReduce job is still composed of three phases: map, shuffle, and reduce (see Figure 5.2). However, Hadoop divides the map and reduce phases into smaller tasks.

Each map phase in Hadoop is divided into five tasks: input format, record reader, mapper, combiner, and partitioner. An *input format* task is in charge of talking to the input data presumably sitting on HDFS, and splitting it into partitions (e.g., by breaking lines at line breaks). Then, a *record reader* task is responsible for translating the split data into the key–value pair records so that they can be processed by the mapper. By default, Hadoop parses files into key–value pairs of line numbers and line contents. However, both input formats and record readers are fully customizable and can be programmed to read custom data including binary files.

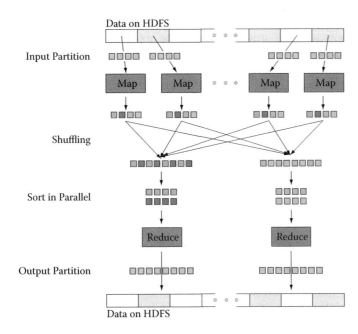

Figure 5.2. Data transfer and communication of a MapReduce job in Hadoop. Data blocks are assigned to several maps, which emit key–value pairs that are shuffled and sorted in parallel. The reduce step emits one or more pairs, with results stored on the HDFS

It is important to note that input formats and record readers only provide data partitioning; they do not move data around computing nodes.

After the records are generated, mappers are spawned—typically on nodes containing the blocks—to run through these records and output zero or more new key–value pairs. A mapper in Hadoop is equivalent to the `map` function of the MapReduce model that we discussed previously. The selection of the key to be output from the mapper will heavily depend on the data processing pipeline and could greatly affect the performance of the framework. Mappers are executed concurrently in Hadoop as long as resources permit.

A combiner task in Hadoop is similar to a function in the MapReduce framework, but it only works locally at each node: it takes output from mappers executed on the same node and produces aggregated values. Combiners are optional but can be used to greatly reduce the amount of data exchange in the shuffle phase; thus, users are encouraged to implement this whenever possible. A common practice is when a `reduce` function is both commutative and associative, and has the same input and output format, one can simply use the `reduce` function as the combiner. Nevertheless, combiners are not guaranteed to be executed by Hadoop, so this should only be treated as a hint. Its execution must not affect the correctness of the program.

A partitioner task is the last process occurring in the map phase on each mapper node, where it hashes the key of each key–value pair output from the mappers or the combiners into bins. By default, the partitioner uses object hash codes and modulus operations to direct a designated reducer to pull data from a map node. Although it is possible to customize the partitioner, it is only advisable to do so when one fully understands the intermediate data distribution as well as the specifications of the cluster. In general, it is better to leave this job to Hadoop.

Each reduce phase in Hadoop is divided into three tasks: reducer, output format, and record writer. The `reducer` task is equivalent to the `reduce` function of the MapReduce model. It basically groups the data produced by the mappers by keys and runs a `reduce` function on each list of grouping values. It outputs zero or more key–value pairs for the output format task, which then translates them into a writable format for the record writer task to serialize on HDFS. By default, Hadoop will separate the key and value with a tab and write separate records on separate lines. However, this behavior is fully customizable. Similarly, the map phase reducers also are executed concurrently in Hadoop.

```
#!/usr/bin/env python
import sys

def parseInput():
    for line in sys.stdin:
        yield line

if __name__=='__main__':
    for line in parseInput():
        fields      = line.strip('\n').split(',')
        awardId     = fields[0]
        domainName = fields[3].split('@')[-1].split('.')[-2:]
        print ('%s\t%s' % (domainName,awardId))
```

Listing 5.2. A Hadoop streaming mapper in Python

5.3.5 Programming language support

Hadoop is written entirely in Java, thus it is best at supporting applications written in Java. However, Hadoop also provides a *streaming API* that allows arbitrary code to be run inside the Hadoop MapReduce framework through the use of UNIX pipes. This means that we can supply a mapper program written in Python or C++ to Hadoop as long as that program reads from the standard input and writes to the standard output. The same mechanism also applies for the combiner and reducer. For example, we can develop from the Python pseudo-code in Listing 5.1 a complete Hadoop streaming mapper (Listing 5.2) and reducer (Listing 5.3).

It should be noted that, in Hadoop streaming, intermediate key-value pairs (the data flowing between mappers and reducers) must be in tab-delimited format, thus we replace the original `yield` command with a `print` formatted with tabs. Although the input format

```
#!/usr/bin/env python
import sys

def parseInput():
    for line in sys.stdin:
        yield line

if __name__=='__main__':
    for line in parseInput():
        (domainName, awardIds) = line.split('\t')
        count = len(set(awardIds))
        print ('%s\t%s' % (domainName, count))
```

Listing 5.3. A Hadoop streaming reducer in Python

and record reader are still customizable in Hadoop streaming, they must be supplied as Java classes. This is one of the most significant limitations of Hadoop for Python developers. They not only have to split their code into separate mapper and reducer programs, but also need to learn Java if they want to work with nontextual data.

5.3.6   Benefits and limitations of Hadoop

- Fault tolerance: By default, HDFS uses checksums to enforce data integrity on its file system and data replication for recovery of potential data losses. Taking advantage of this, Hadoop also maintains fault tolerance of MapReduce jobs by storing data at every step of a MapReduce job to HDFS, including intermediate data from the combiner. Then, the system checks whether a task fails by either looking at its heartbeats (data activities) or whether it has been taking too much time. If a task is deemed to have failed, Hadoop will "kill it" and run it again on a different node. The time limit for the heartbeats and task running duration also may be customized for each job. Although the mechanism is simple, it works well on thousands of machines. It is indeed highly robust because of the simplicity of the model.

- Performance: Hadoop has proven to be a scalable implementation that can run on thousands of cores. However, it is also known for having a relatively high job setup overhead and suboptimal running time. An empty task in Hadoop (i.e., with no mapper or reducer) can take approximately 30 seconds to complete even on a modern cluster. This overhead makes it unsuitable for real-time data or interactive jobs. The problem comes mostly from the fact that Hadoop monitoring processes only live within a job, thus it needs to start and stop these processes each time a job is submitted, which in turns results in this major overhead. Moreover, the brute force approach of maintaining fault tolerance by storing everything on HDFS is expensive, especially for large data sets.

- Hadoop streaming support for non-Java applications: As mentioned previously, non-Java applications may only be integrated with Hadoop through the Hadoop streaming API. However, this API is far from optimal. First, input formats and record readers can only be written in Java, making it impossible to write advanced MapReduce jobs entirely in a different language. Second, Hadoop streaming only communicates with Hadoop through UNIX pipes, and there is no support for data passing within the application using native data structure (e.g., it is necessary to convert Python tuples into strings in the mappers and convert them back into tuples again in reducers).

• Real-time applications: With the current setup, Hadoop only supports batch data processing jobs. This is by design, so it is not exactly a limitation of Hadoop. However, given that increasing numbers of applications are handling real-time massive data sets, the community using MapReduce for real-time processing is constantly growing. Not having support for streaming or real-time data is clearly a disadvantage of Hadoop over other implementations.

• Limited data transformation operations: This is more of a limitation of MapReduce than Hadoop per se. MapReduce only supports two operations, map and reduce; although these operations are sufficient to describe a variety of data processing pipelines, there are classes of applications for which MapReduce is not suitable. Beyond that, developers often find themselves rewriting simple data operations, such as data set joins, finding a min or max, and so on. Sometimes, these tasks require more than one map-and-reduce operation, resulting in multiple MapReduce jobs. This is both cumbersome and inefficient. There are tools to automate this process for Hadoop; however, they are only a layer above, and they are not easy to integrate with existing customized Hadoop applications.

5.4   Other MapReduce Implementations

In addition to Apache Hadoop, other notable MapReduce implementations include MongoDB, GreenplumDB, Disco, Riak, and Spark. MongoDB, Riak, and Greenplum DB are all database systems and thus their MapReduce implementations focus more on the interoperability of MapReduce and the core components, such as MongoDB's aggregation framework, and leave it up to users to customize the MapReduce functionalities for broader tasks. Some of these systems, such as Riak, only parallelize the map phase, and run the reduce phase on the local machine that request the tasks. The main advantage of the three implementations is the ease with which they connect to specific data stores. However, their support for general data processing pipelines is not as extensive as that of Hadoop.

▶ See Chapter 4.

Disco, similar to Hadoop, is designed to support MapReduce in a distributed computing environment, but it is written in Erlang with a Python interface. Thus, for Python developers, Disco might be a better fit. However, it has significantly fewer supporting applications, such as access control and workflow integration, as well as a smaller developing community. This is why the top three big data platforms, Cloudera, Hortonworks, and MapR, still build primarily on Hadoop.

5.5   Apache Spark

Apache Spark is another implementation that aims to support beyond MapReduce. The framework is centered around the concept of resilient distributed data sets and data transformations that can operate on these objects. An innovation in Spark is that the fault tolerance of resilient distributed data sets can be maintained without flushing data onto disks, thus significantly improving the system performance (with a claim of being 100 times faster than Hadoop). Instead, the fault-recovery process is done by replaying a log of data transformations on check-point data. Although this process could require more time than reading data directly from HDFS, it does not occur often and is a fair tradeoff between processing performance and recovery performance.

Beyond map and reduce, Spark also supports various other transformations (Hadoop, nd), including filter, data join, and aggregation. Streaming computation also can be performed in Spark by asking Spark to reserve resources on a cluster to constantly stream data to/from the cluster. However, this streaming method might be resource-intensive (still consuming resources when there is no data coming). Additionally, Spark plays well with the Hadoop ecosystem, particularly with the distributed file system (HDFS) and resource manager (YARN), making it possible to be built on top of current Hadoop applications.

Another advantage of Spark is that it supports Python natively; thus, developers can run Spark in a fraction of the time required for Hadoop. Listing 5.4 provides the full code for the previous example written entirely in Spark. It should be noted that Spark's concept of the `reduceByKey` operator is not the same as Hadoop's, as it is designed to aggregate all elements of a data set into a single element. The closest simulation of Hadoop's MapReduce pattern is a combination of `mapPartitions`, `groupByKey`, and `mapPartitions`, as shown in the next example.

```
import sys
from pyspark import SparkContext
def mapper(lines):
    for line in lines:
        fields      = line.strip('\n').split(',')
        awardId     = fields[0]
        domainName = fields[3].split('@')[-1].split('.')[-2:]
        yield (domainName, awardId)

def reducer(pairs):
    for (domainName, awardIds) in pairs:
        count = len(set(awardIds))
        yield (domainName, count)

if __name__=='__main__':
    hdfsInputPath  = sys.argv[1]
    hdfsOutputFile =  sys.argv[2]
    sc = SparkContext(appName="Counting Awards")
    output = sc.textFile(hdfsInputPath) \
        .mapPartitions(mapper) \
        .groupByKey() \
        .mapPartitions(reducer)

    output.saveAsTextFile(hdfsInputPath)
```

Listing 5.4. Python code for a Spark program that counts the number of awards per institution using MapReduce

Example: Analyzing home mortgage disclosure application data

We use a financial services analysis problem to illustrate the use of Apache Spark.

Mortgage origination data provided by the Consumer Protection Financial Bureau provide insightful details on the financial health of the real estate market. The data, which are a product of the Home Mortgage Disclosure Act (HMDA), highlight key attributes that function as strong indicators of health and lending patterns.

▶ http://www.consumerfin
ance.gov/hmda/learn-more

Lending institutions, as defined by section 1813 in Title 12 of the HMDA, decide on whether to originate or deny mortgage applications based on credit risk. In order to determine this credit risk, lenders must evaluate certain features relative to the applicant, the underlying property, and the location. We want to determine whether census tract clusters could be created based on mortgage application data and whether lending institutions' perception of risk is held constant across the entire US.

For the first step of this process, we study the debt–income ratio for loans originating in different census tracts. This could be achieved simply by computing the debt–income ratio for each loan application and aggregating them for each year by census tract number. A challenge, however, is that the data set provided

Table 5.1. Home Mortgage Disclosure Act data size

Year	Records	File size (gigabytes)
2007	26,605,696	18
2008	17,391,571	12
2009	19,493,492	13
2010	16,348,558	11
2011	14,873,416	9.4
2012	18,691,552	12
2013	17,016,160	11
Total	**130,420,445**	**86.4**

Table 5.2. Home Mortgage Disclosure Act data fields

Index	Attribute	Type
0	Year	Integer
1	State	String
2	County	String
3	Census tract	String
4	Loan amount	Float
5	Applicant income	Float
6	Loan originated	Boolean
. . .	. . .	. . .

by HMDA is quite extensive. In total, HMDA data contain approximately 130 million loan applications between 2007 and 2013. As each record contains 47 attributes, varying in types from continuous variables, such as loan amounts and applicant income, to categorical variables, such as applicant gender, race, loan type, and owner occupancy, the entire data set results in about 86 GB of information. Parsing the data alone could require hours on a single machine if using a naïve approach that scans through the data sequentially. Tables 5.1 and 5.2 highlight the breakdown in size per year and data fields of interest.

Observing the transactional nature of the data, where the aggregation process could be distributed and merged across multiple partitions of the data, we could complete this task in much less time by using Spark. Using a cluster consisting of 1,200 cores, the Spark program in Listing 5.5 requires less than one minute to complete. The substantial performance gain comes less from the large number of processors available and more from the large I/O bandwidth available on the cluster, thanks to the 200 distributed hard disks and fast network interconnects.

5.6   Summary

Analyzing large amounts of data means that it is necessary to both store very large collections of data and perform aggregate computations on those data. This chapter describes an important data storage approach (the Hadoop Distributed File System) and a method for processing large-scale data sets (the MapReduce model, as implemented in both Hadoop and Spark). This model enables not only large-scale data analysis but also provides easy to use implementations for more flexibility for social scientists to work with large amounts of data. These capabilities can both increase analytic throughput and reduce time to insight, speeding up the decision-making process and thus increasing impact.

```
import ast
import sys
from pyspark import SparkContext

def mapper(lines):
    for line in lines:
        fields = ast.literal_eval('(%s)' % line)
        (year, state, county, tract) = fields[:4]
        (amount, income, originated) = fields[4:]

        key = (year, state, county, tract)
        value = (amount, income)

        # Only count originated loans
        if originated:
            yield (key, value)

def sumDebtIncome(debtIncome1, debtIncome2):
    return (debtIncome1[0] + debtIncome2[0], debtIncome1[1] +
    debtIncome2[1])

if __name__=='__main__':
    hdfsInputPath  = sys.argv[1]
    hdfsOutputFile =  sys.argv[2]
    sc = SparkContext(appName="Debt-Income Ratio")
    sumValues = sc.textFile(hdfsInputPath) \
        .mapPartitions(mapper) \
        .reduceByKey(sumDebtIncome)

    # Actually compute the aggregated debt income
    output = sumValues.mapValues(lambda debtIncome: debtIncome[0]/
    debtIncome[1])

    output.saveAsTextFile(hdfsInputPath)
```

Listing 5.5. Python code for a Spark program to aggregate the debt–income ratio for loans originated in different census tracts

5.7 Resources

There is a wealth of online resources describing both Hadoop and Spark. See, for example, the tutorials on the Apache Hadoop and Spark websites. In addition, Lin and Dyer (2010) discuss the use of MapReduce for text analysis.

▶ http://hadoop.apache.org/

▶ https://spark.apache.org/

Part II
Modeling and Analysis

Chapter 6

Information Visualization

M. Adil Yalçın and Catherine Plaisant

This chapter will show you how to use visualization to explore data as well as to communicate results so that data can be turned into interpretable and actionable information. There are many ways of presenting statistical information that convey content in a rigorous manner. The goal of this chapter is to present an introductory overview of effective visualization techniques for a range of data types and tasks, and to explore the foundations and challenges of information visualization at different stages of a project.

6.1 Introduction

One of the most famous discoveries in science—that disease was transmitted through germs, rather than through pollution—is the result of insights derived from a visualization of the location of London cholera deaths near a water pump (Snow, 1855). Information visualization in the 21st century can be used to generate similar insights: detecting financial fraud, understanding the spread of a contagious illness, spotting terrorist activity, or evaluating the economic health of a country. But, the challenge is greater: many (10^2–10^7) items may be manipulated and visualized, often extracted or aggregated from yet larger data sets, or generated by algorithms for analytics.

Visualization tools can organize data in a meaningful way that lowers the cognitive and analytical effort required to make sense of the data and make data-driven decisions. Users can scan, recognize, understand, and recall visually structured representations more rapidly than they can process nonstructured representations. The science of visualization draws on multiple fields, such as perceptual psychology, statistics, and graphic design, to present

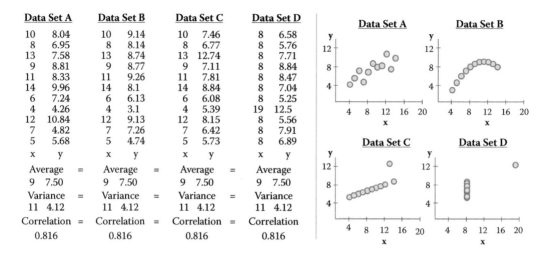

Figure 6.1. Adapted from Anscombe's quartet (Anscombe, 1973)

information, and it utilizes advances in rapid processing and dynamics to design user interfaces that permit powerful interactive visual analysis.

Figure 6.1, "Anscombe's quartet" (Anscombe, 1973), provides a classic example of the value of visualization compared to basic descriptive statistical analysis. The left-side panel includes raw data of four small number-pair data sets (A, B, C, D), which have the same average, median, and standard deviation and have correlation across number pairs. The right-side panel shows these data sets visualized with each point plotted on perpendicular axes (scatterplots), revealing dramatic differences between the data sets, trends, and outliers visually.

In broad terms, visualizations are used either to present results or for analysis and open-ended exploration. This chapter provides an overview of how modern information visualization, or visual data mining, can be used in the context of big data.

6.2   Developing effective visualizations

The effectiveness of a visualization depends on both analysis needs and design goals. Sometimes, questions about the data are known in advance; in other cases, the goal may be to explore new data sets, generate insights, and answer questions that are unknown before starting the analysis. The design, development, and evaluation of a visualization is guided by an understanding of the background and goals of the target audience (see Box 6.1).

▶ See Chapters 2, 3, 4, and 5 for an overview of collecting, merging, storing, and processing data sets.

Box 6.1: Effective visualizations

The development of an effective visualization is a continuous process that generally includes the following activities.

- Specify user needs, tasks, accessibility requirements, and criteria for success.

- Prepare data (clean, transform).

- Design visual representations.

- Design interaction.

- Plan sharing of insights, provenance.

- Prototype/evaluate, including usability testing.

- Deploy (monitor usage, provide user support, manage revision process).

If the goal is to present results, there is a wide spectrum of users and a wide range of options. If the audience is broad, then *infographics* can be developed by graphic designers, as described in classic texts (see Tufte, 2001, 2006; Few, 2009; or the examples compiled by Harrison et al., 2015; Keshif, nd). If, on the other hand, the audience comprises domain experts interested in monitoring the overview status of dynamic processes on a continuous basis, monitoring *dashboards* with no or little interactivity can be used. Examples include the monitoring of sales, or the number of tweets about people, or symptoms of the flu and how they compare to a baseline (Few, 2013). Such dashboards, composed of multiple charts of different operational data, can increase situational awareness so that problems can be noticed and solved early and better decisions can be made with up-to-date information.

Another goal of visualization is to enable *interactive exploratory analysis*. This approach goes beyond a visual snapshot of data for presentation and provides many windows into different parts and relationships within data on demand. Tailor-made solutions can focus on specific querying and navigation tasks given a specific data set. For example, the BabyNameVoyager (http://www.

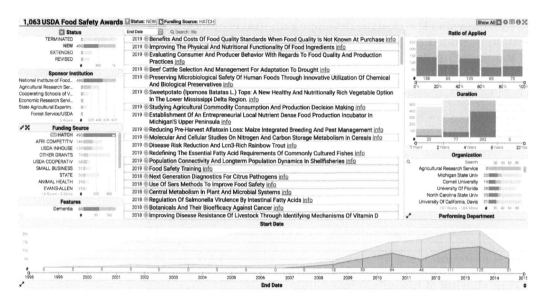

Figure 6.2. Aid Worker Security Incidents Analysis Dashboard (https://gallery.keshif.me/AidWorkerSecurity)

babynamewizard.com/voyager/) lets users type in a name and see a graph of its popularity over the past century. With each letter typed, the page filters baby names starting with the input (such as "Jo" producing results for Joan, Joyce, and John).

In addition, detailed aspects of a dataset can be made explorable by advanced data querying, navigation, and view options. Figure 6.2 shows an interactive dashboard that visualizes the data from the Humanitarian Outcomes' Aid Worker Security Database (https://aidworkersecurity.org/). In this example, the event point locations are clustered on a map, and surrounding charts show trends in attack means, context, location types, region, and country, as well as event date and the number of affected people. This view also presents a breakdown of data by location type, shown using color, including contextual tooltips that provide details on a geographic cluster of points. Additional shortcuts at the top allow navigation to key alternative insights as a storytelling tool.

To create interactive charts and dashboards from new datasets for analysis, products and tools, such as Tableau, PowerBI, Keshif, and others (see Section 6.6), offer a range of chart types with various parameters, as well as visual design environments that allow combining and sharing these charts in potent dashboards. For example, Figure 6.3 shows the charting interface of Tableau on a transaction

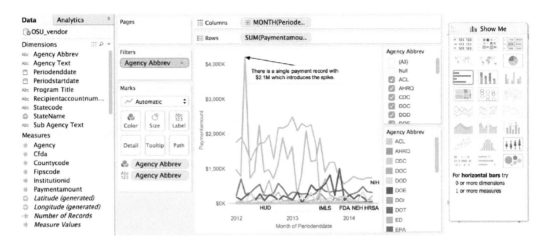

Figure 6.3. Charting interface of Tableau

data set. The left-side panel shows the list of attributes associated with vendor transactions for a given university. The visualization (center) is constructed by placing the month of spending in chart columns, and the sum of payment amount on the chart row, with data encoded using line mark type. Agencies are broken down by color mapping. The agency list, to the right, allows filtering the agencies, which can be used to simplify the chart view. A peak in the line chart is annotated with an explanation of the spike. On the rightmost side, the Show Me panel suggests the applicable chart types potentially appropriate for the selected attributes. This chart can be combined with other charts focusing on other aspects in interactive dashboards. Figure 6.4 shows a treemap (Johnson and Shneiderman, 1991) for agency and sub-agency spending breakdown, combined with a map showing average spending per state. Of the states, Oklahoma is noticable for few but large expenditures. "Mousing-over" Oklahoma reveals details of these expenditures. An additional histogram provides an overview of spending change across three years.

Creating effective visualizations requires careful consideration of many components. Data values may be encoded using one or more visual elements, such as position, length, color, angle, area, and texture (Figure 6.5; see also Cleveland and McGill, 1984; Tufte, 2001). Each of these can be organized in a multitude of ways, discussed in more detail by Munzner (2014). In addition to visual data encoding, units for axes, labels, and legends need to be provided as

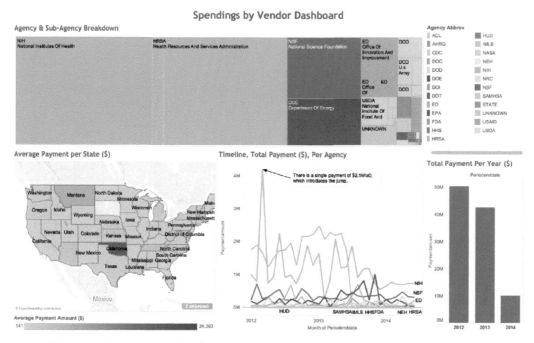

Figure 6.4. A treemap visualization of agency and sub-agency spending breakdown

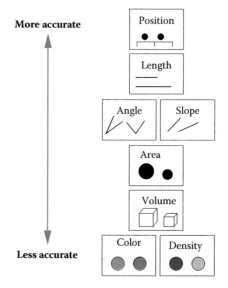

Figure 6.5. Visual elements described by MacKinlay (1986)

well as explanations of the mappings when the design is unconventional. A visually compelling example is the "how to read this data" section of the "A World of Terror" project by Periscopic (https://terror.periscopic.com/).

Annotations or comments can be used to guide viewer attention and describe related insights. Providing attribution and data source, where applicable, is an ethical practice that also enables validating data, and promotes reuse to explore new perspectives.

A short list of guidelines includes: provide immediate feedback upon interaction with the visualization; generate tightly coupled views (i.e., so that selection in one view updates the others); and use a high "data to ink ratio" (Tufte, 2001). Use color carefully and ensure that the visualization is truthful (e.g., watch for perceptual biases or distortion). Avoid use of three-dimensional representations or embellishments, because comparing 3D volumes is perceptually challenging and occlusion is a problem. Labels and legends should be meaningful, novel layouts should be carefully explained, and online visualizations should adapt to different screen sizes. For extended and in-depth discussions, see various textbooks (Tufte, 2001, 2006; Few, 2009; Ward et al., 2010; Kirk, 2012; Munzner, 2014).

We provide a summary of the basic tasks that users typically perform during visual analysis of data in the next section.

6.3   A data-by-tasks taxonomy

We give an overview of visualization approaches for six common data types: multivariate, spatial, temporal, hierarchical, network, and text (Shneiderman and Plaisant, 2015). For each data type listed in this section, we discuss its distinctive properties, the common analytical questions, and examples. Real-life data sets often include multiple data types coming from multiple sources. Even a single data source can include a variety of data types. For example, a single data table of countries (as rows) can have a list of attributes with varying types: the growth rate in the last 10 years (one observation per year, time series data), their current population (single numerical data), the amount of trade with other countries (networked/linked data), and the top 10 exported products (if grouped by industry, hierarchical data).

In addition to common data types, Box 6.2 provides an overview of common tasks for visual data analysis, which can be applied across different data types based on goals and types of visualizations.

Box 6.2: A task categorization for visual data analysis

Select/Query

- Filter to focus on a subset of the data

- Retrieve details of item

- Brush linked selections across multiple charts

- Compare across multiple selections

Navigate

- Scroll along a dimension (1D)

- Pan along two dimensions (2D)

- Zoom along the third dimension (3D)

Derive

- Aggregate item groups and generate characteristics

- Cluster item groups by algorithmic techniques

- Rank items to define ordering

Organize

- Select chart type and data encodings to organize data

- Layout multiple components or panels in the interface

Understand

- Observe distributions

- Compare items and distributions

- Relate items and patterns

Communicate

- Annotate findings

- Share results

- Trace action histories

Interactive visualization design also should consider the devices on which data will be viewed and interacted. Conventionally, visualizations have been designed for mouse and keyboard interaction on desktop computers. However, a wider range of device forms, such as mobile devices with small displays and touch interaction, is becoming common. Creating visualizations for new forms requires special care, although basic design principles such as "less is more" still apply.

6.3.1   Multivariate data

In common tabular data, each record (row) has a list of attributes (columns), whose value is mostly categorical or numerical. The analysis of multivariate data with basic categorical and interval types aims to understand patterns within and across data attributes. Given a larger number of attributes, one of the challenges in data exploration and analytics is to select the attributes and relations on which to focus. Expertise in the data domain can be helpful for targeting relevant attributes.

Multivariate data can be presented in multiple forms of charts depending on the data and relations being explored. One-dimensional (1D) charts present data on a single axis only. An example is a *box plot*, which shows quartile ranges for numerical data. So-called 1.5D charts list the range of possible values on one axis and describe a measurement of data on the other. *Bar charts* are a ubiquitous chart type that can effectively visualize numeric data, e.g., a numeric grade per student or grade average for aggregated student groups by gender. Records also can be grouped over numerical ranges, such as sales price, and bars can show the number of items in each grouping, which generates a *histogram* chart. Two-dimensional (2D) charts plot data along two attributes, such as *scatterplots*. Matrix (grid) charts also can be used to show relations between two attributes. *Heatmaps* visualize each matrix cell using color to represent its value. *Correlation matrices* show the relation between attribute pairs.

To show relations of more than two attributes (3D+), one option is to use additional visual encodings in a single chart, e.g., by adding point size/shape as a data variable in scatterplots. Another option is to use alternative visual designs that can encode multiple relations within a single chart. For example, a *parallel coordinate plot* (Inselberg, 2009) has multiple parallel axes, each one representing an attribute; each record is shown as connected lines passing through the record's values on each attribute. Charts also can show part-of-whole relations using appropriate mappings based on sub-dividing the chart space, such as stacked charts or pie charts.

Finally, another approach to analyzing multidimensional data is to use clustering algorithms to identify similar items. Typically, clusters are represented as a tree structure (see Section 6.3.4). For example, *k*-means clustering starts by users specifying how many clusters to create; the algorithm then places every item into the most appropriate cluster. Surprising relationships and interesting outliers may be identified by these techniques on mechanical analysis algorithms. However, such results may require more effort to interpret.

6.3.2 Spatial data

Spatial data convey a physical context, commonly in a 2D space, such as geographical maps or floor plans. Several of the most prominent examples of information visualization include maps, from the 1861 representation of Napoleon's ill-fated Russian campaign by Minard (popularized by Tufte [2001] and Kraak [2014]) to the inter-active HomeFinder application that introduced the concept of dynamic queries (Ahlberg et al., 1992). The tasks include finding adjacent items, regions containing certain items or with specific characteristics, and paths between items—and performing the basic tasks listed in Box 6.1.

The primary form of visualizing spatial data is *maps*. In *choropleth maps*, color encoding is used to represent each data attribute. *Cartograms* aim to encode the attribute value with the size of regions by distorting the underlying physical space. *Tile grid maps* reduce each spatial area to a uniform size and shape (e.g., a square) so that the color-coded data are easier to observe and compare; these chosen shapes (tiles) can be arranged to approximate and maintain relative physical positions of the regions (DeBelius, 2015; Stanford Visualization Group, nd). Grid maps also make selection of smaller areas (such as small cities or states) easier. *Contour (isopleth) maps* connect areas with similar measurements and color each one separately. *Network maps* aim to show network connectivity between locations, such as flights to/from many regions of the world. In addition, spatial data can be presented with a nonspatial emphasis (e.g., as a hierarchy of continents, countries, and cities, such as by using a treemap chart).

Often, maps are combined with other visualizations. For example, in Figure 6.6, the US Cancer Atlas combines a map showing patterns across states on one attribute with a sortable table providing additional statistical information and a scatterplot that allows users

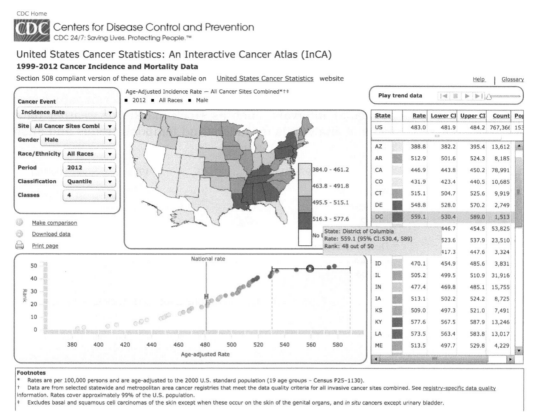

Figure 6.6. The US Cancer Atlas (Centers for Disease Control and Prevention, 2014). Interface based on MacEachren et al. (2008)

to explore correlations between attributes. Figures 6.2 and 6.4 also demonstrate the use of different map designs in the context of larger analytical solutions.

6.3.3 Temporal data

Time is the unique dimension in our physical world that steadily flows forward. While we cannot control time, we frequently record it as a point or an interval. Figures 6.2 and 6.3 exemplify line charts that show trends over multiple years, with each single line representing a subset of the data for cross-comparison of temporal trends. Temporal data also have multiple levels of representation (year, month, day, hour, minute, and so on) with irregularities (leap year, different days per month, etc.). As we measure time based on cyclic events in nature (day/night), our representations are also commonly cyclic. For example, January follows December (first month follows last). This cyclic nature can be captured by circular visual encodings, such as the conventional clock with hour, minute, and second hands.

Time series data (Figures 6.7 and 6.8) describe values measured at regular intervals, such as stock market or weather data. The focus of analysis is to understand temporal trends and anomalies,

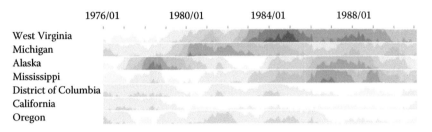

Figure 6.7. Horizon graphs used to display time series

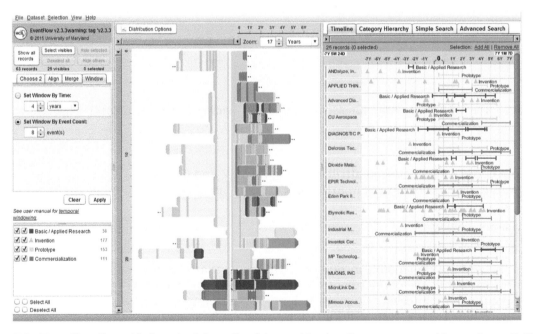

Figure 6.8. EventFlow (https://hcil.umd.edu/eventflow/) is used to visualize sequences of innovation activities by Illinois companies. Created with EventFlow; data sources include NIH, NSF, USPTO, and SBIR. (Image created by C. Scott Dempwolf, used with permission)

querying for specific patterns, or prediction. To show multiple time-series trends across different data categories in a very compact chart area, each trend can be shown with small height using a multi-layered color approach, creating horizon graphs. While perceptually effective after learning to read its encoding, this chart design may not be appropriate for audiences who may lack such training or familiarity.

Another form of temporal analysis is understanding sequences of events. The study of human activity often includes analyzing event sequences. For example, students' records include events such as attending orientation, getting a grade in a class, having an internship, and graduation. In the analysis of event sequences, finding the most common patterns, spotting rare ones, searching for specific sequences, or understanding what leads to particular types of events is important (e.g., what events lead to a student dropping out, precede a medical error, or precipitate a company filing for bankruptcy). Figure 6.8 shows EventFlow used to visualize sequences of innovation activities by Illinois companies. Activity types include research, invention, prototyping, and commercialization. The timeline (right panel) shows the sequence of activities for each company. The overview panel (center) summarizes all of the records aligned by the first prototyping activity of the company. In most of the sequences shown here, the company's first prototype is preceded by two or more patents with a lag of about one year.

6.3.4   Hierarchical data

Often, data are organized in a hierarchical fashion. Each item appears in one grouping (e.g., like a file in a folder), and groups can be grouped to form larger groups (e.g., a folder within a folder), up to the root (e.g., a hard disk). Items, and the relations between items and their grouping, can have their own attributes. For example, the National Science Foundation is organized into directorates and divisions, each with a budget and a number of grant recipients.

Analysis may focus on the structure of the relations, considering questions such as "how deep is the tree," "how many items does this branch have," or "what are the characteristics of one branch compared to another?" In such cases, the most appropriate representation is usually a node-link diagram (Card and Nation, 2002; Plaisant et al., 2002). In Figure 6.9, Spacetree is used to browse a company organizational chart. Because not all of the nodes of the tree fit on the screen, we see an iconic representation of the branches that cannot be displayed, indicating the size of each branch. As the tree

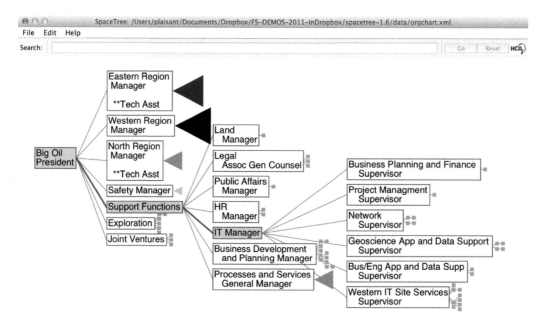

Figure 6.9. SpaceTree (http://www.cs.umd.edu/hcil/spacetree/)

branches are opened or closed, the layout is updated with smooth multiple-step animations to help users remain oriented.

When the structure is less important but the attribute values of the leaf nodes are of primary interest, treemaps, a space-filling approach, are preferable because they can show arbitrarily-sized trees in a fixed rectangular space and map one attribute to the size of each rectangle and another to color. For example, Figure 6.10 shows the Finviz treemap that helps users monitor the stock market. Each stock is shown as a rectangle. The size of the rectangle represents market capitalization, and color indicates whether the stock is going up or down. Treemaps are effective for situational awareness: we can observe that today is a relatively bad day because most stocks are red (i.e., down). Stocks are organized in a hierarchy of industries, allowing users to see that "healthcare technology" is not doing as poorly as most of the other industries. Users also can "zoom" on healthcare to focus on that industry.

6.3.5   Network data

▶ See Chapter 9.

Network data encode relationships between items: for example, social connection patterns (friendships, follows, reposts, etc.), travel

Figure 6.10. The Finviz treemap helps users monitor the stock market (https://www.finviz.com/)

patterns (such as trips between metro stations), and communication patterns (such as emails). The network overviews attempt to reveal the structure of the network, show clusters of related items (e.g., groups of tightly connected people), and allow the path between items to be traced. Analysis also can focus on attributes of the items and the links in between, such as the age of people in communication or the average duration of communications.

Node-link diagrams are the most common representation of network structures and overviews (Figures 6.11 and 6.12) and may use linear (arc), circular, or force-directed layouts for positioning the nodes (items). Matrices or grid layouts are also a valuable way to represent networks (Henry and Fekete, 2006). Hybrid solutions have been proposed, with powerful ordering algorithms to reveal clusters (Hansen et al., 2010). A major challenge in network data exploration is managing larger networks where nodes and edges inevitably overlap by virtue of the underlying network structure, and where aggregation and filtering may be needed before effective overviews can be presented to users.

Figure 6.11 shows the networks of inventors (white) and companies (orange) and their patenting connections (purple lines) in the network visualization NodeXL. Each company and inventor also is connected to a location node (blue = US; yellow = Canada). Green

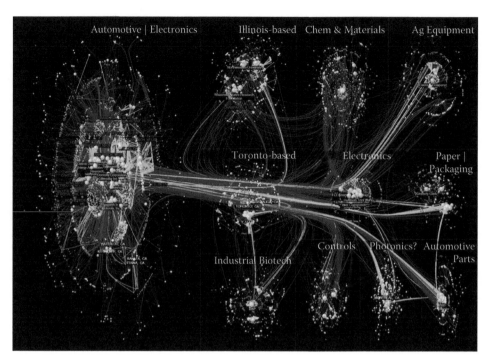

Figure 6.11. NodeXL showing innovation networks of the Great Lakes manufacturing region. Created with NodeXL. Data source: USPTO. (Image created by C. Scott Dempwolf, used with permission)

lines are weak ties based on patenting in the same class and subclass, and they represent potential economic development leads. The largest of the technology clusters are shown using the *group-in-a-box* layout option, which makes the clusters more visible. Note the increasing level of structure moving from the cluster in the lower right to the main cluster in the upper left. NodeXL is designed for interactive network exploration; many controls (not shown in the figure) allow users to zoom on areas of interest or change options. Figure 6.12 shows an example of network visualization on science as a topic used for data presentation in a book and a traveling exhibit. Designed for print media, it includes a clear title and annotations and shows a series of topic clusters at the bottom with a summary of the insights gathered by analysts.

6.3.6   Text data

Usually, text is preprocessed (for word/paragraph counts, sentiment analysis, categorization, etc.) to generate metadata about text

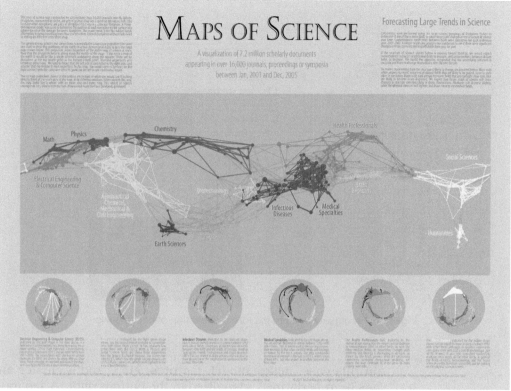

Börner, Katy. *Atlas of Science: Visualizing What We Know.* (2010). The MIT Press. Pg 171.

Figure 6.12. An example from "Maps of Science: Forecasting Large Trends in Science," 2007, The Regents of the University of California, all rights reserved (Börner, 2010)

segments, which then are visualized. Simple visualizations, such as tag clouds, display statistics about word usage in a text collection or can be used to compare two collections or text segments. While visually appealing, they can be easily misinterpreted and often are replaced by word indexes sorted by some count of interest. Specialized visual text analysis tools combine multiple visualizations of data extracted from the text collections, such as matrices to see relations, network diagrams, or parallel coordinates to see entity relationships (e.g., between what, who, where, and when). Timelines can be mapped to the linear dimension of text. Figure 6.13 shows an example using Jigsaw (Stasko et al., 2008) for the exploration of car reviews. Entities have been extracted automatically

▶ See Chapter 8 for text analysis approaches.

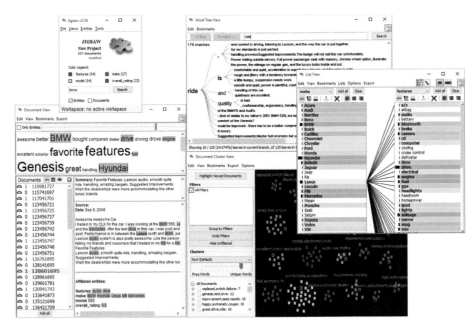

Figure 6.13. Jigsaw used to explore a collection of car reviews

(in this case, make, model, features, etc.), and a cluster analysis has been performed, visualized in the bottom right. A separate view (rightmost) allows analysts to review links between entities. Another view allows traversing word sequences as a tree. Reading original documents is critical, so all of the visualization elements are linked to the corresponding text.

6.4   Challenges

While information visualization is a powerful tool, there are many obstacles to its effective use. We note here four areas of particular concern: scalability, evaluation, visual impairment, and visual literacy.

6.4.1   Scalability

Most visualizations handle relatively small data sets (between one thousand and one hundred thousand, sometimes up to millions, depending on the technique), but scaling visualizations from millions to billions of records does require careful coordination of analytic algorithms to filter data or perform rapid aggregation, effective visual summary designs, and rapid refreshing of displays (Shneiderman, 2008). The visual information-seeking mantra, "Overview first, zoom and filter, then details on demand," remains useful with data at scale. To accommodate one billion records, aggregate markers (which may represent thousands of records) and density plots are useful (Dunne and Shneiderman, 2013). In some cases, the large volume of data can be aggregated meaningfully into a small number of pixels. One example is Google Maps and its visualization of road conditions. A quick glance at the map allows drivers to use a highly aggregated summary of the speed of a large number of vehicles and only a few red pixels are enough to decide when to begin driving on the road.

Although millions of graphic elements may be represented on large screens (Fekete and Plaisant, 2002), perception issues need to be taken into consideration (Yost et al., 2007). Extraction and filtering may be necessary before even attempting to visualize individual records (Wongsuphasawat and Lin, 2014). Preserving interactive rates in querying big data sources is a challenge, with a variety of methods proposed, such as approximations (Fisher et al., 2012) and compact caching of aggregated query results (Lins et al., 2013). Progressive loading and processing will help users review the results as they appear and steer the lengthy data processing (Glueck et al., 2014; Fekete, 2015). Systems are starting to emerge, and strategies to cope with volume and variety of patterns are being described (Shneiderman and Plaisant, 2015).

6.4.2 Evaluation

Human-centric evaluation of visualization techniques can generate qualitative and quantitative assessments of their potential quality, with early studies focusing on the effectiveness of basic visual variables (MacKinlay, 1986). To this day, user studies remain the workhorse of evaluation. In laboratory settings, experiments can demonstrate faster task completion, reduced error rates, or increased user satisfaction. These studies are helpful for comparing visual and interaction designs. For example, studies are reporting on the effects of latency on interaction and understanding (Liu and Heer, 2014), and often they reveal that different visualizations perform better for different tasks (Plaisant et al., 2002; Saket et al., 2014). Evaluations also may aim to measure and study the amount and value of the insights revealed by the use of exploratory visualization tools (Saraiya et al., 2005). Diagnostic usability evaluation remains a cornerstone of user-centered design. Usability studies can be conducted at various stages of the development process to verify that users are able to complete benchmark tasks with adequate speed and accuracy. Comparisons with the technology previously used by target users also may be possible to verify improvements. Metrics need to address the learnability and utility of the system, in addition to performance and user satisfaction (Lam et al., 2012). Usage data logging, user interviews, and surveys also can help identification of potential improvements in visualization and interaction design.

6.4.3   Visual impairment

Color impairment is a common condition that needs to be taken into consideration (Olson and Brewer, 1997). For example, red and green are appealing for their intuitive mapping to positive or negative outcomes (also depending on cultural associations); however, users with red–green color blindness, one of the most common forms, would not be able to differentiate such scales clearly. To assess and assist visual design under different color deficiencies, color simulation tools can be used (see additional resources). The impact of color impairment can be mitigated by careful selection of limited color schemes, using double encoding when appropriate (i.e., using symbols that vary by both shape and color), and allowing users to change or customize color palettes. To accommodate users with low vision, adjustable size and zoom settings can be useful. Users with severe visual impairments may require alternative accessibility-first interface and interaction designs.

6.4.4   Visual literacy

Although the number of people using visualization continues to grow, not everyone is able to accurately interpret graphs and charts. When designing a visualization for a population of users who are expected to make sense of the data without training, it is important to adequately estimate the level of visual literacy of those users. Even simple scatterplots can be overwhelming for some users. Recent work has proposed new methods for assessing visual literacy (Boy et al., 2014), but user testing with representative users in the early stages of design and development will remain necessary to verify that adequate designs are being used. Training is likely to be needed to help analysts when they begin using visual analytics tools. Recorded video demonstrations and online support for question answering are helpful to bring users from novice to expert levels.

6.5 Summary

The use of information visualization is spreading widely, with a growing number of commercial products and additions to statistical packages now available. Careful user testing should be conducted to verify that visual data presentations go beyond the desire for "eye candy" in visualization and to implement designs that have demonstrated benefits for realistic tasks. Visualization is becoming increasingly used by the general public and attention should be given to the goal of universal usability so the widest range of users can access and benefit from new approaches to data presentation and interactive analysis.

6.6 Resources

We have referred to various textbooks throughout this chapter. Tufte's books remain the classics, as inspiring to read as they are instructive (Tufte, 2001, 2006). We also recommend Few's books on information visualization (Few, 2009) and information dashboard design (Few, 2013). See also the book's website for further readings.

Given the wide variety of goals, tasks, and use cases of visualization, many different data visualization tools have been developed that address different needs and appeal to different skill levels. In this chapter, we can only point to a few examples to get started. To generate a wide range of visualizations and dashboards, and to quickly share them online, Tableau and Tableau Public provide a flexible visualization design platform. If a custom design is required and programmers are available, d3 is the de facto low-level library of choice for many web-based visualizations, with its native integration to web standards and flexible methods to convert and manipulate data into visual objects as a JavaScript library. There exist other JavaScript web libraries that offer chart templates (such as Highcharts), or web services that can be used to create a range of charts from given (small) data sets, such as Raw or DataWrapper. To clean, transform, merge, and restructure data sources so that they can be visualized appropriately, tools, such as Trifacta and Alteryx, can be used to create pipelines for data wrangling. For statistical analysis and batch-processing data, programming environments, such as R, or libraries for languages, such as Python (e.g., the Python Plotly library), can be used.

An extended list of visualization tools and books are available at https://gallery.keshif.me/VisTools and https://gallery.keshif.me/VisBooks.

The *Dataset Exploration and Visualization* workbook of Chapter 13 uses `matplotlib` and `seaborn` for creating basic visualizations with Python.

▶ See https://workbooks.coleridgeinitiative.org.

Chapter 7

Machine Learning

Rayid Ghani and Malte Schierholz

This chapter introduces you to the use of machine learning in tackling social science and public policy problems. We cover the end-to-end machine learning process and focus on clustering and classification methods. After reading this chapter, you should have an overview of the components of a machine learning pipeline and methods, and know how to use those in solving social science problems. We have written this chapter to give an intuitive explanation for the methods and to provide a framework and practical tips on how to use them in practice.

7.1 Introduction

You have probably heard of "machine learning" but are not sure exactly what it is, how it differs from traditional statistics, and what you, as social scientists, can do with it. In this chapter, we will demystify machine learning, draw connections to what you already know from statistics and data analysis, and go deeper into some of the novel concepts and methods that have been developed in this field. Although the field originates from computer science, it has been influenced quite significantly by math and statistics in the past 15–20 years. As you will see, many of the concepts you will learn are not entirely new but are simply called something else. For example, you already are familiar with logistic regression (a classification method that falls under the supervised learning framework in machine learning) and cluster analysis (a form of unsupervised learning). You also will learn about new methods that are more exclusively used in machine learning, such as random forests, support vector machines, and neural networks. We will keep formalisms to a minimum and focus on conveying the intuitive information, as well as providing practical tips. Our hope is this chapter will make you comfortable and familiar with machine learning vocabulary, concepts, and processes, and that it allows you to further explore and use these methods and tools in your own research and practice.

7.2   What is machine learning?

When humans improve their skills with experience, they are said to learn. Is it also possible to program computers to do the same? Arthur Samuel, who coined the term *machine learning* in 1959 (Samuel, 1959), is a pioneer in this area, programming a computer to play checkers. Playing against itself and human opponents, Samuel's computer improves its performance with every game. Eventually, after sufficient *training* (and experience), the computer has been shown to be a better player than the human programmer. Today, machine learning has grown significantly beyond learning to play checkers. Machine learning systems have become involved with many tasks, including: learning to drive (and park) autonomous cars; being embedded inside robots; recommending books, products, and movies that are (sometimes) of interest to us; identifying drugs, proteins, and genes that should be investigated further to cure diseases, detect cancer, and other pathologies in x-rays and other types of medical imaging; helping us understand how the human brain learns language; helping identify which voters are persuadable in elections; detecting which students are likely to need extra support to graduate high school on time; and helping solve many more problems. Over the past 20 years, machine learning has become an interdisciplinary field spanning computer science, artificial intelligence, databases, and statistics. At its core, machine learning seeks to design computer systems that improve over time with more experience. In one of the earlier books on machine learning, Tom Mitchell gives a more operational definition, stating that: "A computer program is said to learn from experience E with respect to some class of tasks T and performance measure P, if its performance at tasks in T, as measured by P, improves with experience E" (Mitchell, 1997). We like this definition because it is task-focused and allows us to think of machine learning as a tool used inside a larger system to improve outcomes of importance to us.

Box 7.1: Commercial machine learning examples

- Speech recognition: Speech recognition software uses machine learning algorithms that are built on large amounts of initial training data. Machine learning allows these systems to be tuned and adapt to individual variations in speaking as well as across different domains.

- Autonomous cars: The ongoing development of self-driving cars applies techniques from machine learning. An onboard computer continuously analyzes the incoming video and sensor streams in order to monitor the surroundings. Incoming data are matched with annotated images to recognize objects, such as pedestrians, traffic lights, and potholes. In order to assess the different objects, huge training data sets are required where similar objects already have been identified. This allows the autonomous car to decide which actions to take next.

- Fraud detection: Many public and private organizations face the problem of fraud and abuse. Machine learning systems are widely used to take historical cases of fraud and flag fraudulent transactions as they are occurring. These systems have the benefit of being adaptive and improving with more data over time.

- Personalized ads: Many online stores have personalized recommendations promoting possible products of interest. Based on individual shopping history and what other similar users bought in the past, the website predicts products a user may like and tailors recommendations. Netflix and Amazon are two examples of companies whose recommendation software predicts how a customer would rate a certain movie or product and then suggests items with the highest predicted ratings. Of course, there are some caveats here, because they then adjust the recommendations to maximize profits.

- Face recognition: Surveillance systems, social networking platforms, and imaging software all use face detection and face recognition to first detect faces in images (or video) and then tag them with individuals for various tasks. These systems are trained by giving examples of faces to a machine learning system which then learns to detect new faces and tag known individuals. The bias and fairness chapter will highlight some concerns with these types of systems.

Box 7.2: Social science machine learning examples

Potash et al. (2015) worked with the Chicago Department of Public Health and used random forests (a machine learning classification method) to predict which children were at risk of lead poisoning. Then, this early warning system was used to prioritize lead hazard inspections to detect and remediate lead hazards before they had an adverse effect on the child.

Carton et al. (2016) used a collection of machine learning methods to identify police officers at risk of adverse behavior, such as unjustified use of force or unjustified shootings or sustained complaints, to prioritize preventive interventions such as training and counseling.

Athley and Wagner (2019) used a modification of random forests to estimate heterogeneous treatment effects using a data set from The National Study of Learning Mindsets to evaluate the impact of interventions to improve student achievement.

Voigt et al. (2017) used machine learning methods to analyze footage from body-worn cameras and understand the respectfulness of police officer language toward white and black community members during routine traffic stops.

▶ See Chapter 3.

▶ If you have examples from your own research using the methods we describe in this chapter, please submit a link to the paper (and/or code) here: https://textbook.coleridgeini tiative.org/submitexamples.

Machine learning has grown from the need for systems that were adaptive, scalable, and cost-effective to build and maintain. Now, many tasks are being done using machine learning instead of previous rule-based systems, in which experts would spend considerable time and effort developing and maintaining the rules. The problem with those systems is that they were rigid, not adaptive, difficult to scale, and expensive to maintain. In contrast, machine learning systems have become popular because they can improve the system along all of those dimensions. Box 7.1 mentions several examples where machine learning is being used in commercial applications today. Social scientists are uniquely placed today to take advantage of the same advances in machine learning by having better methods to solve several key problems they are addressing. Box 7.2 describes a few social science and policy problems that are being addressed using machine learning today.

This chapter is not an exhaustive introduction to machine learning. There are many books that have done an excellent job of that (Mitchell, 1997; Hastie et al., 2001; Flach, 2012). Instead, we present a short and accessible introduction to machine learning for social scientists, give an overview of the overall machine learning process, provide an intuitive introduction to machine learning methods, give some practical tips that will be helpful in using these methods, and leave a lot of the statistical theory to machine learning textbooks. As you read more about machine learning in the research literature or the media, you will encounter names of other fields that are related (and practically the same for most social science audiences), such as statistical learning, data mining, and pattern recognition.

7.3 Types of analysis

Many data analysis tasks performed by social scientists can be classified into four types.

- Description: The goal is to describe patterns or groupings in historical data. You already are familiar with descriptive statistics and exploratory data analysis methods, and we will cover more advanced versions of those as we discuss unsupervised learning in this chapter (see Section 7.6.1).

- Detection: The goal here is not necessarily to understand historical behavior but to detect new or emerging anomalies, events, or patterns as they happen. A typical example is early outbreak detection for infectious diseases in order to inform public health officials.

- Prediction: The goal here is to use the same historical data as the description and detection methods, but use it to predict events and behaviors in the future.

- Behavior change (or causal inference): The goal here is to understand the causal relationships in the data in order to influence the outcomes of importance to us.

In this chapter, we will focus primarily on the description and prediction methods, but much work is being performed in developing and using machine learning methods for detection as well as behavior change and causal inference.

7.4 The machine learning process

When solving problems using machine learning methods, it is important to think of the larger data-driven problem-solving process of

▶ See Chapter 3.

▶ http://www.datascience
publicpolicy.org/home/
resources/data-science-
project-scoping-guide/

which these methods are a small part. A typical machine learning problem requires researchers and practitioners to take the following steps.

1. Understand the problem and goal: This sounds obvious but is often nontrivial. Problems typically start as vague descriptions of a goal—improving health outcomes, increasing graduation rates, understanding the effect of a variable X on an outcome Y, etc. It is very important to work with people who understand the domain being studied to discuss and define the problem more specifically. What is the analytical formulation of the metric that you are trying to improve? The Data Science Project Scoping Guide is a good place to start when doing problem scoping for social science or policy problems.

2. Formulate it as a machine learning problem: Is it a classification problem or a regression problem? Is the goal to build a model that generates a ranked list prioritized by risk of an outcome, or is it to detect anomalies as new data come in? Knowing what kinds of tasks machine learning can solve will allow you to map your problem to one or more machine learning settings and give you access to a suite of methods appropriate for that task.

3. Data exploration and preparation: Next, you need to carefully explore the data you have. What additional data do you need or have access to? What variable will you use to match records for integrating different data sources? What variables exist in the data set? Are they continuous or categorical? What about missing values? Can you use the variables in their original form or do you need to alter them in some way?

4. Feature engineering: In machine learning language, what you might know as independent variables or predictors or factors or covariates are called "features." Creating good features is probably the most important step in the machine learning process. This involves doing transformations, creating interaction terms, or aggregating over data points or over time and space.

5. Modeling: Having formulated the problem and created your features, you now have a suite of methods from which to choose. It would be great if there were a single method that always worked best for a specific type of problem, but that would make things too easy. Each method makes a different

assumption about the structure and distribution of the data; and with large amounts of high-dimensional data,* it is difficult to know a priori which assumption will best match the data we have. Typically, in machine learning, you take a collection of methods and try them out to empirically validate which one works the best for your problem. This process not only helps you select the best method for your problem but also helps you understand the structure of your data. We will give an overview of leading methods that are being used today in this chapter.

★ Dimensionality of the data often refers to how many variables we have in the data.

6. Model interpretation: Once we have built the machine learning models, we also want to understand what they are, which predictors have been found to be important and to what extent, what types of entities have been flagged as high risk (and why), where have errors been made, etc. All of these fall under the model interpretation, interpretability, and explainability umbrella which is currently an active area of research in machine learning.

7. Model selection: As you build a large number of possible models, you need a way to select the model that is the "best." This part of the chapter will cover methodology to first test the models on historical data as well as discuss a variety of evaluation metrics. While this chapter will focus primarily on traditionally used metrics, Chapter 11 will expand on this using metrics related to bias and fairness. It is important to note that sometimes the machine learning literature will call this step the "validation" step using historical data, but we want to distinguish it here from validation, which is the next step.

8. Model validation: The next step, after model selection (using historical data), is validation. You should validate on new data as well as designing and running field trials or experiments.

9. Deployment and monitoring: Once you have selected the best model and validated it using historical data as well as a field trial, you are ready to put the model into practice. You still have to keep in mind that new data will be coming in, the world will be changing, and the model might also (need to) change over time. We will not cover too much of those aspects in this chapter, but they are important to keep in mind when putting the machine learning work into practice.

Although each step in this process is critical, a thorough description of each is outside of our scope here. This chapter will focus on models, terms, and techniques that form the core of machine learning.

7.5   Problem formulation: Mapping a problem to machine learning methods

When working on a new problem, one of the first things we need to do is map it to a class of machine learning methods. In general, the problems we will tackle, including the examples above, can be grouped into two major categories.

1. Supervised learning: These are problems where there exists a target variable (continuous or discrete) that we want to predict or use to classify data. Classification, prediction, and regression all fall into this category. More formally, supervised learning methods predict a value Y given input(s) X by learning (or estimating or fitting or training) a function F, where $F(X) = Y$. Here, X is the set of variables (known as *features* in machine learning, or in other fields as *predictors*) provided as input and Y is the target/dependent variable or a *label* (as it is known in machine learning).

   The goal of supervised learning methods is to search for that function F that best estimates or predicts Y. When the output Y is categorical, this is known as *classification*. When Y is a continuous value, this is called *regression*. Sound familiar?

   One key distinction in machine learning is that the goal is not only to find the best function F that can estimate or predict Y for observed outcomes (known Ys) but to find one that best generalizes to new unseen data, often in the future. This distinction makes methods more focused on generalization and less on simply fitting the data we have as best as we can. It is important to note that you do that implicitly when performing regression by not adding increasing numbers of higher-order terms to get better fit statistics. By getting better fit statistics, we *overfit* to the data and the performance on new (unseen) data often goes down. Methods such as the lasso (Tibshirani, 1996) penalize the model for having too many terms by performing what is known as *regularization.*[*]

★ In statistical terms, regularization is an attempt to avoid overfitting the model.

2. Unsupervised learning: These are problems where there does not exist a target variable that we want to predict but we want to understand "natural" groupings or patterns in the data. Clustering is the most common example of this type of analysis where you are given X and want to group similar Xs together. You may have heard of "segmentation," which is used in the marketing world to group similar customers together using clustering techniques. Principal components analysis (PCA) and related methods also fall into the unsupervised learning category.

In between the two extremes of supervised and unsupervised learning, there is a spectrum of methods that have different levels of supervision involved (Figure 7.1). Supervision in this case is the presence of target variables (known in machine learning as *labels*). In unsupervised learning, none of the data points have labels. In supervised learning, all data points have labels. In between, either the percentage of examples with labels can vary or the types of labels can vary. Although this chapter will have a limited discussion of the weakly supervised and semi-supervised methods, it is an active area of research in machine learning; Zhu (2008) provides more details.

7.5.1 Features

Before we address models and methods, we need to turn our raw data into "features." In social science, they are not called features but instead are known as variables or predictors (or covariates if you are doing regression).[*] Feature generation (or engineering, as it is often called) is where a large chunk of the time is spent in

> ★ Good features are what makes machine learning systems effective.

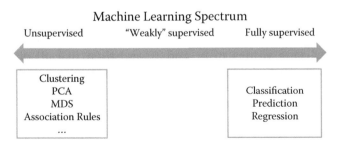

Figure 7.1. Spectrum of machine learning methods from unsupervised to supervised learning

the machine learning process. This is also the phase where previous research and learnings from the domain being tackled can be incorporated into the machine learning process. As social science researchers or practitioners, you have spent a lot of time constructing features, using transformations, dummy variables, and interaction terms. All of that is still required and critical in the machine learning framework. One difference you will need to become comfortable with is that, instead of carefully selecting a few predictors, machine learning systems tend to encourage the creation of *lots* of features and then empirically use holdout data to perform regularization and model selection. It is common to have models that are trained on thousands of features. Of course, it is important to keep in mind that increasing the number of features requires you to have enough data so that you are not overfitting. Commonly used approaches to create features include the following.

- Transformations, such as log, square, and square root.

- Dummy (binary) variables: This is often done by taking categorical variables (such as city) and creating a binary variable for each value (one variable for each city in the data). These are also called indicator variables.

- Discretization: Several methods require features to be discrete instead of continuous. Several approaches exist to convert continuous variables into discrete ones, the most common of which is equal-width binning.

- Aggregation: Aggregate features often constitute the majority of features for a given problem. These aggregations use different aggregation functions (count, min, max, average, standard deviation, etc.), often over varying windows of time and space. For example, given urban data, we would want to calculate the number (and min, max, mean, variance) of crimes within an m-mile radius of an address in the past t months for varying values of m and t, and then use all of them as features in a classification problem. Spatiotemporal aggregation features are going to be extremely important as you build machine learning models.

In general, it is a good idea to have the complexity in features and use a simple model, rather than using more complex models with simple features. Keeping the model simple makes it faster to train and easier to understand and explain.

7.6   Methods

We will start by describing unsupervised learning methods and then discuss supervised learning methods. We focus here on the intuition behind the methods and the algorithm, as well as some practical tips, rather than on the statistical theory that underlies the

---

Box 7.3: Machine learning vocabulary

- Learning: In machine learning, you will notice the term *learning* that will be used in the context of "learning" a model. This is what you probably know as *fitting* or *estimating* a function, or *training* or *building* a model. These terms are all synonyms and are used interchangeably in the machine learning literature.

- Examples: These are data points, rows, observations, or instances.

- Features: These are independent variables, attributes, predictor variables, and explanatory variables.

- Labels: These include the response variable, dependent variable, target variable, or outcomes.

- Underfitting: This occurs when a model is too simple and does not capture the structure of the data well enough.

- Overfitting: This occurs when a model is possibly too complex and models the noise in the data, which can result in poor generalization performance. Using in-sample measures to perform model selection can produce this result.

- Regularization: This is a general method to avoid overfitting by applying additional constraints to the model that is learned. For example, in building logistic regression models, a common approach is to make sure the model weights (coefficients) are, on average, small in magnitude. Two common regularizations are L_1 regularization (used by the lasso), which has a penalty term that encourages the sum of the absolute values of the parameters to be small, and L_2 regularization, which encourages the sum of the squares of the parameters to be small.

methods. We encourage readers to refer to machine learning books listed in Section Resources. Box 7.3 provides brief definitions of several terms we will use in this section.

7.6.1   Unsupervised learning methods

As mentioned previously, unsupervised learning methods are used when we do not have a target variable to estimate or predict but want to understand clusters, groups, or patterns in the data. Often, these methods are used for data exploration, as in the following examples.

1. When faced with a large corpus of text data—for example, email records, congressional bills, speeches, or open-ended free-text survey responses—unsupervised learning methods often are used to understand and comprehend the patterns in our data.

2. Given a data set about students and their behavior over time (academic performance, grades, test scores, attendance, etc.), one might want to understand typical behaviors as well as trajectories of these behaviors over time. Unsupervised learning methods (clustering) can be applied to these data to obtain student "segments" with similar behavior.

3. Given a data set about publications or patents in different fields, we can use unsupervised learning methods (association rules) to determine which disciplines have the most collaboration and which fields have researchers who tend to publish across different fields.

4. Given a set of people who are at high risk of recidivism, clustering can be used to understand different groups of people within the high risk set, to determine intervention programs that may need to be created.

7.6.1.1   Clustering

Clustering is the most common unsupervised learning technique and is used to group data points together that are similar to each other. The goal of clustering methods is to produce results with high intra-cluster (within) similarity and low inter-cluster (between) similarity.

Clustering algorithms typically require a distance (or similarity) metric* to generate clusters. They take a data set and a distance metric (and sometimes additional parameters), and they generate clusters based on that distance metric. The most common

★ Distance metrics are mathematical formulas to calculate the distance between two objects. For example, *Manhattan distance* is the distance a car would drive from one place to another place in a grid-based street system, whereas *Euclidean distance* (in two-dimensional space) is the "straight-line" distance between two points.

distance metric used is Euclidean distance, but other commonly used metrics are Manhattan, Minkowski, Chebyshev, cosine, Hamming, Pearson, and Mahalanobis. Often, domain-specific similarity metrics can be designed for use in specific problems. For example, when performing the record linkage tasks discussed in Chapter 3, you can design a similarity metric that compares two first names and assigns them a high similarity (low distance) if they both map to the same canonical name, so that, for example, Sammy and Sam map to Samuel.

Most clustering algorithms also require the user to specify the number of clusters (or some other parameter that indirectly determines the number of clusters) in advance as a parameter. This is often difficult to do a priori and typically makes clustering an iterative and interactive task. Another aspect of clustering that makes it interactive is often the difficulty in automatically evaluating the quality of the clusters. While various analytical clustering metrics have been developed, the best clustering is task-dependent and thus must be evaluated by the user. There may be different clusterings that can be generated with the same data. You can imagine clustering similar news stories based on the topic content, writing style, or sentiment. The correct set of clusters depends on the user and the task they have. Therefore, typically clustering is used for exploring the data, generating clusters, exploring the clusters, and then rerunning the clustering method with different parameters or modifying the clusters (by splitting or merging the previous set of clusters). Interpreting a cluster can be nontrivial: you can look at the centroid of a cluster, look at frequency distributions of different features (and compare them to the prior distribution of each feature), or you can build a decision tree (a supervised learning method we will cover later in this chapter) where the target variable is the cluster ID that can describe the cluster using the features in your data. A good example of a tool that allows interactive clustering from text data is Ontogen (Fortuna et al., 2007).

7.6.1.2 The k-means clustering

The most commonly used clustering algorithm is called k-means, where k defines the number of clusters. The algorithm works as follows.

1. Select k (the number of clusters you want to generate).

2. Initialize by selecting k points as centroids of the k clusters. This typically is done by selecting k points uniformly at random.

3. Assign each point a cluster according to the nearest centroid.

4. Recalculate cluster centroids based on the assignment in (3) as the mean of all data points belonging to that cluster.

5. Repeat (3) and (4) until convergence.

The algorithm stops when the assignments do not change from one iteration to the next (Figure 7.2). The final set of clusters, however, depends on the starting points. If they are initialized differently, it is possible that different clusters are obtained. One common practical trick is to run k-means several times, each with different (random) starting points. The k-means algorithm is fast, simple, and easy to use, and is often a good first clustering algorithm to try and see if it fits your needs. When the data are of

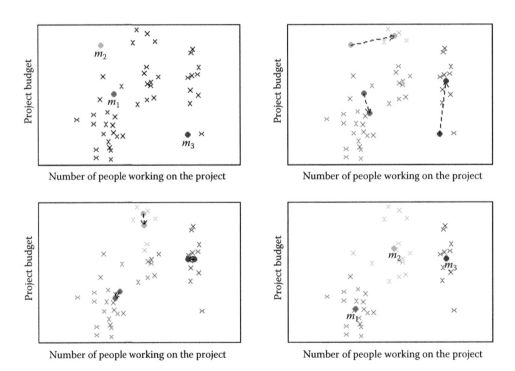

Figure 7.2. Example of k-means clustering with $k = 3$. The upper left panel shows the distribution of the data and the three starting points m_1, m_2, m_3 placed at random. On the upper right, we see what happens in the first iteration. The cluster means move to more central positions in their respective clusters. The lower left panel shows the second iteration. After six iterations, the cluster means have converged to their final destinations and the result is shown in the lower right panel

the form where the mean of the data points cannot be computed, a related method called K-medoids can be used (Park and Jun, 2009).

7.6.1.3 Expectation-maximization clustering

You may be familiar with the expectation-maximization (EM) algorithm in the context of imputing missing data. The EM is a general approach to maximum likelihood in the presence of incomplete data. However, it also is used as a clustering method where the missing data are the clusters to which a data point belongs. Unlike k-means, where each data point is assigned to only one cluster, EM does a soft assignment where each data point gets a probabilistic assignment to various clusters. The EM algorithm iterates until the estimates converge to some (locally) optimal solution.

The EM algorithm is relatively good at handling outliers as well as high-dimensional data, compared to k-means. It also has a few limitations. First, it does not work well with a large number of clusters or when a cluster contains few examples. Also, when the value of k is larger than the number of actual clusters in the data, EM may not yield reasonable results.

7.6.1.4 Mean shift clustering

Mean shift clustering works by finding dense regions in the data by defining a window around each data point and computing the mean of the data points in the window. Then, it shifts the center of the window to the mean and repeats the algorithm until it converges. After each iteration, we can consider that the window shifts to a denser region of the data set. The algorithm proceeds as follows.

1. Fix a window around each data point (based on the bandwidth parameter that defines the size of the window).

2. Compute the mean of data within the window.

3. Shift the window to the mean and repeat until convergence.

Mean shift needs a bandwidth parameter h to be tuned, which influences the convergence rate and the number of clusters. A large h might result in merging distinct clusters. A small h might result in too many clusters. Mean shift might not work well in higher dimensions because the number of local maxima is relatively high and it might converge to a local optimum quickly.

One of the most important differences between mean shift and k-means is that k-means makes two broad assumptions: the number of clusters is already known and the clusters are shaped spherically (or elliptically). Mean shift does not assume anything about the number of clusters (but the value of h indirectly determines that). Also, it can handle arbitrarily shaped clusters.

The k-means algorithm is also sensitive to initializations, whereas mean shift is relatively robust to initializations. Typically, mean shift is run for each point, or sometimes points are selected uniformly randomly. Similarly, k-means is sensitive to outliers, while mean shift is less sensitive. On the other hand, the benefits of mean shift come at a cost—speed. The k-means procedure is fast, whereas classic mean shift is computationally slow but can be easily parallelized.

7.6.1.5 Hierarchical clustering

The clustering methods that we have discussed to this point, often termed *partitioning* methods, produce a flat set of clusters with no hierarchy. Sometimes, we want to generate a hierarchy of clusters, and methods that can do that are of two types.

1. Agglomerative (bottom-up): Start with each point as its own cluster and iteratively merge the closest clusters. The iterations stop either when the clusters are too far apart to be merged (based on a predefined distance criterion) or when there is a sufficient number of clusters (based on a predefined threshold).

2. Divisive (top-down): Start with one cluster and create splits recursively.

Typically, agglomerative clustering is used more often than divisive clustering. One reason is that it is significantly faster, although both of them are typically slower than direct partition methods such as k-means and EM. Another disadvantage of these methods is that they are *greedy*, that is, a data point that is incorrectly assigned to the "wrong" cluster in an earlier split or merge cannot be reassigned again later.

7.6.1.6 Spectral clustering

Figure 7.3 shows the clusters that k-means would generate on the data set in the figure. It is obvious that the clusters produced are not the clusters you would want, and that is one drawback of methods

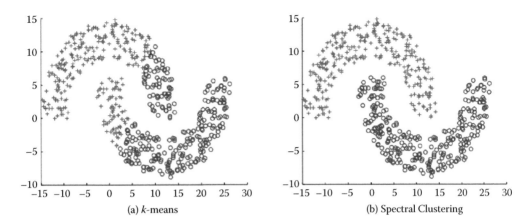

(a) k-means   (b) Spectral Clustering

Figure 7.3. The same data set can produce drastically different clusters: (a) k-means; (b) spectral clustering

such as k-means. Two points that are far away from each other will be put in different clusters even if there are other data points that create a "path" between them. Spectral clustering fixes that problem by clustering data that are connected but not necessarily (what is called) compact or clustered within convex boundaries. Spectral clustering methods work by representing data as a graph (or network), where data points are nodes in the graph and the edges (connections between nodes) represent the similarity between the two data points.

The algorithm works as follows.

1. Compute a similarity matrix from the data. This involves determining a pairwise distance function (using one of the distance functions we described previously).

2. With this matrix, now we can perform graph partitioning, where connected graph components are interpreted as clusters. The graph must be partitioned such that edges connecting different clusters have low weights and edges within the same cluster have high values.

3. Next, we can partition these data represented by the similarity matrix in a variety of ways. One common way is to use the normalized cuts method. Another way is to compute a graph Laplacian from the similarity matrix.

4. Compute the eigenvectors and eigenvalues of the Laplacian.

5. The k eigenvectors are used as proxy data for the original data set, and they are fed into k-means clustering to produce cluster assignments for each original data point.

In general, spectral clustering is much better than k-means in clustering performance but much slower to run in practice. For large-scale problems, k-means is a preferred clustering algorithm to run because of efficiency and speed.

7.6.1.7   Principal components analysis

Principal components analysis is another unsupervised method used for finding patterns and structure in data. In contrast to clustering methods, the output is not a set of clusters but a set of *principal components* that are linear combinations of the original variables. Typically, PCA is used when you have a large number of variables and you want a reduced number that you can analyze. Often, this approach is called *dimensionality reduction*. It generates linearly uncorrelated dimensions that can be used to understand the underlying structure of the data. In mathematical terms, given a set of data on n dimensions, PCA aims to find a linear subspace of dimension d lower than n such that the data points lie mainly on this linear subspace.

The PCA method is related to several other methods of which you already may be aware. Multidimensional scaling, factor analysis, and independent component analysis differ from PCA in the assumptions they make, but they often are used for similar purposes of dimensionality reduction and discovery of the underlying structure in a data set.

7.6.1.8   Association rules

Association rules is a different type of analysis method that originates from the data mining and database community, primarily focused on finding frequent co-occurring associations among a collection of items. This method is sometimes referred to as "market basket analysis," because that was the original application area of association rules. The goal is to find associations of items that occur together more often than you would randomly expect. The classic example (probably a myth) is "men who go to the store to buy diapers will also tend to buy beer at the same time." This type of analysis would be performed by applying association rules to a set of supermarket purchase data. For social scientists, this method can be used on data that contains social services that individuals have received in the past to determine what types of services "co-occur" in people and proactively offer those services to people in need.

Association rules take the form $X_1, X_2, X_3 \Rightarrow Y$ with support S and confidence C, implying that, when a transaction contains items $\{X_1, X_2, X_3\}$ $C\%$ of the time, they also contain item Y and there are at least $S\%$ of transactions where the antecedent is true. This is useful in cases where we want to find patterns that are both *frequent* and *statistically significant*, by specifying thresholds for support S and confidence C.

Support and confidence are useful metrics to generate rules but often are not enough. Another important metric used to generate rules (or reduce the number of spurious patterns generated) is *lift*. Lift is simply estimated by the ratio of the joint probability of two items, x and y, to the product of their individual probabilities: $P(x, y)/[P(x)P(y)]$. If the two items are statistically independent, then $P(x, y) = P(x)P(y)$, corresponding to a lift of 1. Note that anti-correlation yields lift values less than 1, which is also an interesting pattern, corresponding to mutually exclusive items that rarely occur together.

Association rule algorithms work as follows. Given a set of transactions (rows) and items for that transaction:

1. Find all combinations of items in a set of transactions that occur with a specified minimum frequency—these combinations are called *frequent itemsets*; and

2. Generate association rules that express co-occurrence of items within frequent itemsets.

For our purposes, association rule methods are an efficient way to take a *basket* of features (e.g., areas of publication of a researcher, different organizations for which an individual has worked in their career, all of the cities or neighborhoods in which someone may have lived) and find co-occurrence patterns. This may sound trivial, but as data sets and numbers of features grow larger, it becomes computationally expensive and association rule mining algorithms provide a fast and efficient way of proceeding.

7.6.2 Supervised learning

We now turn to the problem of supervised learning, which typically involves methods for classification, prediction, and regression.

Primarily, we will focus on classification methods in this chapter because many of the regression methods in machine learning are relatively similar to methods with which you are already familiar. Remember that classification means predicting a discrete (or categorical) variable. Most of the classification methods that we will cover also can be used for regression (predicting continuous outcomes).

In general, supervised learning methods take as input pairs of data points (X, Y) where X are the predictor variables (features) and Y is the target variable (label). The supervised learning method uses these pairs as *training data* and *learns* a model F, where $F(X) \sim Y$. Then, this model F is used to predict Ys for new data points X. As mentioned previously, the goal is not to build a model that best fits known data but a model that is useful for future predictions and minimizes future generalization error. This is the key goal that differentiates many of the methods that you know from the methods that we will describe next. In order to minimize future error, we want to build models that are not just *overfitting* on past data.

Another goal, often prioritized in the social sciences but for which machine learning methods do not optimize, is obtaining a structural form of the model. Machine learning models for classification can take different structural forms (ranging from linear models, to sets of rules, to more complex nonlinear forms), and it may not always be possible to write them in a compact form as an equation. This does not, however, make them incomprehensible or uninterpretable. Another focus of machine learning models for supervised learning is prediction, and not necessarily causal inference. Some of these models can be used to help with causal inference, but typically they are optimized for prediction tasks. We believe that there are many social science and policy problems where better prediction methods could be extremely beneficial (Kleinberg et al., 2015).

> ▶ The topic of causal inference is addressed in more detail in Chapter 10.

In this chapter, our primary focus is binary classification problems, that is, problems in which the data points are to be classified into one of two categories. Several of the methods that we will cover also can be used for multiclass classification (classifying a data point into one of n categories) or for multi-label classification (classifying a data point into m of n categories where $m \geq 1$). There are also approaches to take multiclass problems and turn them into a set of binary problems; we will briefly discuss them in the next section.

Before we describe supervised learning methods, we will recap a few principles as well as terms that we have used and will be using in the remainder of this chapter.

7.6.2.1   Training a model

After we have finished data exploration, filled in missing values, created predictor variables (features), and decided what our target variable (label) is, then we have pairs of X, Y to begin training (or building) the model.

7.6.2.2   Using the model to score new data

We are building this model so we can predict Y for a new set of Xs— using the model means that we are obtaining new data, generating the same features to obtain the vector X, and then applying the model to produce Y.

One common technique for supervised learning is logistic regression, a method with which you already should be familiar. We will provide an overview of some of the other methods used in machine learning. It is important to remember that, as you use increasingly powerful classification methods, you need more data to *train* the models.

7.6.2.3   The k-nearest neighbor

The method of k-nearest neighbor (k-NN) is one of the simpler classification methods in machine learning. It belongs to a family of models sometimes known as *memory-based models* or *instance-based models*. An example is classified by finding its k nearest neighbors and taking majority vote (or some other aggregation function). We need two key things: a value for k and a distance metric with which to find the k nearest neighbors. Typically, different values of k are used to empirically find the best one. Small values of k lead to predictions having high variance but can capture the local structure of the data. Larger values of k build more global models that are lower in variance but may not capture local structure in the data as well.

Figure 7.4 provides an example for $k = 1, 3, 5$ nearest neighbors. The number of neighbors (k) is a parameter, and the prediction depends significantly on how it is determined. In this example, point B is classified differently if $k = 3$.

Training for k-NN simply means storing the data, making this method useful in applications where data are coming in extremely quickly and a model needs to be updated frequently. All of the work, however, is moved forward to scoring time, because all of the distance calculations will occur when a new data point needs to be classified. There are several optimized methods designed to make

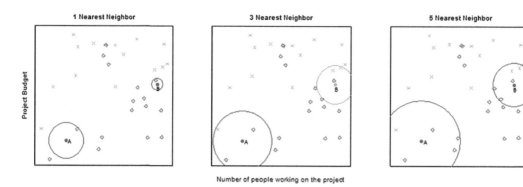

Figure 7.4. Example of k-nearest neighbor with $k = 1, 3, 5$ neighbors. We want to predict the points A and B. The 1-nearest neighbor for both points is red ("Patent not granted"), the 3-nearest neighbor predicts point A (B) to be red (green) with probability 2/3, and the 5-nearest neighbor predicts again both points to be red with probabilities 4/5 and 3/5, respectively

k-NN more efficient that are worth investigating if that is a situation that is applicable to your problem.

In addition to selecting k and an appropriate distance metric, we also have to be careful about the scaling of the features. When distances between two data points are large for one feature and small for a different feature, the method will rely almost exclusively on the first feature to find the closest points. The smaller distances on the second feature are nearly irrelevant to calculate the overall distance. A similar problem occurs when continuous and categorical predictors are used together. To resolve the scaling issues, various options for rescaling exist. For example, a common approach is to center all features at mean 0 and scale them to variance 1.

There are several variations of k-NN. One of these is weighted nearest neighbors, where different features are weighted differently or different examples are weighted based on the distance from the example being classified. The method k-NN also has issues when the data are sparse and have high dimensionality, which means that every point is far away from virtually every other point, and hence pairwise distances tend to be uninformative. This also can occur when many of the features are irrelevant and obscure the relevant features' signal in the distance calculations.

Notice that the nearest-neighbor method can be easily applied to regression problems with a real-valued target variable. In fact, the method is completely oblivious to the type of target variable and potentially can be used to predict text documents, images, and videos, based on the aggregation function after the nearest neighbors are found.

7.6.2.4   Support vector machines

Support vector machines (SVM) are one of the most popular and best-performing classification methods in machine learning today. The mathematics behind SVMs has several prerequisites that are beyond the scope of this book, but we will give you an intuition of how SVMs work, where they work best, and how to use them.

We are all familiar with linear models (e.g., logistic regression) that separate two classes by fitting a line in two dimensions (or a hyperplane in higher dimensions) in the middle (see Figure 7.5). An important decision that linear models have to make is which linear separator we should prefer when there are several we can build.

You can see in Figure 7.5 that multiple lines offer a solution to the problem. Is any of them better than the others? We can intuitively define a criterion to estimate the worth of the lines: a line is bad if it passes too close to the points because it will be noise sensitive and it will not generalize correctly. Therefore, our goal should be to find the line passing as far as possible from all points.

The SVM algorithm is based on finding the hyperplane that maximizes the *margin* of the training data. The training examples that are closest to the hyperplane are called *support vectors* because they are *supporting* the margin (as the margin is only a function of the support vectors).

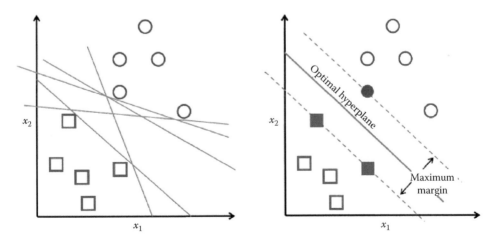

Figure 7.5. Support vector machines

An important concept to learn when working with SVMs is *kernels*. The SVMs are a specific instance of a class of methods called *kernel methods*. To this point, we have only discussed SVMs as linear models. Linear works well in high-dimensional data but sometimes you need nonlinear models, often in cases of low-dimensional data or in image or video data. Unfortunately, traditional methods of generating nonlinear models become computationally expensive because you have to explicitly generate all of the features, such as squares, cubes, and all of the interactions. Kernels are a way to keep the efficiency of the linear machinery but still build models that can capture nonlinearity in the data without creating all of the nonlinear features.

Basically, you may think of kernels as similarity functions and use them to create a linear separation of the data by (implicitly) mapping the data to a higher-dimensional space. Essentially, we take an n-dimensional input vector X, map it into a high-dimensional (infinite-dimensional) feature space, and construct an optimal separating hyperplane in this space. We refer you to relevant papers for more details on SVMs and nonlinear kernels (Schölkopf and Smola, 2001; Shawe-Taylor and Cristianini, 2004). Also, SVMs are related to logistic regression but use a different loss/penalty function (Hastie et al., 2001).

When using SVMs, there are several parameters you must optimize, ranging from the *regularization* parameter C, which determines the tradeoff between minimizing the training error and minimizing model complexity, to more kernel-specific parameters. It is often a good idea to do a grid search to find the optimal parameters. Another tip when using SVMs is to normalize the features; one common approach for this is to normalize each data point to be a vector of unit length.

Linear SVMs are effective in high-dimensional spaces, especially when the space is sparse, such as text classification where the number of data points (perhaps tens of thousands) is often much less than the number of features (one hundred thousand to one million or more). In addition, SVMs are relatively robust when the number of irrelevant features is large (unlike the k-NN approaches mentioned previously) as well as when the class distribution is skewed, that is, when the class of interest is significantly less than 50% of the data.

One disadvantage of SVMs is that they do not directly provide probability estimates. They assign a score based on the distance from the margin. The farther a point is from the margin, the higher the magnitude of the score. This score is good for ranking examples, but obtaining accurate probability estimates entails more work and requires more labeled data to be used to perform probability calibrations.

In addition to classification, there are also variations of SVMs that can be used for regression (Smola and Schölkopf, 2004) and ranking (Chapelle and Keerthi, 2010).

7.6.2.5 Decision trees

Decision trees are yet another set of methods that are helpful for prediction. Typical decision trees learn a set of rules from training data represented as a tree. An exemplary decision tree is shown in Figure 7.6. Each level of a tree *splits* the tree to create a branch using a feature and a value (or range of values). In the example tree, the first split is made on the feature *number of visits in the past year* and the value 4. The second level of the tree now has two splits: one using *average length of visit* with the value 2 days and the other using the value 10 days.

Various algorithms exist to build decision trees, with classification and regression trees (CART), C4.5, and CHAID being the most popular. Each needs to determine the next best feature on which to split. The goal is to find feature splits that can best reduce class impurity in the data, that is, a split that will ideally put all (or as many as possible) positive class examples on one side and all (or as many as possible) negative examples on the other side. One common measure of impurity that comes from information theory is *entropy*, and it is calculated as

$$H(X) = - \sum_x p(x) \log p(x).$$

Entropy is maximum (1) when both classes have equal numbers of examples in a node. It is minimum (0) when all examples are from the same class. At each node in the tree, we can evaluate all of the possible features and select the one that most reduces the entropy given the tree to that point. This expected change in entropy is known as *information gain* and is one of the most common criteria used to create decision trees. Other measures that are used instead of information gain are Gini and chi-squared.

If we continue constructing the tree in this manner, selecting the next best feature on which to split, the resulting tree will be relatively deep and tend to overfit the data. To prevent overfitting, we can either have a stopping criterion or *prune* the tree after it is fully grown. Common stopping criteria include minimum number

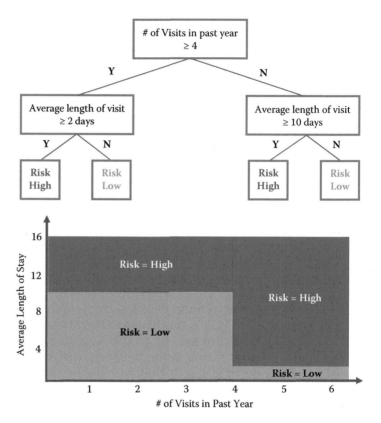

Figure 7.6. An exemplary decision tree. The top figure is the standard representation for trees. The bottom figure offers an alternative view of the same tree. The feature space is partitioned into numerous rectangles, which is another way to view a tree, representing its nonlinear character more explicitly

of data points to have before doing another feature split, maximum depth, and maximum purity. Typical pruning approaches use hold-out data (or cross-validation, which will be discussed later in this chapter) to cut off parts of the tree.

After the tree is built, a new data point is classified by running it through the tree and, once it reaches a terminal node, using some aggregation function to give a prediction (classification or regression). Typical approaches include performing maximum likelihood (if the leaf node contains 10 examples, 8 positive and 2 negative, any data point that gets into that node will obtain an 80% probability of being positive). Trees used for regression often build the tree as we have described but then fit a linear regression model at each leaf node.

Decision trees have several advantages. The interpretation of a tree is straightforward as long as the tree is not too large. Trees can be turned into a set of rules that experts in a particular domain can possibly dig into deeper, validate, and modify. Trees also do not require too much feature engineering. There is no need to create interaction terms because trees can implicitly do that by splitting on two features, one after another.

Unfortunately, along with these benefits comes a set of disadvantages. Decision trees, in general, do not perform well, compared to SVMs, random forests, or logistic regression. They are also unstable: small changes in data can result in very different trees. The lack of stability comes from the fact that small changes in the training data may lead to different splitting points. As a consequence, the whole tree may take a different structure. The suboptimal predictive performance can be seen from the fact that trees partition the predictor space into a few rectangular regions, each one predicting only a single value (see the bottom part of Figure 7.6).

7.6.2.6   Ensemble methods

Combinations of models are generally known as model ensembles. They are among the most powerful techniques in machine learning, often outperforming other methods, although at the cost of increased algorithmic and model complexity.

The intuition behind building ensembles of models is to build several models, each somewhat different. This diversity can come from various sources, such as training models on subsets of the data, training models on subsets of the features, or a combination of these two.

Ensemble methods in machine learning have two things in common. First, they construct multiple diverse predictive models from adapted versions of the training data (most often reweighted or resampled). Second, they combine the predictions of these models in some way, often by simple averaging or voting (possibly weighted).

7.6.2.7   Bagging

Bagging stands for "bootstrap aggregation"*: we first create bootstrap samples from the original data and then aggregate the predictions using models trained on each bootstrap sample. Given a data set of size N, the method works as follows.

★ Bootstrap is a general statistical procedure that draws random samples of the original data with replacement.

1. Create k bootstrap samples (with replacement), each of size N, resulting in k data sets. Only about 63% of the original training examples will be represented in any given bootstrapped set.

2. Train a model on each of the k data sets, resulting in k models.

3. For a new data point X, predict the output using each of the k models.

4. Aggregate the k predictions (typically using average or voting) to obtain the prediction for X.

A nice feature of this method is that any underlying model can be used, but decision trees are often the most commonly used base model. One reason for this is that decision trees are typically high variance and unstable, that is, they can change drastically given small changes in data, and bagging is effective at reducing the variance of the overall model. Another advantage of bagging is that each model can be trained in parallel, making it efficient to scale to large data sets.

7.6.2.8   Boosting

Boosting is another popular ensemble technique, and it often results in improving the base classifier being used. In fact, if your only goal is improving accuracy, you will most likely find that boosting will achieve that. The basic idea is to keep training classifiers iteratively, each iteration focusing on examples that the previous one got wrong. At the end, you have a set of classifiers, each trained on smaller and smaller subsets of the training data. Given a new data point, all of the classifiers predict the target, and a weighted average of those predictions is used to obtain the final prediction, where the weight is proportional to the accuracy of each classifier. The algorithm works as follows.

1. Assign equal weights to every example.

2. For each iteration:

    i. Train classifier on the weighted examples.
    ii. Predict on the training data.
    iii. Calculate error of the classifier on the training data.
    iv. Calculate the new weighting on the examples based on the errors of the classifier.
    v. Reweight examples.

3. Generate a weighted classifier based on the accuracy of each classifier.

One constraint on the classifier used within boosting is that it should be able to handle weighted examples (either directly or by replicating the examples that need to be overweighted). The most common classifiers used in boosting are decision stumps (single-level decision trees), but deeper trees also can work well.

Boosting is a common way to *boost* the performance of a classification method but comes with additional complexity, both in the training time and in interpreting the predictions. A disadvantage of boosting is that it is difficult to parallelize because the next iteration of boosting relies on the results of the previous iteration.

A nice property of boosting is its ability to identify outliers: examples that are either mislabeled in the training data or are inherently ambiguous and difficult to categorize. Because boosting focuses its weight on the examples that are more difficult to classify, the examples with the highest weight often are revealed as outliers. On the other hand, if the number of outliers is large (significant noise in the data), these examples can hurt the performance of boosting by focusing too much on them.

7.6.2.9   Random forests

Given a data set of size N and containing M features, the random forest training algorithm works as follows.

1. Create n bootstrap samples from the original data of size N. Remember, this is similar to the first step in bagging. Increasing n will lead to similar or better results, but also requires more computational resources. Typically, n ranges from 100 to a few thousand but is best determined empirically.

2. For each bootstrap sample, train a decision tree using m features (where m is typically much smaller than M) at each node of the tree. The m features are selected uniformly at random from the M features in the data set, and the decision tree will select the best split among the m features. The value of m is held constant as the forest grows.

3. A new test example/data point is classified by all of the trees, and the final classification is done by majority vote (or another appropriate aggregation method).

Random forests often achieve remarkable results while being simple to use. They can be easily parallelized, making them efficient to run on large data sets, and can handle a large number of features, even with a significant number of missing values. Random forests can become complex, with hundreds or thousands of trees that are relatively deep, so it is difficult to interpret the learned model. At the same time, they provide a nice way to estimate feature importance, providing a sense of what features were important in building the classifier.

Another nice aspect of random forests is the ability to compute a proximity matrix that gives the similarity between every pair of data points. This is calculated by computing the number of times two examples land in the same terminal node. The more that happens, the closer the two examples are. We can use this proximity matrix for clustering, locating outliers, or explaining the predictions for a specific example.

7.6.2.10   Stacking

Stacking is a technique that addresses the task of learning a meta-level classifier to combine the predictions of multiple base-level classifiers. This meta-algorithm is trained to combine the model predictions to form a final set of predictions. This can be used for both regression and classification. The algorithm works as follows.

1. Split the data set into n equal-sized sets: $set_1, set_2, \ldots, set_n$.

2. Train base models on all possible combinations of $n - 1$ sets and, for each model, use it to predict on set_i what has been left out of the training set. This would yield a set of predictions on every data point in the original data set.

3. Then, train a second-stage stacker model on the predicted classes or the predicted probability distribution over the classes from the first-stage (base) model(s).

By using the first-stage predictions as features, a stacker model obtains more information on the problem space than if it were trained in isolation. The technique is similar to cross-validation, an evaluation methodology that we will cover later in this chapter.

7.6.2.11   Neural networks and deep learning

Neural networks are a set of multi-layer classifiers where the outputs of one layer feed into the inputs of the next layer. The layers between the input and output layers are called *hidden layers*, and the more hidden layers a neural network has, the more complex functions it can learn. Popular in the 1980s and early 1990s, neural networks began to "fall out of fashion" because they were slow and expensive to train, even with only one or two hidden layers. Since 2006, a set of techniques has been developed that enable learning in deeper neural networks. These techniques, with access to massive computational resources and large amounts of data, have enabled much deeper (and larger) networks to be trained and, in fact, on many problems their performance has been significantly better than shallow neural networks (with only a single hidden layer). The reason for the better performance is the ability of deep nets to build up a complex hierarchy of concepts, learning multiple levels of representation and abstraction that help to make sense of data such as images, sound, and text.

There are a few different types of neural networks that are popular today.

- Convolutional Neural Networks (**CNN**s): These often are used in detecting objects in images and in conducting an image search, but their applicability goes beyond simply image analysis and they can be used to find patterns in other types of data as well. The CNNs treat input data (such as images) in a spatial manner (in two or three dimensions, for example) and are able to capture spatial dependencies in the data.

- Recurrent Neural Networks (**RNN**s): These are suitable for modeling sequential data that has temporal dependencies. They are trained to generate the next steps in a sequence, such as the next letters in a word, or the next words in a sentence, voice recording, or video. Typically, they are used in translation, speech generation, and time series prediction tasks. A popular variation of RNNs is long short term memory (LSTM) which is used because of its ability and effectiveness in modeling long-range dependencies.

- Generative Adversarial Networks (**GAN**s): These have been shown to be quite adept at generating new realistic images based on other training images. The GANs train two models in parallel. One network (called the generator) is trained to generate data (based on historical examples of previously occurring data, such as images, text, or video). The other network (the discriminator) attempts to classify these generated images as real or synthetic. During training of a GAN, the goal is to generate data that is realistic enough that the discriminator network is fooled to the point that it cannot distinguish the difference between the real and the synthetic input data.

Goodfellow et al. (2016) provide a (mathematical) introduction to deep learning.

Currently, deep neural networks are popular for a certain class of problems and much research is being done on them. It is, however, important to keep in mind that they may often require significantly more data than are available in many problems. In many problems, such as natural language processing, image, and video analysis, there are techniques for starting from a pre-trained neural network model that reduce the need for additional training data. Training deep neural networks also requires a significant amount of computational power, but that is less likely to be an issue for most people today with increased access to computing resources. Typical cases where deep learning has been shown to be effective involve many images, video, and text data. We are in the early stages of development of this class of methods; although there seems to be much potential, we need a better understanding of why they are effective and the problems for which they are well suited.

7.6.3 Binary vs. multiclass classification problems

In the discussion above, we have framed classification problems as binary classification problems with a 0 or 1 output. Many problems, however, have multiple classes, such as classifying companies into their industry codes or predicting whether a student will drop out, transfer, or graduate. Several solutions have been designed to manage the multiclass classification problem.

- Direct multiclass: Use methods that can directly perform multiclass classification. Examples of such methods are k-nearest neighbor, decision trees, and random forests. There are extensions of support vector machines that exist for multiclass classification as well (Crammer and Singer, 2002), but they often can be slow to train.

- Convert to one vs. all (OVA): This is a common approach to solve multiclass classification problems using binary classifiers. Any problem with n classes can be turned into n binary classification problems, where each classifier is trained to distinguish between one versus all other classes. A new example can be classified by combining the predictions from all of the n

classifiers and selecting the class with the highest score. This is a simple and efficient approach, and one that is commonly used, but it suffers from each classification problem possibly having an imbalanced class distribution (due to the negative class being a collection of multiple classes). Another limitation of this approach is that it requires the scores of each classifier to be calibrated so that they are comparable across all of them.

- Convert to pairwise: In this approach, we can create binary classifiers to distinguish between each pair of classes, resulting in $\binom{n}{2}$ binary classifiers. This results in a large number of classifiers, but each classifier usually has a balanced classification problem. A new example is classified by taking the predictions of all of the binary classifiers and using majority voting.

7.6.4   Skewed or imbalanced classification problems

Many problems you will confront will not have uniform (balanced) distributions for both classes. This is often the case with problems in fraud detection, network security, and medical diagnosis where the class of interest is not very common. The same is true in many social science and public policy problems around behavior prediction, such as predicting which students will not graduate on time, which children may be at risk of getting lead poisoning, or which homes are likely to be abandoned in a given city. You will notice that applying standard machine learning methods may result in all of the predictions being for the most frequent category in such situations, making it problematic to detect the infrequent classes. A significant amount of work in machine learning research has been done to address such problems (Chawla, 2005; Kuhn and Johnson, 2013) and we will not cover it in detail here. Common approaches to address class imbalance include oversampling from the minority class and undersampling from the majority class. It is important to keep in mind that the sampling approaches do not need to result in a 1 : 1 ratio. Many supervised learning methods described in this chapter (such as random forests and SVMs) can work well even with a 10 : 1 imbalance. Also, it is critical to make sure that you only resample the training set; keep the distribution of the test set the same as that of the original data because you will not know the class labels of new data in practice and will not be able to resample.

7.6.5   Model interpretability

As social scientists (or good machine learning practitioners), not only do we care about building machine learning models, but we also want to understand what the models "learned," and how to use them to make inferences and decisions. Understanding, or interpreting machine learning models, is a key requirement for most social science and policy problems. There are various reasons for this including:

- Providing information that can help in debugging and improving models

- Increasing trust in the models and hence increasing their adoption by decisionmakers

- Improving the decisions being made using the models by reinforcing the correct predictions and helping the decision-maker override the wrong predictions

- Helping select appropriate interventions based on the explanations

- Providing legal recourse to people being affected by the decisions made using the models.

7.6.5.1   Global vs. individual-level explanations

When thinking about model interpretability, there are two types of interpretability:

- Global: At the overall model level

- Individual: Explaining an individual classification/prediction that is made by a model.

Both of these are important for different reasons. We need global interpretability to help understand the overall model, but we also need explanations for individual classifications when these models are helping a person make decisions about individual cases. A social worker identifying the risk of a client going back to the homeless shelter and determining appropriate interventions to reduce that risk, or a counselor in an employment agency determining how likely an individual is to be long-term unemployed and connecting them with appropriate training programs or job opportunities, needs individual-level explanations of predictions/recommendations that the machine learning model is generating.

7.6.5.1.1 Global interpretability
Each method results in a model that needs to be interpreted in a way that is appropriate for that method. For example, for a decision tree, we may want to view the tree to understand what types of classifications are being made. This can, of course, become cumbersome and difficult if the tree is extremely large. For logistic regression models, we can look at the coefficients and odds-ratios, but it is often difficult to mentally account for different variables controlling for each other. In general, the models discussed above have different ways of exposing their "feature importances"[*] and we often use that as a proxy for global model interpretability.

> ★ Feature importance measures can be calculated for each feature. They are used to identify features that matter most in a given model.

Another way of interpreting a model is to understand how the model scores individual data points. We can take the set of entities that are scored by the model and generate cross-tabs that highlight how the top x% of the scored/predicted entities are different from the rest of the entities. This approach allows us to form an idea of what the model is doing, not in general but on the entities of interest to us, and makes interpretability a little more intuitive and generalizable across different model types.

A different approach that some have taken in this area has been to sparse[*] models, making them easier to interpret. The motivation behind these simple, sparse models is that they are inherently interpretable and do not require the use of additional analysis for humans to understand them. Examples of such work include Caruana et al. (2015), Ustun and Rudin (2016), and Ustun and Rudin (2019). These models may not perform well in every task, so it is important for us to explore the range of models in terms of performance and complexity, and decide what level and type of interpretability we need and how to balance that with the accuracy[*] of those models.

> ★ Models are sparse if only a few features/predictors are used in making predictions whereas most of the available features are irrelevant.

> ★ We are using accuracy as a proxy for different confusion-matrix based performance metrics, such as precision, recall, etc.

7.6.5.1.2 Individual-level explanations
While it is important to understand the models we are building at a global level, in many social science applications, we want to have an explanation for why a data point is classified/predicted a certain way by the model. Much recent work has been done on methods for generating individual-level explanations for predictions made by machine learning models. These fall into two areas: (1) model specific methods—these are used to generate explanations for predictions made by a specific class of methods, such as neural networks or random forests; and (2) model agnostic methods—these can be used to generate explanations for individual predictions made by any type

of model. Examples of this include LIME (Ribeiro et al., 2016), MAPLE (Plumb et al., 2018), and SHAP values (Lundberg and Lee, 2017).

Even so, it is important to keep in mind that these "explanations" typically are not causal, and often are restricted to being a ranked list of "features." One way to think about them is that these features have been most important in assigning this data point the score it has been given by a particular model. This is currently an active area of machine learning research and will hopefully mature into a set of methods and tools useful for social scientists using machine learning to solve problems that require a better and deeper understanding of their predictions.

7.7 Evaluation

The previous section has introduced us to a variety of methods, all with certain pros and cons, and no single method is guaranteed to outperform others for a given problem. This section focuses on evaluation methods, with three primary goals.

1. Model selection: How do we select a method to use in the future? What parameters should we select for that method?

2. Performance estimation: How do we estimate how well our model will perform once it is deployed and applied to new data?

3. Understanding: A deeper understanding of the types of models that work well, and those that do not, can be indicative of the effectiveness and applicability of existing methods and provide a better understanding of the structure of the data and the problem we are tackling.

This section will cover evaluation methodologies as well as commonly used metrics.

7.7.1 Methodology

7.7.1.1 In-sample evaluation

As social scientists, we already evaluate methods on how well they perform in-sample (on the set for which the model was trained). As we mentioned previously in the chapter, the goal of machine learning methods is to generalize to new data, and validating models

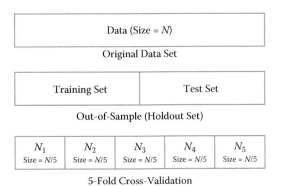

Figure 7.7. Validation methodologies: holdout set and cross-validation

in-sample does not allow us to do that. We focus here on evaluation methodologies that allow us to optimize (as best as we can) for generalization performance. The methods are illustrated in Figure 7.7.

7.7.1.2 Out-of-sample and holdout set

The simplest way to focus on generalization is to *pretend* to generalize to new (unseen) data. One way to do that is to take the original data and randomly split them into two sets: a *training set* and a *test set* (sometimes called the *holdout* or *validation set*). We can decide how much to keep in each set (typically the splits range from 50–50 to 80–20, depending on the size of the data set). Then, we train our models on the training set and classify the data in the test set, allowing us to obtain an estimate of the relative performance of the methods.

One drawback of this approach is that we may be extremely lucky or unlucky with our random split. One way to circumvent the problem is to repeatedly create multiple training and test sets. We can train on TR_1 and test on TE_1, train on TR_2 and test on TE_2, and so on. Then, the performance measures on each test set can give us an estimate of the performance of different methods and how much they vary across different random sets.

7.7.1.3 Cross-validation

Cross-validation is a more sophisticated holdout training and testing procedure that removes some of the shortcomings of the holdout set approach. Cross-validation begins by splitting a labeled data set into k partitions (called folds). Typically, k is set to 5 or 10. Then, cross-validation proceeds by iterating k times. In each iteration, one of the k folds is held out as the test set, while the other $k - 1$ folds are combined and used to train the model. A nice property of cross-validation is that every example is used in one test set for testing the model. Each iteration of cross-validation gives us a performance estimate that then can be aggregated (typically averaged) to generate the overall estimate.

An extreme case of cross-validation is called leave-one-out cross-validation, where given a data set of size N, we create N folds. That means iterating over each data point, holding it out as the test set, and training on the rest of the $N - 1$ examples. This illustrates the benefit of cross-validation by giving us good generalization estimates (by training on as much of the data set as possible) and making sure the model is tested on each data point.

7.7.1.4 Temporal validation

The cross-validation and holdout set approaches described above assume that the data have no time dependencies and that the distribution is stationary over time. This assumption is almost always violated in practice and affects performance estimates for a model.

In most practical problems, we want to use a validation strategy that emulates the way in which our models will be used and provides an accurate performance estimate. We will call this *temporal validation*. For a given point in time t_i, we train our models only on information available to us before t_i to avoid training on data from the "future." We then predict and evaluate on data from t_i to $t_i + d$ and iterate, expanding the training window while keeping the test window size constant at d. Figure 7.8 shows this validation process with $t_i = 2010$ and $d = 1$ year. The test set window d depends on a few factors related to how the model will be deployed to best emulate reality.

1. How far into the future do predictions need to be made? For example, if the set of students who need to be targeted for interventions must be finalized at the beginning of the school year for the entire year, then $d = 1$ year.

2. How often will the model be updated? If the model is being updated daily, then we can move the window by one day at a time to reflect the deployment scenario.

3. How often will the system receive new data? If we receive new data frequently, we can make predictions more frequently.

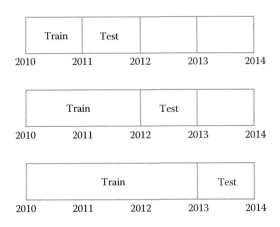

Figure 7.8. Temporal validation

Temporal validation is similar to how time series models are evaluated (also known as backtesting) and should be the validation approach used for most practical problems.

7.7.2 Metrics

The previous subsection has focused on validation methodologies assuming we have an evaluation metric in mind. This section will review commonly used evaluation metrics. You are probably familiar with using R^2, analysis of the residuals, and mean squared error (MSE) to evaluate the quality of regression models. For regression problems, the MSE calculates the average squared differences between predictions $\hat{y}_i$ and true values y_i. When prediction models have smaller MSE, they are better. However, the MSE itself is difficult to interpret because it measures quadratic differences. Instead, the root mean squared error (RMSE) is more intuitive as it is a measure of mean differences on the original scale of the response variable. Yet another alternative is the mean absolute error (MAE), which measures average absolute distances between predictions and true values.

Next, we will describe some additional evaluation metrics commonly used in machine learning for classification. Before we discuss metrics, however, it is important to highlight that machine learning models for classification typically do not predict 0/1 values directly. The SVMs, random forests, and logistic regression all produce a score (which is sometimes a probability) that then is

★ You should never use the predict function in scikit-learn because it assumes a 0.5 threshold.

turned into 0 or 1 based on a user-specific threshold. You might find that certain tools (such as scikit-learn*) use a default value for that threshold (often 0.5), but it is important to know that it is an arbitrary threshold and you should select the threshold based on the data, the model, and the problem you are solving. We will cover that later in this section.

After we have turned the real-valued predictions into 0/1 classification, we can create a *confusion matrix* from these predictions, shown in Figure 7.9. Each data point belongs to either the positive class or the negative class, and for each data point the prediction of the classifier is either correct or incorrect. This is what the four cells of the confusion matrix represent. We can use the confusion matrix to describe several commonly used evaluation metrics.

Accuracy is the ratio of correct predictions (both positive and negative) to all predictions:

$$\text{Accuracy} = \frac{TP + TN}{TP + TN + FP + FN} = \frac{TP + TN}{P + N} = \frac{TP + TN}{P' + N'},$$

where *TP* denotes true positives, *TN* true negatives, *FP* false positives, *FN* false negatives, and other symbols denote row or column totals. Accuracy is the most commonly described evaluation metric for classification but is surprisingly the least useful in practical situations (at least by itself). One problem with accuracy is that it does not give us an idea of *lift* compared to baseline. For example, if we have a classification problem with 95% of the data as positive and 5% as negative, a classifier with 85% is performing worse than

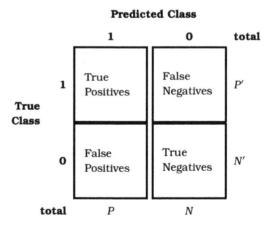

Figure 7.9. A *confusion matrix* created from real-valued predictions

a dumb classifier that predicts positive all of the time (and will have 95% accuracy).

Two additional metrics that often are used are precision and recall, which are defined as follows:

$$\text{Precision} = \frac{TP}{TP + FP} = \frac{TP}{P},$$

$$\text{Recall} = \frac{TP}{TP + FN} = \frac{TP}{P'}.$$

(See also Box 7.3.) Precision measures the accuracy of the classifier when it predicts an example to be positive. It is the ratio of correctly predicted positive examples (TP) to all examples predicted as positive ($TP + FP$). This measure is also called *positive predictive value* in other fields. Recall measures the ability of the classifier to find positive examples. It is the ratio of all of the correctly predicted positive examples (TP) to all of the positive examples in the data ($TP + FN$). This is also called *sensitivity* in other fields.

You might have encountered another metric called *specificity* in other fields. This measure is the true negative rate: the proportion of negatives that are correctly identified.

Another metric that is used is the F_1 score, which is the harmonic mean of precision and recall:

$$F_1 = \frac{2 * \text{Precision} * \text{Recall}}{\text{Precision} + \text{Recall}}.$$

This can be used when you want to balance both precision and recall.

Even so, there often is a tradeoff between precision and recall. By selecting different classification thresholds, we can vary and tune the precision and recall of a given classifier. A highly conservative classifier that only predicts a 1 when it is absolutely certain (e.g., a threshold of 0.9999) will most often be correct when it predicts a 1 (high precision) but will miss most instances of 1 (low recall). At the other extreme, a classifier that gives 1 to every data point (a threshold of 0.0001) will have perfect recall but low precision. Figure 7.10 shows a precision–recall curve that is used to represent the performance of a given classifier.

If we care about optimizing for the entire precision recall space, a useful metric is the *area under the curve* (AUC-PR), which is the area under the precision–recall curve. The AUC-PR must not be confused with AUC-ROC, which is the area under the related receiver operating characteristic (ROC) curve. The ROC curve is created by plotting

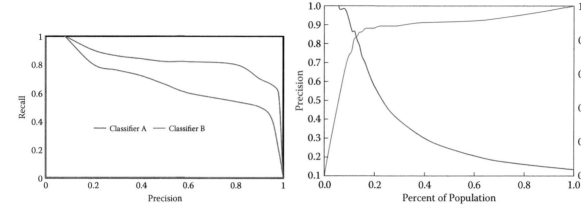

Figure 7.10. Precision–recall curve   **Figure 7.11.** Precision or recall at different thresholds

recall versus (1 – specificity). Both AUCs can be helpful metrics to compare the performance of different methods and the maximum value the AUC can take is 1. If, however, we care about a specific part on the precision–recall curve, we have to look at finer-grained metrics.

Let us consider an example from public health. Most public health agencies conduct inspections of various types to detect health hazard violations (lead hazards, for example). The number of possible places (homes or businesses) to inspect far exceeds the inspection resources typically available. Let us assume further that they can only inspect 5% of all possible places; they would clearly want to prioritize the inspection of places that are most likely to contain the hazard. In this case, the model will score and rank all of the possible inspection places in order of hazard risk. We then would want to know what percentage of the top 5% (the ones that will be inspected) are likely to be hazards, which translates to the precision in the top 5% of the most confidence predictions—precision at 5%, as it is commonly called (see Figure 7.11). *Precision at top k percent* is a common class of metrics widely used in information retrieval and search engine literature, where you want to make sure that the results retrieved at the top of the search results are accurate. More generally, this metric often is used for problems in which the class distribution is skewed and only a small percentage of the examples will be examined manually (inspections, investigations for fraud, etc.). The literature provides many case studies of such applications (Kumar et al., 2010; Lakkaraju et al., 2015; Potash et al., 2015).

One last metric we want to mention is a class of cost-sensitive metrics where different costs (or benefits) can be associated with the different cells in the confusion matrix. To this point, we have implicitly assumed that every correct prediction and every error, whether for the positive class or the negative class, has equal costs and benefits. In many practical problems, that is not the case. For example, we may want to predict whether a patient in a hospital emergency room is likely to go into cardiac arrest in the next six hours. The cost of a false positive in this case is the cost of the intervention (which may be a few extra minutes of a physician's time) while the cost of a false negative could be death. This type of analysis allows us to calculate the expected value of the predictions of a classifier and select the model that optimizes this cost-sensitive metric.

7.8 Practical tips

This section will highlight some practical tips that will be helpful when using machine learning.

7.8.1 Avoiding leakage

Leakage is when your model has access to data at training/building time that it will not have at test/deployment/prediction time. The result is an overoptimistic model that performs much worse when deployed.

The most common forms of leakage occur because of temporal issues—including data from the future in your model because you have it when doing model selection—but there are many other ways leakage is introduced. Here are the most common ones.

The big (and obvious) one

1. Using a proxy for the outcome variable (label) as a feature. This one is often easy to detect because you obtain perfect performance, but it is more nuanced when the proxy is some approximation of the label/outcome variable and the performance increase is more subtle to detect easily.

Doing any transformation or inference using the entire dataset

2. Using the entire data set for imputations. Always do imputation based on your training set only, for each training set.

Including the test set allows information to leak into your models, especially in cases where the world changes in the future (when does it not?!).

3. Using the entire data set for discretizations or normalizations/scaling or many other data-based transformations. Same reason as #2. The range of a variable (e.g., age) can change in the future and knowing that will make your models do/look better than they actually are.

4. Using the entire data set for feature selection. Same reasons as #2 and #3. To play it safe, first split into train and test sets, and then do everything you need to do using that data.

Using information from the future (that will not be available at training or prediction time)

5. Using (proxies/transformation of) future outcomes as features—similar to #1.

6. Doing standard k fold cross-validation when you have temporal data. If you have temporal data (i.e., non-stationary—again, when is it not!), k fold cross-validation will shuffle the data and a training set will (probably) contain data from the future and a test set will (probably) contain data from the past.

7. Using data (as features) that happened before model training time but are not available until later. This is relatively common in cases where there is lag/delay in data collection or access. An event may happen today but not appear in the database until a week, a month, or a year later; while it will be available in the data set you are using to build and select ML models, it will not be available at prediction time in deployment.

8. Using data (as rows) in the training set based on information from the future. Including rows that match certain criteria (in the future) in the training set, such as everyone who receives a social service in the next three months) leaks information to your model via a biased training set.

Humans using knowledge from the future

9. Selecting certain models, features, and other design choices that are based on humans (ML developers, domain experts) knowing what has happened in the future. This is a gray

area—we do want to use all of our domain knowledge to build more effective systems, but sometimes that may not generalize into the future and result in overfitted/over-optimistic models at training time and disappointment once they are deployed.

As a general rule, if you encounter a machine learning model that is performing very well, it probably is because you have made an error that is resulting in leakage. One way to dig deeper is to look at the feature importances of your model to see if the most important feature(s) may be the source of that leakage.

7.8.2 Machine learning pipeline

When working on machine learning projects, it is a good idea to structure your code as a modular pipeline so you can easily try different approaches and methods without major restructuring. The Python workbooks supporting this book will give you an example of a machine learning pipeline. A good pipeline will contain modules for importing data, doing exploration, feature generation, classification, and evaluation. Then, you can instantiate a specific workflow by combining these modules.

► See https://workbooks.coleridgeinitiative.org.

An important component of the machine learning pipeline is comparing different methods. With all of the methods available and all of the hyperparameters involved with them, how do we know which model to use and which hyperparameters to select? And what happens when we add new features to the model or when the data have "temporal drift" and change over time? One simple approach is to have a nested set of loops that loop over all of the methods to which you have access, then enumerate all of the hyperparameters for that method, create a cross-product, and loop over all of them, comparing them across different evaluation metrics and selecting the best one to use going forward. You can even add different feature subsets and time slices to this loop, as the example in the supporting workbooks will show. Triage (http://github.com/dssg/triage) is a good example of a machine learning pipeline that is designed to solve many public policy problems.

7.9 How can social scientists benefit from machine learning?

In this chapter, we have introduced you to some new methods (both unsupervised and supervised), validation methodologies, and evaluation metrics. All of these can benefit social scientists as they tackle problems in research and practice. In this section, we will provide a few specific examples where what you have learned so far can be used to improve some social science tasks.

- Use of better prediction methods and methodology: Traditional statistics and social sciences have not focused much on methods for prediction. Machine learning researchers have spent the past 30 years developing and adapting methods focusing on that task. We believe that there is significant value for social science researchers and practitioners in learning more about those methods, applying them, and even augmenting them (Kleinberg et al., 2015). Two common tasks that can be improved using better prediction methods are generating counterfactuals (essentially a prediction problem) and matching. In addition, holdout sets and cross-validation can be used as a model selection methodology with any existing regression and classification methods, resulting in improved model selection and error estimates.

- Model misspecification: Linear and logistic regressions are common techniques for data analysis in the social sciences. One fundamental assumption within both is that they are additive over parameters. Machine learning provides tools when this assumption is too limiting. Hainmueller and Hazlett (2014), for example, reanalyze data that were originally analyzed with logistic regression and come to substantially different conclusions. They argue that their analysis, which is more flexible and based on supervised learning methodology, provides three additional insights when compared to the original model. First, predictive performance is similar or better, although they do not need an extensive search to find the final model specification as it was done in the original analysis. Second, their model allows them to calculate average marginal effects that are mostly similar to the original analysis. However, for one covariate they find a substantially different result, which is due to model misspecification in the original model. Finally, the reanalysis also discovers interactions that were missed in the original publication.

- Better text analysis: Text is everywhere, but unfortunately humans are slow and expensive in analyzing text data. Thus, computers are needed to analyze large collections of text. Machine learning methods can help make this process more efficient. Feldman and Sanger (2006) provide an overview of different automatic methods for text analysis. Grimmer and Stewart (2013) give examples that are more specific for social scientists, and Chapter 8 provides more details on this topic.

- Adaptive surveys: Some survey questions have a large number of possible answer categories. For example, international job classifications describe more than 500 occupational categories, and it is prohibitive to ask all categories during the survey. Instead, respondents answer an open-ended question about their job and machine learning algorithms can use the verbatim answers to suggest small sets of plausible answer options. Then, the respondents can select the option that is the best description for their occupation, thereby saving the costs for coding after the interview (Schierholz et al., 2018).

- Estimating heterogeneous treatment effects: A standard approach to causal inference is the assignment of different treatments (e.g., medicines) to the units of interest (e.g., patients). Then, researchers usually calculate the average treatment effect—the average difference in outcomes for both groups. It is also of interest if treatment effects differ for various subgroups (e.g., is a medicine more effective for younger people?). Traditional subgroup analysis has been criticized and challenged by various machine learning techniques (Green and Kern, 2012; Imai et al., 2013).

- Variable selection: Although there are many methods for variable selection, regularized methods, such as the lasso, are highly effective and efficient when confronting large amounts of data. Varian (2014) goes into more detail and discusses other methods from machine learning that can be useful for variable selection. In addition, we can find interactions between pairs of variables (to feed into other models) using random forests, by looking at variables that co-occur in the same tree, and by calculating the strength of the interaction as a function of in how many trees they co-occur, how high they occur in the trees, and how far apart they are in a given tree.

7.10 Advanced topics

This has been a short but intense introduction to machine learning, and we have omitted several important topics that are useful and interesting for you to know and that are being actively researched in the machine learning community. We mention them here so you know what they are, but we will not describe them in detail. They include the following.

- Semi-supervised learning: For this, a combination of labeled and unlabeled data are used for training, given a set of assumptions. Such methods are useful when labeling data is costly and where unlabeled (not manually labeled/tagged data) can help improve the machine learning models. See the MIT Press edited volume (Chapelle et al., 2006) for explanations and examples.

- Recommender systems: These are commonly used by audio and video services, such as YouTube to generate playlists, or by online shops, such as Amazon to suggest additional products a customer might wish to buy. More generally, recommender systems aim to predict the preferences a user might have. One strategy is to recommend products that have similar characteristics to the ones already selected by the same user (independent of others). Another strategy recommends a product if other persons with a similar profile have selected the same product in the past.

- Active learning: Here, a set of machine learning algorithms query the user or some other information source to obtain labels for data points that are most beneficial for the machine learning models. This is in contrast to the standard machine learning process where we often select data points to label/tag randomly. Active learning approaches to selecting data points to label have been shown to reduce the effort needed to train machine learning models.

- Reinforcement learning: The "supervised" machine learning methods we have covered in this chapter are "one-shot" and take data points and labels as inputs. Reinforcement learning is a different machine learning paradigm where the machine learning program takes a series of actions/decisions and obtains delayed feedback (reward or penalty) when performing a task. The goal of reinforcement learning is to determine the next best action to take in order to maximize long-term performance. This has been applied to scenarios such as playing games (checkers, chess, backgammon, etc.) and in robotics (Sutton and Barto, 2018).

7.11   Summary

Machine learning is an active research field, and in this chapter we have given you an overview of how the work developed in this field can be used by social scientists. We have covered the overall machine learning process, methods, evaluation approaches and metrics, and some practical tips, as well as how all of this can benefit social scientists. The material described in this chapter is a snapshot of a fast-changing field, and, as we are seeing increasing collaborations between machine learning researchers and social scientists, the hope and expectation is that the next few years will bring advances that will allow us to tackle social and policy problems much more effectively using new types of data and improved methods.

7.12   Resources

We provide a machine learning "cheat sheet" to reference at https://textbook.coleridgeinitiative.org/mlcheatsheet.

Literature for further reading that also explains most topics from this chapter in greater depth:

- *Data Science for Business* (Provost and Fawcett, 2013) is a good practical handbook for using machine learning to solve real-world problems.

- *The Elements of Statistical Learning* (Hastie et al., 2001) is a classic and is available online for free.

- *An Introduction to Statistical Learning* (James et al., 2013), from the same authors, includes less mathematics and is more approachable. It is also available online.

- *Machine Learning* (Mitchell, 1997) is a classic introduction to some of the methods and gives a good motivation underlying them.

- "Top 10 Algorithms in Data Mining" (Wu et al., 2008).

Software:

- Python (with libraries such as `scikit-learn`, `pandas`, and more).

- R has many relevant packages.

▶ https://cran.r-project.org/web/views/MachineLearning.html

- Cloud-based: AzureML, Amazon ML, Google.

- Free: KNIME, Rapidminer, Weka (mostly for research use).

- Commercial: IBM Modeler, SAS Enterprise Miner, Matlab.

Many excellent courses are available online (Zygmunt, 2013).

Major conferences in this area include the International Conference on Machine Learning, the Annual Conference on Neural Information Processing Systems (NeurIPS), and the ACM International Conference on Knowledge Discovery and Data Mining (KDD).

Chapter 8

Text Analysis

Evgeny Klochikhin and Jordan Boyd-Graber

This chapter provides an overview of how social scientists can utilize text data using computational data analysis methods. We cover the types of analysis that can be done with text data (search, topic detection, classification, etc.) and provide an overview of how to perform these analyses, the social science tasks for which they are useful, and how to evaluate the results produced. We also provide pointers to some tools that are commonly used for conducting text analysis.

8.1   Understanding human-generated text

As social scientists, often we are handling text data from a variety of sources: open-ended survey responses, phone call transcriptions, social media data, notes from electronic health records, news articles, and research publications. A challenge we face when managing these types of data is how to efficiently analyze as we would analyze traditional tabular data.

For example, when analyzing survey responses or electronic health records data, both of which contain narrative text (from the respondents and medical practitioners, respectively), the text data often is ignored or selectively read by the analysts (manually) and used anecdotally. Text analysis techniques described in this chapter allow you to use all of the data available (structured and unstructured) and incorporate large amounts of text data in your analysis.

8.2   How is text data different than "structured" data?

Generally, we are comfortable analyzing "structured data" that is organized as rows and columns. Text data, also known as unstructured data,[*] are more difficult to analyze using traditional data analysis tools because it is not retrieved as a set of rows and columns, but instead consists of characters, words, sentences, and paragraphs. In traditional "structured" data, a human has already decided what constitutes a row (e.g., a person), what constitutes a column (e.g., their age, sex, address), and the relationship between them. We have discussed this in Chapter 4 where we create a data model for a given domain. When handling text data, we must create the tabular ourselves.

While creating that tabular structure, we must manage the reality that human language is complex and nuanced, which makes automatically analyzing it difficult. We often make simplifying assumptions: we assume our input is clean text; we ignore humor (Halevy et al., 2009) and deception (Ott et al., 2011; Niculae et al., 2015); and we assume "standard" English (Kong et al., 2014). Text data also often reflects human observations that are exceptions to regular processes, e.g., the ubiquitous "other" or the "anything else you want to tell us" field in questionnaires. Recognizing this complexity, the goal of text analysis is to efficiently extract important information from large amounts of text, and use it for/in our analysis as we would use tabular data.

> ★ This term is used often, but it is a fallacy. Text contains much structure—the structure of chapters, paragraphs, sentences, and syntax within a sentence (Marcus et al., 1993) allows you, the reader, to understand what we are writing here. Nevertheless, you will see the term "unstructured data" used to refer to text or, in some cases, to other forms of non-tabular data, such as images and videos.

> ▶ See Chapter 7 for a discussion of speech recognition, which can turn spoken language into text.

8.3   What can we do with text data?

There are many types of analysis that we can do with text data. Table 8.1 lists a summary of these types of analysis.

For this chapter, we will focus on two types of use cases containing text data that social scientists will address.

> ▶ If you have examples from your own research using the methods we describe in this chapter, please submit a link to the paper (and/or code) here: https://textbook.coleridgeini tiative.org/submitexamples.

1. Content understanding: We have some text "corpus," e.g., open-ended survey responses, news articles, or research publications, and our goal is to understand the content—patterns, themes, trends—of that data. This often involves methods from unsupervised machine learning (as discussed in the previous chapter). The analysis can be combined with tabular data that might accompany the text. For example, the survey responses

Table 8.1. Text analysis examples

Type of analysis	Description	Examples
Search	Finding relevant content based on some information need, often specified as a set of keywords/phrases but can be more structured.	Our use of these techniques in systematic literature reviews to facilitate the discovery and retrieval of relevant publications related to early grade reading in Latin America and the Caribbean.
Topic detection/clustering	Used to explore and understand what types of words, phrases, and topics exist in text data.	Given thousands of e-mails from a corporation, characterizing the broad themes that are prominent in the firm's communication.
Classification	Used to classify text content into one or more predefined categories.	Given SMS messages from a disaster region, deciding whether the sender needs medical assistance, food, or shelter (Yates and Paquette, 2010).
Sentiment analysis	Detection of sentiment or opinions at different levels of granularity—document, paragraph/sentence, or entity (person, organization, etc.) level.	Using machine learning to analyze the flow and topic segmentation of political debates and behaviors (Nguyen et al., 2012, 2015) and to assign automated tags to documents (Tuarob et al., 2013).
Word clustering/synonyms	Finding groups of words that are similar to each other. Depending on the problem need, similarity can be defined as strictly synonyms or aliases (e.g., IBM and Big Blue being synonyms in a specific context).	In a search engine, when a user searches for "Russian astronaut" and search results also contain "Soviet cosmonaut" (Zeng et al., 2012).
Named entity linking	Recognition, tagging, and extraction of named entities (typically of type Person, Location, Organization) from text data. Typically limited to proper nouns.	Given an e-mail, automatically linking all of the names to their corresponding Wikipedia page (Ferragina and Scaiella, 2010).
General extraction	Recognition, tagging, and extraction of specific classes of words/phrases that may be entities, events, relationships between entities, etc.	Automatically detecting words as types of events (e.g., holiday, party, graduation) and classifying them into types (e.g., related to sports, politics, or religion) from tweets (Ritter et al., 2012).
Visualization	Visualization of text data and/or visual mash-ups combining text with other forms of data (e.g., maps or networks).	Given grants funded by the NIH, creating a visualization to find areas where directorates could collaborate with each other (**?**).
Summarization	Summarization of a document (or a set of documents), either as a set of important keywords, or important sentences extracted from the text, or new sentences generated to produce a summary.	Using topic modeling to produce category-sensitive text summaries and annotations on large-scale document collections (Wang et al., 2009b).
Translation	Automatic translation of text from one language to another.	Looking at reactions to a political event in newspapers of different countries in different languages.

also may have structured information supplied by the respondent, or the news article or research publication may have meta-data that can be augmented with information generated from the text analysis.

2. Content classification: The second use case is less focused on "discovery" and "understanding new content" and instead focuses on efficiently classifying content into a pre-defined set of categories. The text data are similar to the previous use case, but the task is different, and often can be a follow-up task to the previous use case. We might have news articles about politics that we need to automatically classify into issue areas that are being discussed, such as healthcare, education, foreign policy, etc. Another example is analyzing research publications that we need to classify into topics or research areas. This falls into supervised learning in the machine learning framework that we discussed in the previous chapter.

8.4 How to analyze text

Text analysis, especially as related to the clustering and classification use cases, requires us to build an analysis pipeline that processes data through a series of steps.

- Initial processing: We take raw text data (word documents, HTML content scraped from webpages, etc.) and run it through some initial processing where the goal is to clean the text (managing content that is redundant or dirty, such as cleaning up HTML if processing data from web pages), turning sentences or documents into words or phrases, or removing words that are not considered useful for a specific analysis.

- Adding linguistic features: This step is only needed when the problem requires deeper linguistic analysis. For example, when trying to understand the structure of a sentence, we can augment the raw words with their part-of-speech tags using a part-of-speech tagger for deeper analysis. Or, we can use a statistical parser to generate a "parse tree" that shows relationships between different components of a sentence.

- Converting the enriched text to a matrix: After we have cleaned the text data and split them into sentences, phrases, words, and their corresponding linguistic attributes, the goal of this step is to make decisions that turn our "document" into a matrix. The key decisions we have to make in this step are (1) defining what a row is, (2) defining what a column is, and (3) determining what we put as the value for a cell in a given row and column.

- Analysis: After we have a matrix, then we can apply the methods we covered in the previous chapter (e.g., clustering and classification) as well as any other data analysis methods available to us. Later in this chapter, we will go deeper into applying these methods to text data as well as describing new methods that are specifically designed for text analysis.

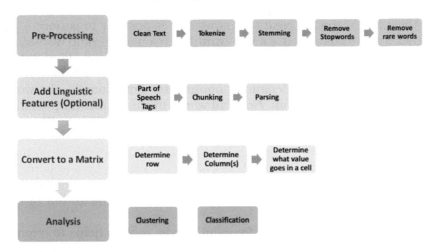

8.4.1   Initial processing

The first important step in working with text data is cleaning and processing. Textual data are often messy and unstructured, which makes many researchers and practitioners overlook their value. Depending on the source, cleaning and processing these data can require varying amounts of effort but typically involve a set of established techniques.

▶ Cleaning and processing are discussed extensively in Chapter 3.

8.4.1.1   Tokenization

The first step in processing text is deciding what terms and phrases are meaningful. Tokenization separates sentences and terms from each other. The Natural Language Toolkit (NLTK) (Bird et al., 2009) provides simple reference implementations of standard natural language processing algorithms such as tokenization—for example, sentences are separated from each other using punctuation such as period, question mark, or exclamation mark. However, this does not cover all cases such as quotes, abbreviations, or informal communication on social media. While separating sentences in a single language is difficult enough, some documents "code-switch" (Molina et al., 2016), combining multiple languages in a single document. These complexities are best addressed through data-driven machine learning frameworks (Kiss and Strunk, 2006).

8.4.1.2   Stop words

After the tokens are clearly separated, it is possible to perform further text processing at a more granular token level. Stop words are a category of words that have limited semantic meaning (and hence utility) regardless of the document content. Such words can be prepositions, articles, common nouns, etc. For example, the word "the" accounts for about 7% of all words in the Brown Corpus, and "to" and "of" are more than 3% each (Malmkjær, 2002). We may choose to remove stopwords if we think that they will not be useful in our analysis. For example, words such as "the," "is," "or" may not be useful if the task is to classify news articles by article topic. On the other hand, they may provide useful information if the task is to classify a document into a genre or identify the author of the document.

In addition to removing frequent words, it often helps to remove words that only appear a few times. These words—names, misspellings, or rare technical terms—are also unlikely to bear significant contextual meaning. Similar to stop words, these tokens usually are disregarded in further modeling either by the design of the method or by manual removal from the corpus before the actual analysis.

8.4.1.3   The N-grams

Even so, individual words are sometimes not the correct unit of analysis. For example, blindly removing stop words can obscure important phrases, such as "systems of innovation," "cease and desist," or "commander in chief." Identifying these N-grams requires looking for statistical patterns to discover phrases that often appear together in fixed patterns (Dunning, 1993). Typically, these combinations of phrases are called *collocations*, as their overall meaning is more than the sum of their parts. These N-grams can be created over any unit of analysis, such as sequences of characters (called character n-grams) or sequences of phonemes that are used in speech recognition.

8.4.1.4 Stemming and lemmatization

Text normalization is another important aspect of preprocessing textual data. Given the complexity of natural language, words can take multiple forms dependent on the syntactic structure with limited change of their original meaning. For example, the word "system" morphologically has a plural "systems" or an adjective "systematic." All of these words are semantically similar and—for many tasks— should be treated the same. For example, if a document has the word "system" occurring three times, "systems" once, and "systematic" twice, one can assume that the word "system" with similar meaning and morphological structure can cover all instances and that variance should be reduced to "system" with six instances.

Generally, the process for text normalization is implemented using established lemmatization and stemming algorithms. A *lemma* is the original dictionary form of a word. For example, "go," "went," and "goes" will all have the lemma "go." The stem is a central part of a given word bearing its primary semantic meaning and uniting a group of similar lexical units. For example, the words "order" and "ordering" will have the same stem "ord." Morphy (a lemmatizer provided by the electronic dictionary WordNet), Lancaster Stemmer, and Snowball Stemmer are common tools used to derive lemmas and stems for tokens, and all have implementations in the NLTK (Bird et al., 2009).

8.4.2 Linguistic analysis

To this point, we have treated words as tokens without regard to the meaning of the word or the way it is used, or even what language from which the word is derived. There are several techniques in text analysis that are language-specific that go deeper into the syntax of the document, paragraph, and sentence structure to extract linguistic characteristics of the document.

8.4.2.1 Part-of-speech tagging

When the examples x are individual words and the labels y represent the grammatical function of a word (e.g., whether a word is a noun, verb, or adjective), the task is called part-of-speech tagging. This level of analysis can be useful for discovering simple patterns in text: distinguishing between when "hit" is used as a noun (a Hollywood hit) and when "hit" is used as a verb (the car hit the guard rail).

Unlike document classification, the examples x are not independent: knowing whether the previous word was an adjective makes it far more likely that the next word will be a noun rather than a verb. Thus, the classification algorithms need to incorporate structure into the decisions. Two traditional algorithms for this problem are hidden Markov models (Rabiner, 1989) and conditional random fields (Lafferty et al., 2001), but more complicated models have greater accuracy (Plank et al., 2016).

8.4.2.2 Order matters

All text-processing steps are critical to successful analysis. Some of them bear more importance than others, depending on the specific application, research questions, and properties of the corpus. Having all of these tools ready is imperative to producing a clean input for subsequent modeling and analysis. Some simple rules should be followed to prevent typical errors. For example, stop words should not be removed before performing n-gram indexing, and a stemmer should not be used where data are complex and require accounting for all possible forms and meanings of words. Reviewing interim results at every stage of the process can be helpful.

8.4.3 Turning text data into a matrix: How much is a word worth?

The processing stages described above provide us with the columns in our matrix. Now, we have to decide how to assign values in that column for each word or phrase. In text analysis, we typically refer to them as tokens (where a token can be a word or a phrase). One simple approach would be to give each column a binary 0 or 1 value—if this token occurs in a document, we assign that cell a value of 1 and 0 otherwise. Another approach would be to assign it the value of how many times this token occurs in that document (often known as frequency of that term or token). This is essentially a way to define the importance or value of this token in this document. Not all words are worth the same; in an article about sociology, "social" may be less important or informative than "inequality." Appropriately weighting and calibrating words is important for both human

▶ Term weighting is an example of feature engineering discussed in Chapter 7.

and machine consumers of text data: humans do not want to see "the" as the most frequent word of every document in summaries, and classification algorithms benefit from knowing which features are actually important to making a decision.

Weighting words requires balancing how often a word appears in a local context (e.g., a document) with how much it appears overall in the document collection. Term frequency–inverse document frequency (TFIDF) (Salton, 1968) is a weighting scheme to explicitly balance these factors and prioritize the most meaningful words. The TFIDF model takes into account both the term frequency of a given token and its document frequency (Box 8.1) so that, if a highly frequent word also appears in almost all documents, its meaning for the specific context of the corpus is negligible. Stop words are a good example when highly frequent words also bear limited meaning because they appear in virtually all documents of a given corpus.

Box 8.1: The TFIDF

For every token t and every document d in the corpus D of size $|D| = N$, TFIDF is calculated as

$$tfidf(t, d, D) = tf(t, d) \times idf(t, D),$$

where term frequency is either a simple count,

$$tf(t, d) = f(t, d),$$

or a more balanced quantity,

$$tf(t, d) = 0.5 + \frac{0.5 \times f(t, d)}{\max\{f(t, d) : t \in d\}},$$

and inverse document frequency is

$$idf(t, D) = \log \frac{N}{|\{d \in D : t \in d\}|}.$$

8.4.4 Analysis

Now that we have a matrix with documents as rows, words/phrases as columns, and perhaps the TFIDF score of the value of words in the document, we are ready to run machine learning methods on this data. We will not recap all of the methods and evaluation methodologies already covered in Chapter 7 here, but they can all be used with text data.

We will focus on three types of analysis: finding similar documents, clustering, and classification. For each type of analysis, we will focus on what it allows us to do, for what types of tasks social scientists will find it useful, and how to evaluate the results of the analysis.

8.4.4.1   Use case: Finding similar documents

One task that may interest social scientists is finding documents that are similar to a document they are analyzing. This is a routine task during literature review, where we may have a paper and we are interested in finding similar papers, or in disciplines such as law, where lawyers looking at a case file want to find all prior cases related to the case being reviewed. The key challenge here is to define what makes two documents similar—what similarity metrics should we use to calculate this similarity? Two commonly used metrics are cosine similarity and Kullback–Leibler divergence (Kullback and Leibler, 1951).

Cosine similarity is a popular measure in text analysis. Given two documents d_a and d_b, this measure first turns the documents into vectors (each dimension of the vector can be a word or phrase) $\vec{t_a}$ and $\vec{t_b}$, respectively, and uses the cosine similarity (the cosine of the angle between the two vectors) as a measure of their similarity. This is defined as

$$SIM_C(\vec{t_a}, \vec{t_b}) = \frac{\vec{t_a} \cdot \vec{t_b}}{|\vec{t_a}| \times |\vec{t_b}|}.$$

Kullback–Leibler (KL) divergence is a measure that allows us to compare probability distributions in general and is often used to compare two documents represented as vectors. Given two term vectors $\vec{t_a}$ and $\vec{t_b}$, the KL divergence from vector $\vec{t_a}$ to $\vec{t_b}$ is

$$D_{KL}(\vec{t_a} \| \vec{t_b}) = \sum_{t=1}^{m} w_{t,a} \times \log\left(\frac{w_{t,a}}{w_{t,b}}\right),$$

where $w_{t,a}$ and $w_{t,b}$ are term weights in the two vectors, respectively, for terms $t = 1, \ldots, m$.

Then, an averaged KL divergence metric is defined as

$$D_{AvgKL}(\vec{t_a} \| \vec{t_b}) = \sum_{t=1}^{m} (\pi_1 \times D(w_{t,a} \| w_t) + \pi_2 \times D(w_{t,b} \| w_t)),$$

where
$$\pi_1 = \frac{w_{t,a}}{w_{t,a} + w_{t,b}}, \quad \pi_2 = \frac{w_{t,b}}{w_{t,a} + w_{t,b}},$$
and $w_t = \pi_1 \times w_{t,a} + \pi_2 \times w_{t,b}$ (Huang, 2008).

A Python-based `scikit-learn` library provides an implementation of these measures as well as other machine learning models and approaches.

8.4.4.2 Example: Measuring similarity between documents

National Science Foundation (NSF) awards are not labeled by scientific field—they are labeled by program. This administrative classification is not always useful to assess the effects of certain funding mechanisms on disciplines and scientific communities. A common need is to understand how awards are similar to each other even if they have been funded by different programs. Cosine similarity allows us to do just that.

8.4.4.3 Example code

The Python `numpy` module is a powerful library of tools for efficient linear algebra computation. Among other things, it can be used to compute the cosine similarity of two documents represented by numeric vectors, as described above. The `gensim` module, often used as a Python-based topic modeling implementation, can be used to produce vector space representations of textual data.

8.4.4.4 Augmenting similarity calculations with external knowledge repositories

Frequently, it is the case that two, especially short, documents do not have any words in common but are still similar. In such cases, cosine similarity or KL divergence do not help us with the similarity calculations without augmenting the data with additional information. Often, external data resources that provide relationships between words, documents, or concepts present in specific domains can be used to achieve that. Established corpora, such as the Brown Corpus and Lancaster–Oslo–Bergen Corpus, are one type of such preprocessed repositories.

Wikipedia and WordNet are examples of another type of lexical and semantic resources that are dynamic in nature and that can provide a valuable basis for consistent and salient information retrieval and clustering. These repositories have the innate

hierarchy, or ontology, of words (and concepts) that are explicitly
linked to each other either by inter-document links (Wikipedia) or
by the inherent structure of the repository (WordNet). In Wikipedia,
concepts thus can be considered as titles of individual Wikipedia
pages and the contents of these pages can be considered as their
extended semantic representation.

Information retrieval techniques build on these advantages of
WordNet and Wikipedia. For example, Meij et al. (2009) map search
queries to the DBpedia ontology (derived from Wikipedia topics and
their relationships); they find that this mapping enriches the search
queries with additional context and concept relationships. One way
of using these ontologies is to retrieve a predefined list of Wikipedia
pages that would match a specific taxonomy. For example, scientific
disciplines are an established way of tagging documents—some are
in physics, others in chemistry, engineering, or computer science.
If a user retrieves four Wikipedia pages on "Physics," "Chemistry,"
"Engineering," and "Computer Science," they can be further mapped
to a given set of scientific documents to label and classify them, such
as a corpus of award abstracts from the NSF.

Personalized PageRank is a similarity system that can help with
the task. This system uses WordNet to assess semantic relationships and relevance between a search query (document *d*) and possible results (the most similar Wikipedia article or articles). This
system has been applied to text categorization (Navigli et al., 2011)
by comparing documents to *semantic model vectors* of Wikipedia
pages constructed using WordNet. These vectors account for the
term frequency and their relative importance given their place in
the WordNet hierarchy, so that the overall *wiki* vector is defined as

$$SMV_{wiki}(s) = \sum_{w \in Synonyms(s)} \frac{tf_{wiki}(w)}{|Synsets(w)|},$$

where *w* is a token (word) within *wiki*, *s* is a WordNet synset (a
set of synonyms that share a common meaning) that is associated
with every token *w* in WordNet hierarchy, *Synonyms(s)* is the set of
words (i.e., synonyms) in the synset *s*, $tf_{wiki}(w)$ is the term frequency
of the word *w* in the Wikipedia article *wiki*, and *Synsets(w)* is the
set of synsets for the word *w*.

The overall probability of a candidate document *d* (e.g., an NSF
award abstract or a PhD dissertation abstract) matching the target
query, or in our case a Wikipedia article *wiki*, is

$$wiki_{BEST} = \sum_{w_t \in d} \max_{s \in Synsets(w_t)} SMV_{wiki}(s),$$

where $Synsets(w_t)$ is the set of synsets for the word w_t in the target document (e.g., NSF award abstract) and $SMV_{wiki}(s)$ is the semantic model vector of a Wikipedia page, as defined above.

8.4.4.5   Evaluating "find similar" methods

When developing methods to find similar documents, we want to ensure that we find all relevant documents that are similar to the document under consideration, and we want to ensure we do not find any non-relevant documents. Chapter 7 already has mentioned the importance of precision and recall for evaluating the results of machine learning models (Box 8.2 provides a reminder of the formulae). The same metrics can be used to evaluate the two goals we have in finding relevant and similar documents.

Box 8.2: Precision and recall

Precision computes the type I errors—*false positives* (retrieved documents that are not relevant)—and is formally defined as

$$\text{Precision} = \frac{|\{\text{relevant documents}\} \cap \{\text{retrieved documents}\}|}{|\{\text{retrieved documents}\}|}.$$

Recall accounts for type II errors—*false negatives* (relevant documents that were not retrieved)—and is defined as

$$\text{Recall} = \frac{|\{\text{relevant documents}\} \cap \{\text{retrieved documents}\}|}{|\{\text{relevant documents}\}|}.$$

We assume that a user has three sets of documents, $D_a = \{d_{a1}, d_{a2}, \ldots, d_n\}$, $D_b = \{d_{b1}, d_{b2}, \ldots, d_k\}$, and $D_c = \{d_{c1}, d_{c2}, \ldots, d_l\}$. All three sets are clearly tagged with a disciplinary label: D_a are computer science documents, D_b are physics, and D_c are chemistry.

The user also has a different set of documents—Wikipedia pages on "Computer Science," "Chemistry," and "Physics." Knowing that all documents in D_a, D_b, and D_c have clear disciplinary assignments, let us map the given Wikipedia pages to all documents within those three sets. For example, the Wikipedia-based query on "Computer Science" should return all computer science documents and none in physics or chemistry. So, if the query based on the "Computer Science" Wikipedia page returns only 50% of all computer science documents, then 50% of the relevant documents are lost: the recall is 0.5.

On the other hand, if the same "Computer Science" query returns 50% of all computer science documents but also 20% of the physics documents and 50% of the chemistry documents, then all of the physics and chemistry documents returned are false positives. Assuming that all document sets are of equal size, so that $|D_a| = 10$, $|D_b| = 10$, and $|D_c| = 10$, then the precision is $\frac{5}{12} = 0.42$.

8.4.4.6   The F score

The *F score* or F_1 *score* combines precision and recall. In formal terms, the F score is the harmonic mean of precision and recall:

$$F_1 = 2 \cdot \frac{\text{Precision} \cdot \text{Recall}}{\text{Precision} + \text{Recall}}.$$

In terms of type I and type II errors:

$$F_\beta = \frac{(1 + \beta^2) \cdot \text{true positive}}{(1 + \beta^2) \cdot \text{true positive} + \beta^2 \cdot \text{false negative} + \text{false positive}},$$

where β is the balance between precision and recall. Thus, F_2 puts more emphasis on the recall measure and $F_{0.5}$ puts more emphasis on precision.

8.4.4.7   Examples

Some examples from our recent work can demonstrate how Wikipedia-based labeling and labeling via Latent Dirichlet allocation (LDA, which will be discussed in a subsequent section) (Ramage et al., 2009; Nguyen et al., 2014) cope with the task of document classification and labeling in the scientific domain. See Table 8.2.

8.4.4.8   Use case: Clustering

Another task social scientists will perform is finding themes, topics, and patterns in a text data set, such as open-ended survey responses, news articles, or publications. Given open-ended responses from a survey on how people feel about a certain issue, we may be interested in finding the common themes that occur in these responses. Clustering methods are designed to do exactly that. With text data, clustering will be used to explore what topics and concepts are present in a new corpus (collection of documents). It is important to note that, if we already have a

Table 8.2. Wikipedia articles as potential labels generated by n-gram indexing of NSF awards

Abstract excerpt	ProQuest subject category	Labeled LDA	Wikipedia-based labeling
Reconfigurable computing platform for small-scale resource-constrained robot. Specific applications often require robots of small size for reasons such as costs, access, and stealth. Smallscale robots impose constraints on resources such as power or space for modules. . .	Engineering, Electronics and Electrical; Engineering, Robotics	Motor controller	Robotics, Robot, Field-programmable gate array
Genetic mechanisms of thalamic nuclei specification and the influence of thalamocortical axons in regulating neocortical area formation. Sensory information from the periphery is essential for all animal species to learn, adapt, and survive in their environment. The thalamus, a critical structure in the diencephalon, receives sensory information. . .	Biology, Neurobiology	HSD2 neurons	Sonic hedgehog, Induced stem cell, Nervous system
Poetry 'n acts: The cultural politics of twentieth century American poets' theater. This study focuses on the disciplinary blind spot that obscures the productive overlap between poetry and dramatic theater and prevents us from seeing the cultural work that this combination can perform. . .	Literature, American; Theater	Audience	Counter-culture of the 1960s, Novel, Modernism

pre-specified set of categories and documents that are tagged with those categories, and the goal is to tag new documents, then we would use classification methods instead of clustering methods. As addressed in the previous chapter, clustering is a form of unsupervised learning where the goal is exploration and understanding of the data.

As we have discussed, unsupervised analysis of large text corpora without extensive investment of time provides additional opportunities for social scientists and policymakers to gain insights into policy and research questions through text analysis. The clustering methods described in Chapter 7, such as k-means clustering, can be used for text data as well after the text has been converted to a matrix as described previously. Next, we will describe topic modeling, which provides us with another clustering approach specifically designed for text data.

8.4.5 Topic modeling

Topic modeling is an approach that describes *topics* that constitute the high-level themes of a text corpus. Topic modeling can be described as an *information discovery* process: describing what "concepts" are present in a corpus. We refer to them as "concepts" or "topics" using quotation marks because they typically will be represented as a probability distribution over the words (that the topic modeling method groups together) which may or may not be semantically coherent as a traditional topic to social scientists.

As topic modeling is a broad subfield of natural language processing and machine learning, we will restrict our focus to a single method called Latent Dirichlet allocation (LDA) (Blei et al., 2003). The LDA is a fully Bayesian extension of probabilistic latent semantic indexing (Hofmann, 1999), itself a probabilistic extension of latent semantic analysis (Landauer and Dumais, 1997). Blei and Lafferty (2009) provide a more detailed discussion of the history of topic models.

The LDA, like all topic models, assumes there are topics that form the building blocks of a corpus. Topics are distributions over words and often are shown as a ranked list of words, with the highest probability words at the top of the list (see Figure 8.1). However, we do not know what the topics are a priori; the goal is to discover what they are (which we will discuss next).

In addition to assuming that there exist some number of topics that explain a corpus, LDA also assumes that each document in a corpus can be explained by a small number of topics. For example, taking the example topics from Figure 8.1, the document titled "Red Light, Green Light: A 2-Tone L.E.D to Simplify Screens" would be about Topic 1, which appears to be about technology. However, a document such as "Forget the Bootleg, Just Download the

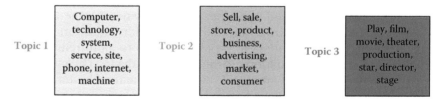

Figure 8.1. Topics are distributions over words. Here are three example topics learned by latent Dirichlet allocation from a model with 50 topics discovered from the *New York Times* (Sandhaus, 2008). Topic 1 appears to be about technology, Topic 2 about business, and Topic 3 about the arts

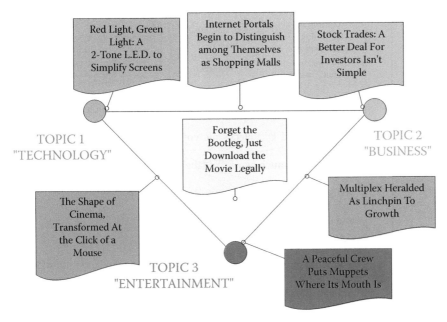

Figure 8.2. Allocations of documents to topics

Movie Legally" would require all three of the topics. The set of topics that are used by a document is called the document's *allocation*
(Figure 8.2). This terminology explains the name *latent Dirichlet allocation*: each document has an allocation over latent topics governed
by a Dirichlet distribution.

8.4.5.1   Inferring "topics" from raw text

Algorithmically, the problem can be viewed as: given a corpus and
an integer k as input, provide the k topics that best describe the
document collection; a process called *posterior inference*. The most
common algorithm for solving this problem is a technique called
Gibbs sampling (Geman and Geman, 1990).

Gibbs sampling works at the word level to discover the topics
that best describe a document collection. Each word is associated
with a single topic, explaining why that word has appeared in a
document. For example, consider the sentence, "Hollywood studios
are preparing to let people download and buy electronic copies of
movies over the Internet." Each word in this sentence is associated
with a topic: "Hollywood" might be associated with an arts topic,
"buy" with a business topic, and "Internet" with a technology topic
(Figure 8.3).

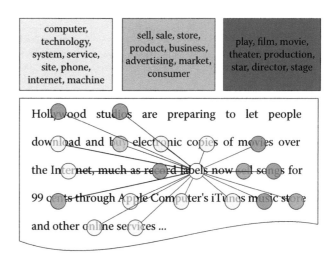

Figure 8.3. Each word is associated with a topic. Gibbs sampling inference iteratively resamples the topic assignments for each word to discover the most likely topic assignments that explain the document collection

This is where we should arrive eventually. However, we do not know this at the start. So, initially we can assign words to topics randomly. This will result in poor topics, but we can make those topics better. We improve these topics by taking each word, pretending that we do not know the topic, and selecting a new topic for the word.

You should want your topic model to (not) do two things: you do not want it to use many topics in a document and you do not want it to use many words in a topic. Therefore, the algorithm will keep track of how many times a document d has used a topic k, $N_{d,k}$, and how many times a topic k has used a word w, $V_{k,w}$. For notational convenience, it also will be useful to keep track of marginal counts of how many words are in a document,

$$N_{d,\cdot} \equiv \sum_k N_{d,k},$$

and how many words are associated with a topic,

$$V_{k,\cdot} \equiv \sum_w V_{k,w}.$$

The algorithm removes the counts for a word from $N_{d,k}$ and $V_{k,w}$ and then changes the topic of a word (hopefully to a better topic than the one it had previously). Through many thousands of iterations of this process, the algorithm can find topics that are coherent, useful, and characterize the data well.

The two goals of topic modeling—balancing document allocations to topics and topics' distribution over words—come together in an equation that multiplies them together. A good topic will be both common in a document and explain a word's appearance well.

Example: Gibbs sampling for topic models

The topic assignment $z_{d,n}$ of word n in document d is proportional to

$$p(z_{d,n} = k) \propto \left(\underbrace{\frac{N_{d,k} + a}{N_{d,\cdot} + Ka}}_{\text{how much doc likes the topic}} \right) \left(\underbrace{\frac{V_{k,w_{d,n}} + \beta}{V_{k,\cdot} + V\beta}}_{\text{how much topic likes the word}} \right),$$

where a and β are smoothing factors that prevent a topic from having zero probability if a topic does not use a word or a document does not use a topic (Wallach et al., 2009). Recall that we do not include the token that we are sampling in the counts for N or V.

For an explicit example, assume that we have three documents with the following topic assignments:

- Document 1: $^A\text{dog}_3$ $^B\text{cat}_2$ $^C\text{cat}_3$ $^D\text{pig}_1$,
- Document 2: $^E\text{hamburger}_2$ $^F\text{dog}_3$ $^G\text{hamburger}_1$,
- Document 3: $^H\text{iron}_1$ $^I\text{iron}_3$ $^J\text{pig}_2$ $^K\text{iron}_2$.

If we want to sample token B (the first instance of "cat" in document 1), we compute the conditional probability for each of the three topics ($z = 1, 2, 3$):

$$p(z_B = 1) = \frac{1 + 1.000}{3 + 3.000} \times \frac{0 + 1.000}{3 + 5.000} = 0.333 \times 0.125 = 0.042,$$

$$p(z_B = 2) = \frac{0 + 1.000}{3 + 3.000} \times \frac{0 + 1.000}{3 + 5.000} = 0.167 \times 0.125 = 0.021,$$

$$p(z_B = 3) = \frac{2 + 1.000}{3 + 3.000} \times \frac{1 + 1.000}{4 + 5.000} = 0.500 \times 0.222 = 0.111.$$

To reiterate, we do not include token B in these counts: in computing these conditional probabilities, we consider topic 2 as never appearing in the document and "cat" as never appearing in topic 2. However, "cat" does appear in topic 3 (token C), so it has a higher probability than the other topics. After renormalizing, our conditional probabilities are $(0.24, 0.12, 0.64)$. Then, we sample the new assignment of token B to be topic 3 two times out of three. Griffiths and Steyvers (2004) provide more details on the derivation of this equation.

8.4.5.2   Applications of topic models

Topic modeling is most often used for topic exploration, allowing users to understand the contents of large text corpora. Thus, topic models have been used, for example, to understand what the National Institutes of Health funds (Talley et al., 2011), to compare and contrast what was discussed in the North and South in the Civil War (Nelson, 2010), and to understand how individuals code in large programming projects (Maskeri et al., 2008).

Topic models also can be used as features to more elaborate algorithms, such as machine translation (Hu et al., 2014), detecting objects in images (Wang et al., 2009a), or identifying political polarization (Paul and Girju, 2010). Boyd-Graber et al. (2017) summarize applications of topic models in the humanities, information retrieval, and social sciences.

Blei and McAuliffe (2007) apply topic models to classification and regression tasks, such as sentiment analysis. As discussed in the previous chapter, such methods require a feature-based representation of the data. An advantage of using topic models is that the distribution over topics itself can serve as a feature.

For example, to predict whether a legislator will vote on a bill, Gerrish and Blei (2012) learn a topic model that encodes each bill (proposed piece of legislation) as a vector. To predict how a legislator will vote on a bill, the model takes a dot product between the bill's distribution over topics and a legislator's ideology vector. The higher the score, the more compatible they are and the more likely the legislator is to vote on the bill. Conversely, the lower the score, the less likely it is the legislator will vote on the bill.

This formulation should remind you of logistic regression; however, the features are learned automatically rather than the feature engineering approach described in the last chapter.

Use case: Document classification   The previous section focuses on the task of finding topics and themes in a new text data set. In many cases, we already know a set of topics—this could be the set of topics or research fields as described by the Social Science Research Network or the set of sections (local news, international, sports, finance, etc.) in a news publication. The task we often face is to automatically categorize new documents into an existing set of categories. In text analysis, this is called text classification or categorization, and it uses supervised learning techniques from machine learning as described in Chapter 7.

Text classification typically requires two things: a set of categories into which we want documents to be categorized (each document can belong to one or more categories) and a set of documents annotated/tagged with one or more categories from step 1.

For example, if we want to classify Twitter or Facebook posts as being about health or finance, a classification method would take a small number of posts, manually tagged as belonging to either health or finance, and train a classification model. Then, this model can be used to automatically classify new posts as belonging to either health or finance.

All of the classification (supervised learning) methods discussed in Chapter 7 can be used here once the text data has been processed and converted to a matrix. Neural networks (Iyyer et al., 2015), naïve Bayes (Lewis, 1998), and support vector machines (SVM) (Zhu et al., 2013) are some of the commonly used methods applied to text data.

Example: Using text to categorize scientific fields

The National Center for Science and Engineering Statistics, the US statistical agency charged with collecting statistics on science and engineering, uses a rule-based system to manually create categories of science; these are used to categorize research as "physics" or "economics" (Mortensen and Bloch, 2005; Organisation of Economic Co-operation and Development, 2004). In a rule-based system there is no ready response to the question, "how much do we spend on climate change, food safety, or biofuels," because existing rules have not created such categories. Text analysis techniques can be used to provide such detail without manual collation. For example, data about research awards from public sources and about people funded on research grants from UMETRICS can be linked with data about their subsequent publications and related student dissertations from ProQuest. Both award and dissertation data are text documents that can be used to characterize what research has been done, provide information about which projects are similar within or across institutions, and potentially identify new fields of study (Talley et al., 2011).

Applications

Spam detection

One simple but ubiquitous example of document classification is spam detection: an email is either an unwanted advertisement (spam) or it is not. Document classification techniques, such as naïve Bayes (Lewis, 1998), touch essentially every email sent worldwide, making email functional even though most emails are spam.

Sentiment analysis

Instead of being what a document is about, a label y could reveal the speaker. A recent subfield of natural language processing uses machine learning to reveal the internal state of speakers based on what they say about a subject (Pang and Lee, 2008). For example, given an example of sentence x, can we determine whether the speaker is a liberal or a conservative? Is the speaker happy or sad?

Simple approaches use dictionaries and word counting methods (Pennebaker and Francis, 1999), but more nuanced approaches make use of *domain*-specific information to make better predictions. One uses different approaches to praise a toaster than to praise an air conditioner (Blitzer et al., 2007); liberals and conservatives each frame health care differently from how they frame energy policy (Nguyen et al., 2013).

Evaluating text classification methods

The metrics used to evaluate text classification methods are the same as those used in supervised learning, as described in Chapter 7. The most commonly used metrics include accuracy, precision, recall, AUC, and F_1 score.

8.5 Word embeddings and deep learning

In discussing topic models, we have learned a vector that summarizes the content of each document. This is useful for applications where you can use a single short vector to summarize a document for a downstream machine learning application. However, modern research does not stop there—it learns vector representations of everything from documents down to sentences and words.

First, let us consider this from a high-level perspective. The goal of representation learning (Bengio et al., 2013) is to take an input and transform it into a vector that computers can understand. Similar inputs should be close together in vector space; e.g., "dog" and "poodle" should have similar vectors, while "dog" and "chainsaw" should not.

A well-known technique for word representation is word2vec (Mikolov et al., 2013). Using an objective function similar to logistic regression, it predicts, given a word, whether another word will appear in the same context. For example, the dot product for "dog" and "pet," "dog" and "leash," and "dog" and "wag" will be high, but those for "dog" and "rectitude," "dog" and "examine," and "dog" and "cloudy" will be lower. Training a model to do this for all of the words in English will produce vector representations for "dog" and "poodle" that are quite close together.

This model has been well adopted throughout natural language processing (Ward, 2017). Downloading word2vec vectors for words and using them as features in your machine learning pipeline (e.g., for document classification by averaging the words in the document) will likely improve a supervised classification task.

Even so, word representations are not the end of the story. A word only makes sense in the context of the sentence in which it appears, e.g., "I deposited my check at the bank" versus "The airplane went into a bank of clouds." A single word per vector does not capture these subtle effects. More recent models named after Muppets attempt to capture broader relationships between words within sentences to create *contextualized representations*. [Note: the story behind the Muppet names is long and uninteresting so we will not recount it here.]

Both ELMO (Peters et al., 2018) and BERT (Devlin et al., 2019) use deep learning to take word vectors (via word2vec) to create representations that make sense given a word's *context*. These are also useful features to use in supervised machine learning contexts if higher accuracy is your goal.

However, these techniques are not always the best tools for social scientists. They are not always interpretable—it is often difficult to discern why you received the answer you did (Ribeiro et al., 2016), and slightly changing the input to the models can dramatically change the results (Feng et al., 2018). Given that our typical goal is understanding our data, it is probably better to start first with the simpler (and faster methods) we have mentioned, so that you understand your data first.

8.6   Text analysis tools

We are fortunate to have access to a set of powerful open source text analysis tools. We describe several here.

8.6.1   The natural language toolkit

The NLTK is a commonly used natural language toolkit that provides a large number of relevant solutions for text analysis. It is Python-based and can be easily integrated into data processing and analytical scripts by a simple `import nltk` (or similar for any one of its submodules).

The NLTK includes a set of tokenizers, stemmers, lemmatizers, and other natural language processing tools typically applied in text analysis and machine learning. For example, a user can extract tokens from a document *doc* by running the command `tokens = nltk.word_tokenize(doc).`

Useful text corpora also are present in the NLTK distribution. For example, the stop words list can be retrieved by running the command `stops = nltk.corpus.stopwords.words(language)`. These stop words are available for several languages within NTLK, including English, French, and Spanish.

Similarly, the Brown Corpus or WordNet can be called by running `from nltk.corpus import wordnet/brown`. After the corpora are loaded, their various properties can be explored and used in text analysis; for example, `dogsyn = wordnet.synsets('dog')` will return a list of WordNet synsets related to the word "dog."

Term frequency distribution and *n*-gram indexing are other techniques implemented in NLTK. For example, a user can compute frequency distribution of individual terms within a document *doc* by running a command in Python: `fdist = nltk.FreqDist(text)`. This command returns a dictionary of all tokens with associated frequency within *doc*.

The *N*-gram indexing is implemented as a chain-linked collocations algorithm that takes into account the probability of any given two, three, or more words appearing together in the entire corpus. In general, *n*-grams can be discovered as easily as running `bigrams = nltk.bigrams(text)`. However, a more sophisticated approach is needed to discover statistically significant word collocations, as we show in Listing Bigrams.

▶ http://www.nltk.org

Bird et al. (2009) provide a detailed description of NLTK tools and techniques. See also the official NLTK website.

8.6.2   Stanford CoreNLP

While NLTK's emphasis is on simple reference implementations, Stanford's CoreNLP (Manning et al., 2014) is focused on fast

```
def bigram_finder(texts):
  # NLTK bigrams from a corpus of documents separated by new line
  tokens_list = nltk.word_tokenize(re.sub("\n"," ",texts))
  bgm = nltk.collocations.BigramAssocMeasures()
  finder = nltk.collocations.BigramCollocationFinder.from_words(
    tokens_list)
  scored = finder.score_ngrams( bgm.likelihood_ratio )

  # Group bigrams by first word in bigram.
  prefix_keys = collections.defaultdict(list)
  for key, scores in scored:
      prefix_keys[key[0]].append((key[1], scores))

  # Sort keyed bigrams by strongest association.
  for key in prefix_keys:
      prefix_keys[key].sort(key = lambda x: -x[1])
```

Listing 8.1. Python code to find bigrams using NLTK

implementations of cutting-edge algorithms, particularly for syntactic analysis (e.g., determining the subject of a sentence).

▶ https://stanfordnlp.github.io/CoreNLP/

8.6.3   The MALLET

For probabilistic models of text, the MAchine Learning for LanguagE Toolkit (MALLET) (McCallum, 2002), often strikes the right balance between usefulness and usability. It is written to be fast and efficient but with sufficient documentation and easy interfaces that enable it to be used by novices. It offers fast popular implementations of conditional random fields (for part-of-speech tagging), text classification, and topic modeling.

8.6.3.1   Spacy.io

While NLTK is a Python toolkit optimized for teaching NLP concepts to students, Spacy.io [http://spacy.io] is optimized for practical application. It is fast and contains many models for well-trodden tasks (classification, parsing, finding entities in sentences, etc.). It also has pre-trained models (including word and sentence representations) that can help practitioners quickly build competitive models.

8.6.3.2   Pytorch

For the truly adventurous who want to build their own deep learning models for text, PyTorch [http://pytorch.org] offers the flexibility to go from word vectors to complete deep representations of sentences.

8.7 Summary

Many of the new sources of data that are of interest to social scientists appear as text: tweets, Facebook posts, corporate emails, and the news of the day. However, the meaning of these documents is buried beneath the ambiguities and noisiness of the informal inconsistent ways by which humans communicate with each other and traditional data analysis methods do not work with text data directly. Despite attempts to formalize the meaning of text data through asking users to tag people, apply metadata, or to create structured representations, these attempts to manually curate meaning are often incomplete, inconsistent, or both.

These aspects make text data difficult to work with, but also a rewarding object of study. Unlocking the meaning of a piece of text helps bring machines closer to human-level intelligence—as language is one of the most quintessentially human activities—and helps overloaded information professionals do their jobs more effectively: understand large corpora, find the right documents, or automate repetitive tasks. And, as an added bonus, the better computers become at understanding natural language, the easier it is for information professionals to communicate their needs: one day using computers to grapple with big data may be as natural as sitting down for a conversation over coffee with a knowledgeable and trusted friend.

8.8 Resources

Text analysis is one of the more complex tasks in big data analysis. Because it is unstructured, text (and natural language overall) requires significant processing and cleaning before we can engage in interesting analysis and learning. In this chapter, we have referenced several resources that can be helpful in mastering text mining techniques.

▶ http://www.nltk.org

- The Natural Language Toolkit is one of the most popular Python-based tools for natural language processing. It has a variety of methods and examples that are easily accessible online. The book by Bird et al. (2009), available online, contains multiple examples and tips on how to use NLTK. This is a great package to use if you want to *understand* these models.

- A paper by Anna Huang (2008) provides a brief overview of the key similarity measures for text document clustering discussed in this chapter, including their strengths and weaknesses in different contexts.

- Materials at the MALLET website (McCallum, 2002) can be specialized for the unprepared reader but are helpful when looking for specific solutions with topic modeling and machine classification using this toolkit.

- We provide an example of how to run topic modeling using MALLET on textual data from the National Science Foundation and Norwegian Research Council award abstracts.

  ▶ http://www.umiacs.umd. edu/~jbg/lda_demo

- If you do not care about understanding and only want models that are easy to use and fast, spaCy [https://spacy.io/] has a useful minimal core of models for the average user. Indeed, spaCy is the most useful toolkit for the preprocessing steps of dataset preparation.

- For more advanced models (classification, tagging, etc.), the AllenNLP toolkit [https://allennlp.org/] is useful if you want to run state of the art models and only adjust them slightly.

- Text corpora—sets of multiple similar documents, each called a *corpus*—can be very helpful. For example, the Brown University Standard Corpus of Present-Day American English, or simply the Brown Corpus (Francis and Kucera, 1979), is a collection of processed documents from works published in the United States in 1961. The Brown Corpus represents a historical milestone: it was a machine-readable collection of 1 million words across 15 balanced genres with each word tagged with its part of speech (e.g., noun, verb, preposition). The British National Corpus (University of Oxford, 2006) has repeated that process for British English at a larger scale. The Penn Treebank (Marcus et al., 1993) provides additional information: in addition to part-of-speech annotation, it provides *syntactic* annotation. For example, what is the object of the sentence, "The man bought the hat?" These standard corpora serve as training data to train the classifiers and machine learning techniques to automatically analyze text (Halevy et al., 2009).

- The *Text Analysis* workbook in Chapter 13 provides an introduction to topic modeling with Python.

  ▶ See https://workbooks. coleridgeinitiative.org.

Chapter 9

Networks: The Basics

Jason Owen-Smith

Social scientists are typically interested in describing the activities of individuals and organizations (such as households and firms) in a variety of economic and social contexts. The frame within which data has been collected typically will have been generated from tax or other programmatic sources. The new types of data permit new units of analysis—particularly network analysis—largely enabled by advances in mathematical graph theory. This chapter provides an overview of how social scientists can use network theory to generate measurable representations of patterns of relationships connecting entities. The value of the new framework is not only in constructing different right-hand-side variables but also in studying an entirely new unit of analysis that lies somewhere between the largely atomistic actors that occupy the markets of neo-classical theory and the tightly managed hierarchies that are the traditional object of inquiry of sociologists and organizational theorists.

9.1   Introduction

Social scientists have studied networks for many years. In fact, much of the theory behind network analysis comes from the social sciences where we have studied relationships between people, groups, and organizations (Moreno, 1934). What is different today is the scale of the data available to us to perform this analysis. Instead of studying a group of 25 participants in a karate club (Zachary, 1977), we now have data about hundreds of millions of people communicating with each other through social media channels, or hundreds of thousands of employees in a large multinational organization collaborating on projects. This increased scale requires us to explore new methods of answering the same questions which remain of interest to us, while it also opens up avenues to answering new questions that could not be answered previously.

▶ If you have examples from your own research using the methods we describe in this chapter, please submit a link to the paper (and/or code) here: https://textbook.coleridge initiative.org/submitexam ples.

Box 9.1: Network analysis examples

- Example 1: Cascading information. By using network analysis, researchers were able to characterize how information travels and recurs in patterns, exhibiting multiple bursts of popularity (Ugander et al., 2011).

- Example 2: Large-scale social networks. Using data from a variety of sources to determine how people are connected, researchers were able to observe how communities form and social interactions change over time (Stopczynski et al., 2014).

- Example 3: Facebook social graph. A study of the Facebook friendship network was able to characterize how connected people were on the social networking site. By studying clustering and friendship preferences, the researchers were able to see how communities form and explore the "six degrees of separation" phenomenon on Facebook (Cheng et al., 2016).

This chapter provides a basic introduction to the analysis of large networks for social science research and practice. We describe how to use data from existing social networks as well as how to turn "non-network" data into a network to perform further analysis. Then, we describe different measures that can be calculated to understand the properties of the network being analyzed, show different network visualization techniques, and discuss social science questions that these network measures and visualizations can help us answer.

We use the comparison of the collaboration networks of two research-intensive universities to show how to perform network analysis but the same approach generalizes to other types of problems. The collaboration networks and a grant co-employment network for a large public university examined in this chapter are derived from data produced by the multi-university Committee on Institutional Cooperation (CIC)'s UMETRICS project (Lane et al., 2015). The snippets of code that are provided are from the `igraph` package for network analysis as implemented in Python.

9.2   What are networks?

Networks are measurable representations of relationships connecting entities. What this means is that there are two fundamental questions to ask of any network representation. First, what are the nodes, or entities, that are connected? Second, what are the relationships (ties or edges) connecting the nodes? Once we have the representation, then we can analyze the underlying data and relationships through the measures and methods described in this chapter. This is of great interest because a significant amount of research in social sciences demonstrates that networks are essential to understanding behaviors and outcomes at both the individual and the organizational level.

Networks offer not only another convenient set of right-hand-side variables but an entirely new unit of analysis that lies somewhere between the largely atomistic actors that occupy the markets of neo-classical theory and the tightly managed hierarchies that are the traditional object of inquiry of sociologists and organizational theorists. As Walter W. Powell (2003) states in a description of buyer supplier networks of small Italian firms: "when the entangling of obligation and reputation reaches a point where the actions of the parties are interdependent, but there is no common ownership or legal framework ... such a transaction is neither a market exchange nor a hierarchical governance structure, but a separate, different mode of exchange."

Existing as they do between the uncoordinated actions of independent individuals and coordinated work of organizations, networks offer a unique level of analysis for the study of scientific and creative teams (Wuchty et al., 2007), collaborations (Kabo et al., 2015), and clusters of entrepreneurial firms (Owen-Smith and Powell, 2004).

The following sections will introduce you to this approach to studying innovation and discovery, focusing on examples drawn from high-technology industries and particularly from the scientific collaborations among grant-employed researchers at UMET-RICS universities. We make particular use of a network that connects individual researchers to grants that paid their salaries in 2012 for a large public university. The grants network for university A includes information on 9,206 individuals who were employed on 3,389 research grants from federal science agencies in a single year. The web of partnerships that emerges from university scientists' decentralized efforts to build effective collaborations and teams generates a distinctive social infrastructure for cutting-edge science.

While this chapter focuses on social networks, the techniques described here have been used to examine the structure of networks such as the World Wide Web, the national railway route map of the US, the food web of an ecosystem, or the neuronal network of a particular species of animal.

9.3   Structure for this chapter

The chapter first introduces the most common structures for large
network data, briefly introduces three key social "mechanisms of
action" by which social networks are thought to have their effects,
and then presents a series of basic measures that can be used to
quantify characteristics of entire networks and the relative position
individual people or organizations hold in the differentiated social
structure created by networks.

Taken together, these measures offer an excellent starting point
for examining how global network structures create opportunities
and challenges for the people in them, for comparing and explaining
the productivity of collaborations and teams, and for making sense
of the differences between organizations, industries, and markets
that hinge on the pattern of relationships connecting their participants.

Understanding the productivity and effects of university research
thus requires an effort to measure and characterize the networks
on which it depends. As suggested, those networks influence outcomes in three ways: first, they distinguish among individuals; second, they differentiate among teams; and third, they help to distinguish among research-performing universities. Most research-
intensive institutions have departments and programs that cover
similar arrays of topics and areas of study. What distinguishes
them from one another is not the topics they cover but the ways
in which their distinctive collaboration networks lead them to have
quite different scientific capabilities.

9.4   Turning data into a network

Networks are comprised of *nodes*, which represent entities that can
be connected to one another, and of ties that represent the relationships connecting nodes. When ties are undirected (representing
a relationship between the nodes that is not directional), they are
called *edges*. When they are directed (as when I lend money to you
and you do or do not reciprocate), they are called *arcs*. Nodes, edges,
and arcs can, in principle, be anything: patents and citations, web
pages and hypertext links, scientists and collaborations, teenagers
and intimate relationships, nations and international trade agreements. The very flexibility of network approaches means that the
first step toward doing a network analysis is to turn our data into

a graph by clearly defining what counts as a node and what counts as a tie. In traditional social network analysis studies, there is a natural representation of the data as a network. People are often nodes, and some type of communication between them form the "ties." While this seems like an easy move, it often requires deep thought. For instance, an interest in innovation and discovery could take several forms. We could be interested in how universities differ in their capacity to respond to new requests for proposals (a macro question that would require the comparison of full networks across campuses). We could wonder what sorts of training arrangements lead to the best outcomes for graduate students (a more micro-level question that requires us to identify individual positions in larger networks). Or, we could ask what team structure is likely to lead to more or less radical discoveries (a decidedly meso-level question that requires we identify substructures and measure their features).

Each of these is a network question that relies on the ways in which people are connected to one another. The first challenge of measurement is to identify the nodes (what is being connected) and the ties (the relationships that matter) in order to construct the relevant networks. The next is to collect and structure the data in a fashion that is sufficient for analysis. Finally, measurement and visualization decisions must be made.

9.4.1 Types of networks

Network ties can be directed (flowing from one node to another) or undirected. In either case, they can be binary (indicating the presence or absence of a tie) or valued (allowing for relationships of different types or strengths). Network data can be represented in matrices or as lists of edges and arcs. All of these types of relationships can connect one type of node (what is commonly called *one-mode* network data) or multiple types of nodes (what is called *two-mode* or affiliation data). Varied data structures correspond to different classes of network data. The simplest form of network data represents instances where the same kinds of nodes are connected by undirected ties (edges) that are binary. An example of this type of data is a network where nodes are firms and ties indicate the presence of a strategic alliance connecting them (Powell et al., 2005). This network would be represented as a square symmetric matrix or a list of edges connecting nodes. Figure 9.1 summarizes this simple data structure, highlighting the idea that network data of this form can be represented either as a matrix or as an edge list. If this data was representing acquisitions, we could turn it into a

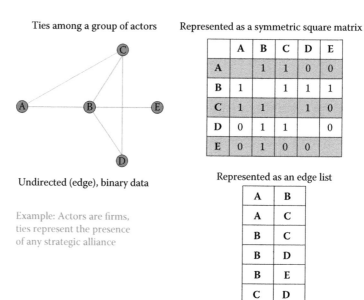

Ties among a group of actors   Represented as a symmetric square matrix

	A	B	C	D	E
A		1	1	0	0
B	1		1	1	1
C	1	1		1	0
D	0	1	1		0
E	0	1	0	0	

Undirected (edge), binary data

Example: Actors are firms,
ties represent the presence
of any strategic alliance

Represented as an edge list

A	B
A	C
B	C
B	D
B	E
C	D

Figure 9.1. Undirected, binary, one-mode network data

directed graph where the edge would be directed from the acquiring firm to the acquired firm.

A much more complicated network would be one that is both directed and valued. One example might be a network of nations connected by flows of international trade. Goods and services flow from one nation to another and the value of those goods and services (or their volume) represents ties of different strengths. When networks connecting one class of nodes (in this case nations) are directed and valued, they can be represented as asymmetric valued matrices or lists of arcs with associated values. (See Figure 9.2 for an example.)

Many studies of small- to medium-sized social networks rely on one-mode data. Large-scale social network data of this type are relatively rare, but one-mode data of this sort are relatively common in relationships among other types of nodes, such as web pages, citations connecting patents, or publications. Nevertheless, much "big" social network analysis is conducted using two-mode data. The UMETRICS employee data set is a two-mode network that connects people (research employees) to the grants that pay their wages. These two types of nodes can be represented as a rectangular matrix that is either valued or binary. It is relatively rare

Ties among a group of actors   Represented as an asymmetric square matrix

	A	B	C	D	E
A		0	1	0	0
B	3		3	1	0
C	4	0		0	0
D	0	0	0		5
E	0	2	0	5	

Directed (arc), valued data

Example: Nodes are faculty,
ties represent the number of
payments from one principal
investigator's grant to another's

Represented as an arc list

A C 1
B A 3
B C 3
B D 1
C A 4
D E 5
E B 2
E D 5

Figure 9.2. Directed, valued, one-mode network data

to analyze untransformed two-mode network data. Instead, most
analyses take advantage of the fact that such networks are *dual*
(White et al., 1976). In other words, a two-mode network connecting grants and people can be conceptualized (and analyzed) as two
one-mode networks, or *projections.**

> ★ Key insight: A two-mode
> network can be conceptual-
> ized and analyzed as two
> one-mode networks, or pro-
> jections.

9.4.2   Inducing one-mode networks from two-mode data

The most important technique in large-scale social network analysis is that of inducing one-mode, or unipartite, networks (e.g.,
employee × employee relationships) from two-mode, or bipartite,
data. But the ubiquity and potential value of two-mode data can
come at a cost. Not all affiliations are equally likely to represent
real meaningful relationships. While it seems plausible to assume
that two individuals paid by the same grant have interactions that
reasonably pertain to the work funded by the grant, this need not
be the case.

For example, consider the two-mode grant × person network for
university A. I have used SQL to create a representation of this network that is readable by a freeware network visualization program

called Pajek (Batagelj and Mrvar, 1998). In this format, a network is represented as two lists: a *vertex list* that lists the nodes in the graph and an *edge list* that lists the connections between those nodes. In our grant × person network, we have two types of nodes, people and grants, and one kind of edge, used to represent wage payments from grants to individuals.

I present a brief snippet of the resulting network file in what follows, showing first the initial 10 elements of the vertex list and then the initial 10 elements of the edge list, presented in two columns for compactness. (The complete file comprises information on 9,206 employees and 3,389 grants, for a total of 12,595 vertices and 15,255 edges. The employees come first in the vertex list, and so the 10 rows shown all represent employees.) Each vertex is represented by a vertex number–label pair and each edge by a pair of vertices plus an optional value. Thus, the first entry in the edge list (1 10419) specifies that the vertex with identifier 1 (which happens to be the first element of the vertex list, which has value "00100679") is connected to the vertex with identifier 10419 by an edge with value 1, indicating that employee "00100679" is paid by the grant described by vertex 10419.

*Vertices	12595 9206	*Edges	
1	"00100679"	1	10419
2	"00107462"	2	10422
3	"00109569"	3	9855
4	"00145355"	3	9873
5	"00153190"	4	9891
6	"00163131"	7	10432
7	"00170348"	7	12226
8	"00172339"	8	10419
9	"00176582"	9	11574
10	"00203529"	10	11196

*Grant-Person-Network.

The network excerpted here is two-mode because it represents relationships between two different classes of nodes, grants and people. In order to use data of this form to address questions about patterns of collaboration on UMETRICS campuses, we first must transform it to represent collaborative relationships.

A person-by-person projection of the original two-mode network assumes that ties exist between people when they are paid by the same grant. By the same token, a grant-by-grant projection of the original two-mode network assumes that ties exist between grants when they pay the same people. Transforming two-mode data into

Ties linking nodes of two different types   Represented by a rectangular matrix

Example: Researchers and grants

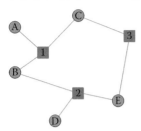

	1	2	3
A	1	0	0
B	1	1	0
C	1	0	1
D	0	1	0
E	0	1	1

$\mathbf{X} = g \times h$ matrix
$\mathbf{X}' = h \times g$ transpose

XX' yields a $g \times g$ symmetrical matrix that represents co-employment ties among individuals

X'X yields an $h \times h$ symmetrical matrix that represents grants connected by people

Figure 9.3. Two-mode affiliation data

one-mode projections is a relatively simple matter. If **X** is a rectangular matrix, $p \times g$, then a one-mode projection, $p \times p$, can be obtained by multiplying **X** by its transpose **X'**. Figure 9.3 summarizes this transformation.

In the following snippet of code, I use the `igraph` package in Python to read in a Pajek file and then transform the original two-mode network into two separate projections. Because my focus in this discussion is on relationships among people, I then move on to work exclusively with the employee-by-employee projection. However, every technique that I describe also can be used with the grant-by-grant projection, which provides a different view of how federally funded research is put together by collaborative relationships on campus.

```
from igraph import *

# Read the graph
g = Graph.Read_Pajek("public_a_2m.net")

# Look at result
summary(g)


# IGRAPH U-WT 12595 15252 --
```

```
# + attr: color (v), id (v), shape (v), type (v), x (v), y (v), z
    (v), weight (e)
# ...
# ...

# Transform to get 1M projection
pr_g_proj1, pr_g_proj2= g.bipartite_projection()

# Look at results
summary(pr_g_proj1)

# IGRAPH U-WT 9206 65040 --
# + attr: color (v), id (v), shape (v), type (v), x (v), y (v), z
    (v), weight (e)

summary(pr_g_proj2)
# IGRAPH U-WT 3389 12510 --
# + attr: color (v), id (v), shape (v), type (v), x (v), y (v), z
    (v), weight (e)

# pr_g_proj1 is the employeeXemployee projection, n=9,206 nodes
# Rename to emp for use in future calculations

emp=pr_g_proj1
```

Now, we can work with the graph `emp`, which represents the collaborative network of federally funded research on this campus. Care must be taken when inducing one-mode network projections from two-mode network data, because not all affiliations provide equally compelling evidence of actual social relationships. While assuming that people who are paid by the same research grants are collaborating on the same project seems plausible, it might be less realistic to assume that all students who take the same university classes have meaningful relationships. For the remainder of this chapter, the examples I discuss are based on UMETRICS employee data rendered as a one-mode person-by-person projection of the original two-mode person-by-grants data. In constructing these networks, I assume that a tie exists between two university research employees when they are paid any wages from the same grant during the same year. Other time frames or thresholds might be used to define ties if appropriate for particular analyses.[*]

> ★ Key insight: Care must be taken when inducing one-mode network projections from two-mode network data because not all affiliations provide equally compelling evidence of actual social relationships.

9.5 Network measures

The power of networks lies in their unique flexibility and ability to address many phenomena at multiple levels of analysis. But, harnessing that power requires calculating measures that take into

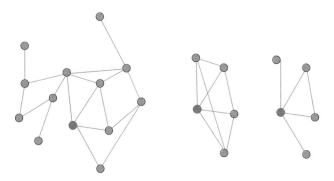

Figure 9.4. Reachability and indirect ties

account the overall structure of relationships represented in a given network. The key insight of structural analysis is that outcomes for any individual or group are a function of the complete pattern of connections among them. In other words, the explanatory power of networks is driven as much by the pathways that *indirectly* connect nodes as by the particular relationships that *directly* link members of a given dyad. Indirect ties create reachability in a network.[*]

> ★ Key insight: Structural analysis of outcomes for any individual or group are a function of the complete pattern of connections among them.

9.5.1   Reachability

Two nodes are said to be reachable when they are connected by an unbroken chain of relationships through other nodes. For instance, two people who have never met may nonetheless be able to reach each other through a common acquaintance who is positioned to broker an introduction (Obstfeld, 2005) or the transfer of information and resources (Burt, 2004). It is the reachability that networks create that makes them so important for understanding the work of science and innovation.

Consider Figure 9.4, which presents three schematic networks. In each, one focal node, ego, is colored orange. Each ego has four alters, but the fact that each has connections to four other nodes masks important differences in their structural positions. Those differences involve the number of other nodes they can reach through the network and the extent to which the other nodes in the network are connected to each other. The orange node (ego) in each network has four partners, but their positions are far from equivalent. Centrality measures on full network data can tease out the differences. The networks also vary in their gross characteristics. Those differences, too, are measurable.[*]

> ★ Key insight: Much of the power of networks (and their systemic features) is due to indirect ties that create reachability. Two nodes can reach each other if they are connected by an unbroken chain of relationships; often these are called indirect ties.

Networks in which more of the possible connections among nodes are realized are denser and more cohesive than networks in which fewer potential connections are realized. Consider the two smaller networks in Figure 9.4, each of which is comprised of five nodes. Only five ties connect those nodes in the network on the far right of the figure. One smaller subset of that network, the triangle connecting ego and two alters at the center of the image, represents a more cohesively connected subset of the networks. In contrast, eight of the nine ties that are possible connect the five nodes in the middle figure; no subset of those nodes is clearly more interconnected than any other. Although these kinds of differences may seem trivial, they have implications for the orange nodes, and for the functioning of the networks as a whole. Structural differences between the positions of nodes, the presence and characteristics of cohesive "communities" within larger networks (Girvan and Newman, 2002), and many important properties of entire structures can be quantified using different classes of network measures. Newman (2010) provides the most recent and most comprehensive look at measures and algorithms for network research.

The most essential element to be able to understand about larger scale networks is the pattern of indirect connections among nodes. What is most important about the structure of networks is not necessarily the ties that link particular pairs of nodes to one another. Instead, it is the chains of indirect connections that make networks function as a system and thus make them worthwhile as new levels of analysis for understanding social and other dynamics.

9.5.2 Whole-network measures

The basic terms needed to characterize whole networks are relatively simple. It is useful to know the size (in terms of nodes and ties) of each network you study. This is true both for the purposes of being able to generally gauge the size and connectivity of an entire network and because many of the measures that one might calculate using such networks should be standardized for analytic use. While the list of possible network measures is long, a few commonly used indices offer useful insights into the structure and implications of entire network structures.

9.5.2.1 Components and reachability

As we have discussed, a key feature of networks is reachability. The reachability of participants in a network is determined by their membership in what network theorists call *components*, subsets of

larger networks where every member of a group is indirectly connected to every other. If you imagine a standard node and line drawing of a network, a component is a portion of the network where you
can trace paths between every pair of nodes without ever having to
lift your pen.

Most large networks have a single dominant component that
typically includes 50% to 90% of its participants as well as many
smaller components and isolated nodes that are disconnected from
the larger portion of the network. Because the path length centrality measures described below can only be computed on connected
subsets of networks, it is typical to analyze the largest component of
any given network. Thus, any description of a network or any effort
to compare networks should report the number of components and
the percentage of nodes reachable through the largest component.
In the following code snippet, I identify the weakly connected components of the employee network, emp.

```
# Add component membership
emp.vs["membership"] = emp.clusters(mode="weak").membership

# Add component size
emp.vs["csize"] = [emp.clusters(mode="weak").sizes()[i] for i in
    emp.clusters(mode="weak").membership]

# Identify the main component
# Get indices of max clusters
maxSize = max(emp.clusters(mode="weak").sizes())
emp.vs["largestcomp"] = [1 if maxSize == x else 0 for x in emp.vs[
    "csize"]]

# Add component membership

emp.vs["membership"] = emp.clusters(mode="weak").membership
```

The main component of a network is commonly analyzed and
visualized because the graph-theoretic distance among unconnected
nodes is infinite, which renders calculation of many common network measures impossible without strong assumptions about how
far apart unconnected nodes actually are. While some researchers
replace infinite path lengths with a value that is one plus the longest
path, called the network's diameter, observed in a given structure, it
is also common to simply analyze the largest connected component
of the network.

9.5.2.2   Path length

One of the most robust and reliable descriptive statistics about
an entire network is the average path length, l_G, among nodes.

Networks with shorter average path lengths have structures that may make it easier for information or resources to flow among members in the network. Longer path lengths, by comparison, are associated with greater difficulty in the diffusion and transmission of information or resources. Let g be the number of nodes or vertices in a network. Then,

$$l_G = \frac{1}{g(g-1)} \sum_{i \neq j} d(n_i, n_j),$$

where $d(n_i, n_j)$ is the path length between n_i and n_j. Typically, the path length is defined as the length of the shortest path[*] between two nodes.

★ A *shortest path* is a path that requires the fewest steps, taking into account values of ties if applicable. In networks with unvalued ties, most pairs have several of those. In networks with valued ties, the shortest path may not be the one with the fewest vertices.

As with other measures based on reachability, it is most common to report the average path length for the largest connected component of the network because the graph-theoretic distance between two unconnected nodes is infinite. In an electronic network, such as the World Wide Web, a shorter path length means that any two pages can be reached through fewer hyperlink clicks.

The next snippet of code identifies the distribution of shortest path lengths among all pairs of nodes in a network and the average path length. I also include a line of code that calculates the network distance among all nodes and returns a matrix of those distances. That matrix (saved as empdist) can be used to calculate additional measures or to visualize the graph-theoretic proximities among nodes.

```
# Calculate distances and construct distance table

dfreq=emp.path_length_hist(directed=False)
print(dfreq)

# N = 12506433, mean +- sd: 5.0302 +- 1.7830
# Each * represents 51657 items
# [ 1,  2): * (65040)
# [ 2,  3): ********* (487402)
# [ 3,  4): ******************************** (1831349)
# [ 4,  5): ****************************************************************
    **** (2996157)
# [ 5,  6): *********************************************************
    (2733204)
# [ 6,  7): *************************************** (1984295)
# [ 7,  8): ************************ (1267465)
# [ 8,  9): ************ (649638)
# [ 9, 10): ***** (286475)
# [10, 11): ** (125695)
# [11, 12): * (52702)
# [12, 13):  (18821)
```

```
# [13, 14):    (5944)
# [14, 15):    (1682)
# [15, 16):    (403)
# [16, 17):    (128)
# [17, 18):    (28)
# [18, 19):    (5)
print(dfreq.unconnected)
# 29864182

print(emp.average_path_length(directed=False))
#[1] 5.030207

empdist= emp.shortest_paths()
```

These measures provide a few key insights into the employee network we have been considering. First, the average pair of nodes that are connected by indirect paths are slightly more than five steps from one another. Second, however, many node pairs in this network ($unconnected = 29,864,182$) are unconnected and thus unreachable to each other. Figure 9.5 presents a histogram of the distribution of path lengths in the network. It represents the numeric values returned by the `distance.table` command in this code snippet. In this case, the diameter of the network is 18 and five pairs of nodes are reachable at this distance, but the largest group of dyads is reachable ($N = 2,996,157$ dyads) at distance 4. In short, nearly 3 million pairs of nodes are collaborators of collaborators of collaborators of collaborators.

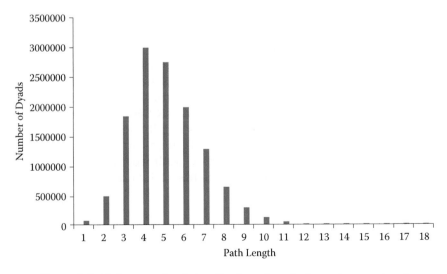

Figure 9.5. Histogram of path lengths for university A employee network

9.5.2.3 Degree distribution

Another powerful way to describe and compare networks is to look at the distribution of centralities across nodes. While any of the centrality measures described above could be summarized in terms of their distribution, it is most common to plot the degree distribution of large networks. Degree distributions commonly have extremely long tails. The implication of this pattern is that most nodes have a small number of ties (typically one or two) and that a small percentage of nodes account for the majority of a network's connectivity and reachability. Typically, degree distributions are so skewed that it is common practice to plot degree against the percentage of nodes with that degree score on a log–log scale.

High-degree nodes are often particularly important actors. In the UMETRICS networks that are employee × employee projections of employee × grant networks, for instance, the nodes with the highest degree seem likely to include high-profile faculty—the investigators on larger institutional grants such as National Institutes of Health-funded Clinical and Translational Science Awards and National Science Foundation-funded Science and Technology Centers, and perhaps staff whose particular skills are in demand (and paid for) by multiple research teams. For instance, the head technician in a core microscopy facility or a laboratory manager who serves multiple groups might appear highly central in the degree distribution of a UMETRICS network.

Most importantly, the degree distribution is commonly taken to provide insight into the dynamics by which a network was created. Highly skewed degree distributions often represent scale-free networks (Barabási and Albert, 1999; Newman, 2005; Powell et al., 2005), which grow in part through a process called *preferential attachment*, where new nodes entering the network are more likely to attach to already prominent participants. In the kinds of scientific collaboration networks that UMETRICS represents, a scale-free degree distribution might arise as faculty new to an institution attempt to enroll more established colleagues on grants as coinvestigators. In the comparison exercise outlined next, I plot degree distributions for the main components of two different university networks.

9.5.2.4 Clustering coefficient

The third commonly used whole-network measure captures the extent to which a network is cohesive, with many nodes interconnected. In networks that are more cohesively clustered, there are fewer opportunities for individuals to assume the kinds of brokering roles that we will discuss in the context of betweenness centrality. Less cohesive networks, with lower levels of clustering, are potentially more conducive to brokerage and the kinds of innovation that accompany it.

However, the challenge of innovation and discovery is both the moment of invention, the "aha!" of a good new idea, and the often complicated, uncertain, and collaborative work that is required to turn an initial insight into a scientific finding. While less clustered, open networks are more likely to create opportunities for brokers to develop fresh ideas; more cohesive and clustered networks support the kinds of repeated interactions, trust, and integration that are necessary to do uncertain and difficult collaborative work.

While it is possible to generate a global measure of cohesiveness in networks, which is generically the number of closed triangles (groups of three nodes all connected to one another) as a proportion of the number of possible triads, it is more common to take a local measure of connectivity and average it across all nodes in a network. This local connectivity measure more closely approximates the notion of cohesion around nodes that is at the heart of studies of networks as a means to coordinate difficult and risky work. The next code snippet calculates both the global clustering coefficient and a vector of node-specific clustering coefficients whose average represents the local measure for the employee × employee network projection of the university A UMETRICS data.

```
# Calculate clustering coefficients
emp.transitivity_undirected()
# 0.7241

local_clust=emp.transitivity_local_undirected(mode="zero")
# (isolates="zero" sets clustering to zero rather than undefined)

import pandas as pd
print(pd.Series(local_clust).describe())
# count    9206.000000
# mean        0.625161
# std         0.429687
# min         0.000000
# 25%         0.000000
# 50%         0.857143
# 75%         1.000000
# max         1.000000
#-------------------------------------------------#
```

Together, these summary statistics—number of nodes, average path length, distribution of path lengths, degree distribution, and the clustering coefficient—offer a robust set of measures to examine and compare whole networks. It is also possible to distinguish among the positions nodes hold in a particular network. Some of the most powerful centrality measures also rely on the idea of indirect ties.[*]

9.5.2.5   Centrality measures

This class of measures is the most common way to distinguish between the positions individual nodes hold in networks. There are many different measures of centrality that capture different aspects of network positions, but they fall into three general types. The most basic and intuitive measure of centrality, *degree centrality*, simply counts the number of ties that a node has. In a binary undirected network, this measure resolves into the number of unique alters to which each node is connected. In mathematical terms, it is the row or column sum of the adjacency matrix that characterizes a network. Degree centrality, $C_D(n_i)$, represents a clear measure of the prominence or visibility of a node. Let

$$C_D(n_i) = \sum_j x_{ij}.$$

The degree of a node is limited by the size of the network in which it is embedded. In a network of g nodes, the maximum degree of any node is $g-1$. The two orange nodes in the small networks presented in Figure 9.4 have the maximum degree possible (4). In contrast, the orange node in the larger 13-node network in that figure has the same number of alters but the possible number of partners is three times as large (12). For this reason, it is problematic to compare raw degree centrality measures across networks of different sizes. Thus, it is common to normalize degree by the maximum value defined by $g-1$:

$$C_D'(n_i) = \frac{\sum_j x_{ij}}{g-1}.$$

While the normalized degree centrality of the two orange nodes of the smaller networks in Figure 9.4 is 1.0, the normalized value for the node in the large network of 13 nodes is 0.33. Despite the fact that the highlighted nodes in the two smaller networks have the same degree centrality, the pattern of indirect ties connecting their alters means they occupy meaningfully different positions. There are a number of degree-based centrality measures that take more of the structural information from a complete network into consideration by using a variety of methods to account not only for the number of partners a particular ego might have but also for the prominence of those partners. Two well-known examples are eigenvector centrality and page rank (see Newman, 2010; Chapters 7 and 8).

Consider two additional measures that capture aspects of centrality that have more to do with the indirect ties that increase reachability. Both make explicit use of the idea that reachability is the source of many of the important social and economic benefits of salutary network positions, but they do so with different substantive emphases. Both of these approaches rely on the idea of a network geodesic, the shortest path connecting any pair of actors. Because these measures rely on reachability, they are only useful when applied to components. When nodes have no ties (degree 0) they are called *isolates*. The geodesic distances are infinite and thus path-based centrality measures cannot be calculated. This is a shortcoming of these measures, which can only be used on connected subsets of graphs where each node has at least one tie to another and all are indirectly connected.

Closeness centrality C_C is based on the idea that networks position some individuals closer to or farther away from other participants. The primary idea is that shorter network paths between actors increase the likelihood of communication and with it the ability to coordinate complicated activities. Let $d(n_i, n_j)$ represent the number of network steps in the geodesic path connecting two nodes i and j. As d increases, the network distance between a pair of nodes grows. Thus, a standard measure of closeness is the inverse of the sum of distances between any given node and all of the others that are reachable in a network:

$$C_C(n_i) = \frac{1}{\sum_{j=1}^{g} d(n_i, n_j)}.$$

The maximum of closeness centrality occurs when a node is directly connected to every possible partner in the network. As with degree centrality, closeness depends on the number of nodes in a network. Thus, it is necessary to standardize the measure to allow comparisons across multiple networks:

$$C_C'(n_i) = \frac{g - 1}{\sum_{j=1}^{g} d(n_i, n_j)}.$$

Similar to closeness centrality, betweenness centrality C_B relies on the concept of geodesic paths to capture nuanced differences of the positions of nodes in a connected network. Where closeness assumes that communication and the flow of information increase with proximity, betweenness captures the idea of brokerage made famous by Burt (1993). Here, too, the idea is that flows of information and resources pass between nodes that are not directly connected through indirect paths. The key to the idea of brokerage is that such paths pass through nodes that can interdict, or otherwise profit from their position "in between" unconnected alters. This idea has been particularly important in network studies of innovation (Owen-Smith and Powell, 2003; Burt, 2004), where flows of information through strategic alliances among firms or social networks connecting individuals significantly influence explanations of why some organizations or individuals are better able to develop creative ideas than others.

To calculate betweenness as originally specified, two strong assumptions are required (Freeman, 1979). First, one must assume that, when people (or organizations) search for new information through their networks, they are capable of identifying the shortest path to what they seek. When multiple paths of equal length exist, we assume that each path is equally likely to be used. Newman (2005) describes an alternative betweenness measure based on random paths through a network, rather than shortest paths, that relaxes these assumptions. For now, let g_{jk} equal the number of geodesic paths linking any two actors. Then, $1/g_{jk}$ is the probability that any given path will be followed on a particular node's search for information or resources in a network. In order to calculate the betweenness score of a particular actor i, it is then necessary to determine how many of the geodesic paths connecting j to k include i. That quantity is $g_{jk}(n_i)$. With these (unrealistic) assumptions in place, we calculate $C_B(n_i)$ as

$$C_B(n_i) = \sum_{j<k} g_{jk}^{(n_i)}/g_{jk}.$$

Here, too, the maximum value depends on the size of the network. For example, $C_B(n_i) = 1$ if i sits on every geodesic path in the network. While this is only likely to occur in small star-shaped networks, it is still common to standardize the measure. Instead of conceptualizing network size in terms of the number of nodes, however, this measure requires that we consider the number of possible pairs of actors (excluding ego) in a structure. When there are g nodes, that quantity is $(g-1)(g-2)/2$ and the standardized betweenness measure is

$$C'_B(n_i) = \frac{C_B(n_i)}{(g-1)(g-2)/2}.$$

Centrality measures of various sorts are the most commonly used means to examine network effects at the level of individual participants in a network. In the context of UMETRICS, such indices might be applied to examine the differential scientific or career success of graduate students as a function of their positions in the larger networks of their universities. In such an analysis, care must be taken to use the standardized measures as university collaboration networks can vary dramatically in size and structure. Describing and accounting for such variations and the possibility of analyses conducted at the level of entire networks or subsets of networks, such as teams and labs, requires a different set of measures. The code snippet presented next calculates each of these measures for the university A employee network we have been examining.

```
# Calculate centrality measures
emp.vs["degree"]=emp.degree()
emp.vs["close"]=emp.closeness(vertices=emp.vs)
emp.vs["btc"]=emp.betweenness(vertices=emp.vs, directed=False)
```

9.6 Case study: Comparing collaboration networks

Consider Figure 9.6, which presents visualizations of the main component of two university networks. Both of these representations are drawn from a single year (2012) of UMETRICS data. Nodes represent people, and ties reflect the fact that those individuals were paid with the same federal grant in the same year. The images are scaled so that the physical location of any node is a function of its position in the overall pattern of relationships in the network. The size and color of nodes represent their betweenness centrality. Larger darker nodes are better positioned to play the role of brokers in the network. A complete review of the many approaches to network visualization and their dangers in the absence of descriptive statistics, such as those presented here, is beyond the scope of this chapter, but consider the guidelines presented in Chapter 6 on information visualization as well as useful discussions by Powell et al. (2005) and Healy and Moody (2014).

Consider the two images. University A is a major public institution with a significant medical school. Similarly, university B is a public institution, but it lacks a medical school and primarily is known for engineering. The two networks manifest some interesting and suggestive differences. Note first that the network on the left (university A) appears much more tightly connected. There is a dense center and there are fewer very large nodes whose positions bridge less well-connected clusters. Similarly, the network on the right (university B) seems at a glance to be characterized by a number of densely interconnected groups that are pulled together by ties through high-degree brokers. One part of this may involve the size and structure of university A's medical school, whose significant NIH funding dominates the network. In contrast, university B's engineering-dominated research portfolio seems to be arranged around clusters of researchers working on similar topic areas and lacks the dominant central core apparent in university A's image.

The implications of these kinds of university-level differences are only starting to be realized, and the UMETRICS data offer great possibilities for exactly this kind of study. These networks, in essence, represent the local social capacity to respond to new problems and to develop scientific findings. Two otherwise similar institutions might have quite different capabilities based on the structure and composition of their collaboration networks.

The intuitions suggested by Figure 9.6 also can be checked against some of the measures we have described. Figure 9.7, for

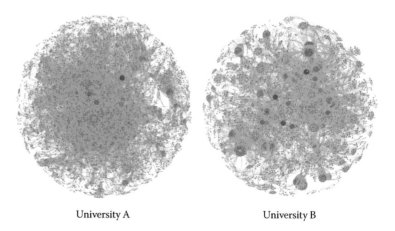

University A   University B

Figure 9.6. The main component of two university networks

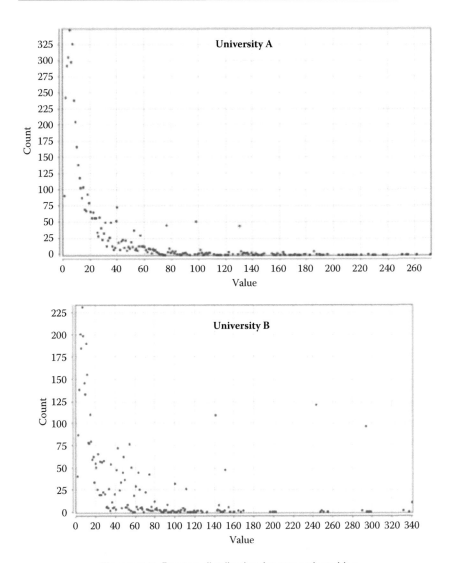

Figure 9.7. Degree distribution for two universities

instance, presents degree distributions for each of the two networks. Figure 9.8 presents the histogram of path lengths for each network.

It is evident from Figure 9.7 that they are quite different in character. University A's network follows a more classic skewed distribution of the sort that often is associated with the kinds of power-law degree distributions common to scale-free networks. In contrast,

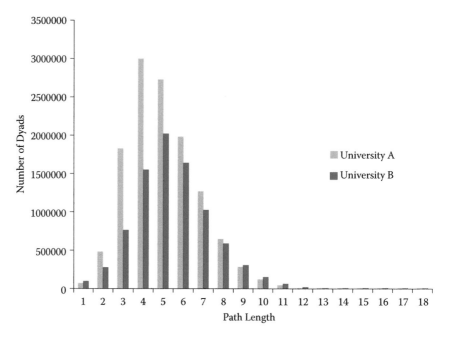

Figure 9.8. Distribution of path lengths for universities A and B

university B's distribution has some interesting features. First, the left side of the distribution is more dispersed than it is for university A, suggesting that there are many nodes with moderate degree. These nodes also may have high betweenness centrality if their ties allow them to span different subgroups within the networks. Of course, this might reflect the fact that each cluster also has members that are more locally prominent. Finally, consider the few instances on the right-side end of the distribution where there are relatively large numbers of people with surprisingly high degree. I suspect these are the result of large training grants or center grants that employ many people. A quirk of relying on one-mode projections of two-mode data is that every person associated with a particular grant is connected to every other. More work needs to be done to validate these hypotheses, but for now it suffices to say that the degree distribution of the networks is consistent with the intuition we obtained from the images—they are significantly different.

The path length histogram presented in Figure 9.8 suggests a similar pattern. While the average distance among any pair of connected nodes in both networks is relatively similar (see Table 9.1),

Table 9.1. Descriptive statistics for the main components of two university networks

	University A	University B
Nodes	4,999	4,144
Edges (total)	57,756	91,970
% nodes in main component	68.67%	67.34%
Diameter	18	18
Average degree	11.554	44.387
Clustering coefficient	0.855	0.913
Density	0.005	0.011
Average path length	5.034	5.463

university B has a larger number of unconnected nodes and university A has a greater concentration of more closely connected dyads. The preponderance of shorter paths in this network also could be a result of a few larger grants that connect many pairs of nodes at unit distance and thus shorten overall path lengths.

But, how do the descriptive statistics "shake out?" Table 9.1 presents the basic descriptive statistics we have discussed for each network. University A's network includes 855 more nodes than university B's, a difference of about 20%. In contrast, there are far fewer edges connecting university A's research employees than connecting university B's, a difference that appears particularly starkly in the much higher density of university B's network. Part of the explanation can be found in the average degree of nodes in each network. As the degree distributions presented in Figure 9.6 suggest, the average researcher at university B is much more highly connected to others than is the case at university A. The difference is stark and quite likely involves the presence of larger grants that employ many individuals.

Both schools have a low average path length (around 5), suggesting that no member of the network is more than five acquaintances away from any other. Similarly, the diameter of both networks is 18, which means that on each campus the most distant pair of nodes is separated by only 18 steps. University A's slightly lower path length may be accounted for by the centralizing effect of its large medical school grant infrastructure. Finally, consider the clustering coefficient. This measure approaches 1 as it becomes more likely that two partners to a third node will themselves be connected. The likelihood that collaborators of collaborators will collaborate is high on both campuses, but substantially higher at university B.

9.7 Summary

This chapter has provided a brief overview of the basics of networks and how to conduct large-scale network analysis. While network measures can produce new and exciting ways to characterize social dynamics, they also are important levels of analysis in their own right. Concepts such as reachability, cohesion, brokerage, and reciprocity are important for a variety of reasons—they can be used to describe networks in terms of their composition and community structure. This chapter provides a classic example of how well social science meets data science. Social science is needed to identify the nodes (what is being connected) and the ties (the relationships that matter) in order to construct the relevant networks. Computer science is necessary to collect and structure the data in a manner that is sufficient for analysis. The combination of data science and social science is key to making the correct measurement and visualization decisions.

9.8 Resources

For more information about network analysis *in general*, the International Network for Social Network Analysis (http://www.insna.org/) is a large interdisciplinary association dedicated to network analysis. It publishes a traditional academic journal, *Social Networks*, an online journal, *Journal of Social Structure*, and a short-format journal, *Connections*, all dedicated to social network analysis. Its several listservs offer vibrant international forums for discussion of network issues and questions. Finally, its annual meetings include numerous opportunities for intensive workshops and training for both beginning and advanced analysts.

A new journal, *Network Science* (http://journals.cambridge.org/action/displayJournal?jid=NWS), published by Cambridge University Press and edited by a team of interdisciplinary network scholars, is a good venue to follow for cutting-edge articles on computational network methods and for substantive findings from a wide range of social, natural, and information science applications.

There are some good software packages available. *Pajek* (http://mrvar.fdv.uni-lj.si/pajek/) is a freeware package for network analysis and visualization. It is routinely updated and has a vibrant user group. Pajek is exceptionally flexible for large networks and has a number of utilities that allow import of its relatively simple file types into other programs and packages for network analysis. *Gephi* (https://gephi.org/) is another freeware package that supports large-scale network visualization. Although I find it less flexible than Pajek, it offers strong support for compelling visualizations.

Stanford Network Analysis Platform (SNAP) (<snap.stanford.edu>) is a general purpose library for network analysis and graph mining. It scales to very large networks, efficiently manipulates large graphs, calculates structural properties, and generates regular and random graphs.

Network Workbench (http://nwb.cns.iu.edu/) is a freeware package that supports extensive analysis and visualization of networks. This package also includes numerous shared data sets from many different fields that can used to test and hone your network analytic skills.

iGraph (http://igraph.org/redirect.html) is my preferred package for network analysis. Implementations are available in R, Python, and C libraries. The examples in this chapter have been coded in iGraph for Python.

Nexus is a growing repository for network data sets that includes some classic data dating back to the origins of social science network research as well as more recent data from some of the best-known publications and authors in network science.

The *Networks* workbook in Chapter 13 provides an introduction to network analysis and visualizations.

▶ See https://workbooks.coleridgeinitiative.org.

Part III
Inference and Ethics

Chapter 10

Data Quality and Inference Errors

Paul P. Biemer

This chapter addresses inference and the errors associated with big data. Social scientists know only too well the cost associated with bad data—we have highlighted both the classic *Literary Digest* example and the more recent Google Flu Trends problems in Chapter 1. Although the consequences are well understood, the new types of data are so large and complex that their properties often cannot be studied in traditional ways. In addition, the data-generating function is such that the data are often selective, incomplete, and erroneous. Without proper data hygiene, the errors can quickly compound. This chapter provides, for the first time, a systematic way to think about the error framework in a big data setting.

10.1 Introduction

Chapters 7 and 11 discuss how analysis errors can lead to bad inferences and suboptimal decision making. In fact, the whole workflow we depict in Chapter 1—and the decisions made along the way—can contribute to errors. In this chapter, we will focus on frameworks that help to detect errors in our data, highlight in general how errors can lead to incorrect inferences, and discuss some strategies to mitigate the inference risk from errors.

The massive amounts of high-dimensional and unstructured data that have recently become available to social scientists, such as data from social media platforms and microdata from administrative data sources, bring both new opportunities and new challenges. Many of the problems with these types of data are well known (see, for example, the AAPOR report by Japec et al., 2015): these data often have selection bias, are incomplete, and are erroneous. As the data are processed and analyzed, new errors can be introduced in downstream operations.

These new sources of data are typically aggregated from disparate sources at various points in time and integrated to form data sets for further analysis. The processing pipeline involve linking records together, transforming them to form new attributes (or variables), documenting the actions taken (although sometimes inadequately), and interpreting the newly created features of the data. These activities may introduce new errors into the data set: errors that may be either *variable* (i.e., errors that create random noise resulting in poor reliability) or *systematic* (i.e., errors that tend to be directional, thus exacerbating biases). Using these new sources of data in statistically valid ways is increasingly challenging in this environment; however, it is important for social scientists to be aware of the error risks and the potential effects of these errors on inferences and decision making. The massiveness, high dimensionality, and accelerating pace of data, combined with the risks of variable and systematic data errors, requires new robust approaches to data analysis.

The core issue that often is the cause of these errors is that such data may not be generated from instruments and methods designed to produce valid and reliable data for scientific analysis and discovery. Rather, this is data that are being repurposed for uses not originally intended. It has been referred to as "found" data or "data exhaust" because it is generated for purposes that often do not align with those of the data analyst. In addition to inadvertent errors, there are also errors from mischief in the data generation process; for example, automated systems have been written to generate bogus content in social media that is indistinguishable from legitimate or authentic data. Social scientists using this data must be keenly aware of these limitations and should take the necessary steps to understand and hopefully mitigate the effects of hidden errors on their results.

10.2   The total error paradigm

We now provide a framework for describing, mitigating, and interpreting the errors in essentially any data set, be it structured or unstructured, massive or small, static or dynamic. This framework has been referred to as the total error framework or paradigm. We begin by reviewing the traditional paradigm, acknowledging its limitations for truly large and diverse data sets, and we suggest how this framework can be extended to encompass the new error structures that we have described.

10.2.1 The traditional model

Managing the risks that errors introduce in big data analysis can be facilitated through a better understanding of the sources and nature of those errors. Such knowledge is gained through in-depth understanding of the data-generating mechanism, the data processing/ transformation infrastructure, and the approaches used to create a specific data set or the estimates derived from it. For survey data, this knowledge is embodied in the well-known *total survey error* (TSE) framework that identifies all of the major sources of error contributing to data validity and estimator accuracy (Biemer and Lyberg, 2003; Groves, 2004; Biemer, 2010). The TSE framework attempts to describe the nature of the error sources and what they may suggest about how the errors could affect inference. The framework parses the total error into bias and variance components that, in turn, may be further subdivided into subcomponents that map the specific types of errors to unique components of the total mean squared error. It should be noted that, while our discussion on issues regarding inference has quantitative analyses in mind, some of the issues discussed here are also of interest to more qualitative uses of big data.

For surveys, the TSE framework provides useful insights regarding how data generating, reformatting, and file preparation processes affect estimation and inference, and it suggests methods for either reducing the errors at their source or adjusting for their effects in the final products to produce inferences of higher quality.

The traditional TSE framework is quite general in that it can be applied to essentially any data set that conforms to the format shown in Figure 10.1. However, in most practical situations, it is quite limited because it makes no attempt to describe how the processes that generated the data, maybe have contributed to what

Record #	V_1	V_2	...	V_P
1				
2				
...				
N				

Figure 10.1. A typical rectangular data file format

could be construed as data errors. In some cases, these processes
constitute a "black box," and the best approach is to attempt to
evaluate the quality of the end product. For survey data, the TSE
framework provides a relatively complete description of the error-
generating processes for survey data and survey frames (Biemer,
2010). But as of this writing, little effort has been devoted to enu-
merating the error sources, the error-generating processes for big
data, and the effect of these errors on some common methods of
data analysis (for an exception see Amaya et al., 2020). Related
articles include three recent papers that discuss some of the issues
associated with integrating multiple data sets for official statistics,
including the effects of integration on data uncertainty (see Zhang,
2012; Holmberg and Bycroft, 2017; Reid et al., 2017). There also
has been some effort to describe these processes for population registers and administrative data (Wallgren and Wallgren, 2007). In
addition, Hsieh and Murphy (2017) develop an error model expressly
for Twitter data.

10.2.1.1 Types of errors

Many administrative data sets have a simple tabular structure, as
do survey sampling frames, population registers, and accounting
spreadsheets. Figure 10.1 is a representation of tabular data as an
array consisting of rows (records) and columns (variables), with their
size denoted by N and p, respectively. The rows typically represent
units or elements of our target population, the columns represent
characteristics, variables (or features) of the row elements, and the
cells correspond to values of the column features for elements on
the rows.

The total error for this data set may be expressed by the following
heuristic formula:

Total error = Row error + Column error + Cell error.

10.2.1.1.1 Row error   For the situations considered in this chapter,
the row errors may be of three types.

- Omissions: Some rows are missing, which implies that elements in the target population are not represented in the file.

- Duplications: Some population elements occupy more than
one row.

- Erroneous inclusions: Some rows contain elements or entities
that are not part of the target population.

10.2.1.1.1.1 Omissions: For survey sample data sets, omissions include members of the target population that are either inadvertently or deliberately absent from the frame, as well as nonsampled frame members. For other types of data, the selectivity of the capture mechanism is a common cause of omissions. For example, a data set consisting of people who have conducted a Google search in the past week can be used to make inferences about that specific population; but, if our goal is to make inferences about the larger population of Internet users, this data set will exclude people who have not conducted a Google search. This selection bias can lead to inference errors if the people who did not conduct a Google search are different from those who did.

Therefore, such exclusions can be viewed as a source of selectivity bias if inference is to be made about an even larger set of people, such as the general population. For one, persons who do not have access to the Internet are excluded from the data set. These exclusions may be biasing in that persons with Internet access may have quite different demographic characteristics from persons who do not have Internet access (Dutwin and Buskirk, 2017). The selectivity of big data capture is similar to frame noncoverage in survey sampling and can bias inferences when researchers fail to consider it and compensate for it in their analyses.

Example: Google searches

As an example, in the United States, the word "Jewish" is included in 3.2 times more Google searches than "Mormon" (Stephens-Davidowitz and Varian, 2015). This does not mean that the Jewish population is 3.2 times larger than the Mormon population. Other possible explanations could be that Jewish people use the Internet in higher proportions, have more questions that require using the word "Jewish," or there could be more searches for "Jewish food" than "Mormon food." Thus, Google search data are more useful for relative comparisons than for estimating absolute levels.

A well-known formula in the survey literature provides a useful expression for the so-called *coverage bias* in the mean of some variable, V. Denote the mean by $\bar{V}$, and let $\bar{V}_T$ denote the (possibly hypothetical because it may not be observable) mean of the target population of N_T elements, including the $N_T - N$ elements that are missing from the observed data set. Then, the bias due to this *noncoverage* is $B_{NC} = \bar{V} - \bar{V}_T = (1 - N/N_T)(\bar{V}_C - \bar{V}_{NC})$, where $\bar{V}_C$ is the mean of the *covered* elements (i.e., the elements in the observed data set) and $\bar{V}_{NC}$ is the mean of the $N_T - N$ *noncovered* elements. Thus, we see that, to the extent that the difference between the covered and noncovered elements is large or the fraction of missing elements $(1 - N/N_T)$ is large, the bias in the descriptive statistic also will be large. As in survey research, often we can only speculate about the sizes of these two components of bias. Nevertheless, speculation is useful for understanding and interpreting the results of data analysis and cautioning ourselves regarding the risks of false inference.

10.2.1.1.1.2 Duplication: We also can expect that big data sets, such as a data set containing Google searches during the previous week, could have the same person represented many times. People who have conducted many searches during the data capture period would be disproportionately represented relative to those who have conducted fewer searchers. If the rows of the data set correspond to tweets in a Twitter feed, duplication can arise when the same tweet is retweeted or when some persons are quite active in tweeting while others observe but tweet much less frequently. Whether such duplications should be regarded as "errors" depends upon the goals of the analysis.

For example, if inference is to be made to a population of persons, persons who tweet multiple times on a topic would be over-represented. If inference is to be made to the population of tweets, including retweets, then such duplication does not bias inference. This is also common in domains such as healthcare or human services where certain people have more interactions with the systems (medical appointments, consumption of social services, etc.) and can be over-represented when performing analysis at an individual interaction level.

When it is a problem, it still may not be possible to identify duplications in the data. Failing to account for them could generate duplication biases in the analysis. If these unwanted duplications can be identified, they can be removed from the data file (i.e., deduplication). Alternatively, if a certain number of rows, e.g., d, correspond to the same population unit, those row values can be weighted by $1/d$ to correct the estimates for the duplications.

10.2.1.1.1.3 Erroneous inclusions: Erroneous inclusions also can create biases. For example, Google searches or tweets may not be generated by a person but rather by a computer either maliciously or as part of an information-gathering or publicity-generating routine. Similarly, some rows may not satisfy the criteria for inclusion in an analysis—for example, an analysis by age or gender includes some row elements not satisfying the criteria. If the criteria can be applied accurately, the rows violating the criteria can be excluded prior to analysis. However, with big data, some out-of-scope elements still may be included as a result of missing or erroneous information, and these inclusions will bias inference.

10.2.1.2 Column error

The most common type of column error in survey data analysis is caused by inaccurate or erroneous labeling of the column data—an example of metadata error. In the TSE framework, this is referred to as a *specification* error. For example, a business register may include a column labeled "number of employees," defined as the number of persons in the company who received a payroll check in the preceding month. Even so, the column may contain the number of persons on the payroll whether or not they received a check in the prior month, thus including, for example, persons on leave without pay.

When analyzing a more diverse set of data sources, such errors could occur because of the complexities involved in producing a data set. For example, data generated from an individual tweet may undergo a number of transformations before it is included in the analysis data set. This transformative process can be quite complex, involving parsing phrases, identifying words, and classifying them as to subject matter and then perhaps further classifying them as either positive or negative expressions about some phenomenon such as the economy or a political figure. There is considerable risk of the resulting variables being either inaccurately defined or misinterpreted by the data analyst.

Example: Specification error with Twitter data

As an example, consider a Twitter data set where the rows correspond to tweets and one of the columns supposedly includes an indicator of whether the tweet contained one of the following keywords: marijuana, pot, cannabis, weed, hemp, ganja, or THC. Instead, the indicator actually corresponds to whether the tweet contained a shorter list of words, e.g., either marijuana or pot. The mislabeled column is an example of specification error which could be a biasing factor in an analysis. For example, estimates of marijuana use based upon the indicator could be underestimates.

10.2.1.3 Cell errors

Finally, cell errors can be of three types: content error, specification error, or missing data.

10.2.1.3.1 Content error
A content error occurs when the value in a cell satisfies the column definition but still deviates from the true value, whether or not the true value is known. For example, the value satisfies the definition of "number of employees" but is outdated because it does not agree with the current number of employees. Errors in sensitive data, such as drug use, prior arrests, and sexual misconduct, may be deliberate. Thus, content errors may be the result of the measurement process, a transcription error, a data processing error (e.g., keying, coding, editing), an imputation error, or some other cause.

10.2.1.3.2 Specification error
A specification error is the same as described for a column error but applied to a cell. For example, the column is correctly defined and labeled; however, a few companies have provided values that, although otherwise highly accurate, are nevertheless inconsistent with the required definition.

10.2.1.3.3 Missing data
Missing data, as the name implies, are simply empty cells. As described in Kreuter and Peng (2014), data sets derived from big data are notoriously affected by all three types of cell error, particularly missing or incomplete data, perhaps because that is the most obvious deficiency.

10.2.1.3.4 Missing data can take two forms
The two forms of missing data are missing information in a cell of a data matrix (referred to as *item missingness*) or missing rows (referred to as *unit missingness*), with the former being readily observable whereas the latter can be completely hidden from the analyst. Much is known from the survey research literature about how both types of missingness affect data analysis (see, for example, Rubin, 1976; Little and Rubin, 2014). Rubin (1976) introduces the term *missing completely at random (MCAR)* to describe data where the data that are available (e.g., the rows of a data set) can be considered as a simple random sample of the inferential population (i.e., the population to which inferences from the data analysis will be made). Because the data set represents the population, MCAR data provide results that are generalizable to this population.

A second possibility also exists for the reasons why data are missing. For example, students who have high absenteeism may be missing because they were ill on the day of the test. They may otherwise be average performers on the test so, in this case, it has little to do with how they would score. Thus, the values are missing for reasons related to another variable, health, that may be available in the data set and completely observed. Students with poor health tend to be missing test scores, regardless of those student's performance on the test. Rubin (1976) uses the term *missing at random (MAR)* to describe data that are missing for reasons related to completely observed variables in the data set. It is possible to compensate for this type of missingness in statistical inferences by modeling the missing data mechanism.

However, most often, missing data may be related to factors that are not represented in the data set and, thus, the missing data mechanism cannot be adequately modeled. For example, there may be a tendency for test scores to be missing from school administrative data files for students who are poor academic performers. Rubin calls this form of missingness *nonignorable*. With nonignorable missing data, the reasons for the missing observations depend on the values that are missing. When we suspect a nonignorable missing data mechanism, we need to use procedures much more complex than will be described here. Little and Rubin (2014) and Schafer (1997) discuss methods that can be used for nonignorable missing data. Ruling out a nonignorable response mechanism can simplify the analysis considerably.

In practice, it is quite difficult to obtain empirical evidence about whether or not the data are MCAR or MAR. Understanding the data generation process is invaluable for specifying models that appropriately represent the missing data mechanism and that then will be successful in compensating for missing data in an analysis. (Note that Schafer and Graham (2002) provide a more thorough discussion of this issue.)

One strategy for ensuring that the missing data mechanism can be successfully modeled is to have available on the data set many variables that may be causally related to missing data. For example, features such as personal income are subject to high item missingness, and often the missingness is related to income. However, less sensitive surrogate variables, such as years of education or type of employment, may be less subject to missingness. The statistical relationship between income and other income-related variables increases the chance that information lost in missing variables is supplemented by other completely observed variables. Model-based methods use the multivariate relationship between variables to handle the missing data. Thus, the more informative the data set, the more measures we have on important constructs, the more successfully we can compensate for missing data using model-based approaches.

In the next section, we consider the impact of errors on some forms of analysis that are common in the big data literature. We will limit the focus on the effects of content errors on data analysis. However, there are numerous resources available for studying and mitigating the effects of missing data on analysis, such as Little and Rubin (2014), Schafer (1997), and Allison (2001).

10.3 Example: Google Flu Trends

▶ See the discussion in Section 1.3.

A well-known example of the risks of bad inference is provided by the Google Flu Trends series that uses Google searches on flu symptoms, remedies, and other related keywords to provide near-real-time estimates of flu activity in the US and 24 other countries. Compared to CDC data, the Google Flu Trends provides remarkably accurate indicators of flu incidence in the US between 2009 and 2011. However, for the 2012–2013 flu seasons, the Google Flu Trends estimates are almost double the CDC's (Butler, 2013). Lazer et al. (2014) cite two causes of this error: big data hubris and algorithm dynamics.

Hubris occurs when the big data researcher believes that the volume of the data compensates for any of its deficiencies, thus obviating the need for traditional scientific analytic approaches. As Lazer et al. (2014) note, big data hubris fails to recognize that "quantity of data does not mean that one can ignore foundational issues of measurement and construct validity and reliability."

Algorithm dynamics refers to properties of algorithms that allow them to adapt and "learn" as the processes generating the data change over time. Although explanations vary, the fact remains that Google Flu Trends estimates are too high and by considerable margins for 100 out of 108 weeks starting in July 2012. Lazer et al. (2014) also blame "blue team dynamics," which arises when the data-generating engine is modified in such a way that the formerly highly predictive search terms eventually fail to work. For example, when a Google user searches on "fever" or "cough," Google's other programs start recommending searches for flu symptoms and treatments—the very search terms the algorithm uses to predict flu. Thus, flu-related searches artificially spike as a result of changes to the algorithm and the impact these changes have on user Pbehavior.

In survey research, this is similar to the measurement biases induced by interviewers who suggest to respondents who are coughing that they might have flu, then ask the same respondents if they think they might have flu.

Algorithm dynamic issues are not limited to Google. Platforms such as Twitter and Facebook are also frequently modified to improve the user experience. A key lesson provided by Google Flu Trends is that successful analyses using big data today may fail to produce good results tomorrow. All of these platforms change their methodologies more or less frequently, with ambiguous results for any kind of long-term study unless highly nuanced methods are routinely used. Recommendation engines often exacerbate effects in a certain direction, but these effects are difficult to determine. Furthermore, other sources of error may affect Google Flu Trends to an unknown extent. For example, selectivity may be an important issue because the demographics of people with Internet access are quite different from the demographic characteristics related to flu incidence (Thompson et al., 2006). Thus, the "at risk" population for influenza and the implied population based on Google searches do not correspond. This illustrates only one type of representativeness issue that often plagues big data analysis. In general, it is an issue that algorithms are not (publicly) measured for accuracy, because they are often proprietary. Google Flu Trends is special in that it publicly failed. From what we have seen, most models fail privately and often without anyone noticing.

10.4 Errors in data analysis

The total error framework described above focuses on different types of errors in the data that can lead to incorrect inference. In addition to direct inference errors because of errors in the data, our analysis also can be incorrect because of these data errors. This section goes deeper into these common types of analysis errors when analyzing a diverse set of data sources. We begin by exploring errors that can occur under the assumption of accurate data, and then we consider errors in three common types of analysis when data are not accurate: classification, correlation, and regression.

10.4.1 Analysis errors despite accurate data

Data deficiencies represent only one set of challenges for the big data analyst. Even if data are correct, other challenges can arise solely as a result of the massive size, rapid generation, and vast dimensionality of the data (Meng, 2018). Fan et al. (2014) identify three issues—noise accumulation, spurious correlations, and incidental endogeneity—which will be discussed in this section. These issues should concern social scientists even if the data could be regarded as infallible. Content errors, missing data, and other data deficiencies will only exacerbate these problems.

10.4.2 Noise accumulation

To illustrate noise accumulation, Fan et al. (2014) consider the following scenario. Suppose an analyst is interested in classifying individuals into two categories, C_1 and C_2, based upon the values of 1,000 variables in a big data set. Suppose further that, unknown to the researcher, the mean value for persons in C_1 is 0 on all 1,000 variables while persons in C_2 have a mean of 3 on the first 10 variables and 0 on all other variables. Because we are assuming the data are error-free, a classification rule based upon the first $m \leq 10$ variables performs quite well, with little classification error. However, as more variables are included in the rule, classification error increases because the uninformative variables (i.e., the 990 variables having no discriminating power) eventually overwhelm the informative signals (i.e., the first 10 variables). In the Fan et al. (2014) example, when $m > 200$, the accumulated noise exceeds the signal embedded in the first 10 variables and the classification rule becomes equivalent to a coin-flip classification rule.

10.4.3 Spurious correlations

High dimensionality also can introduce coincidental (or *spurious*) correlations in that many unrelated variables may be highly correlated simply by chance, resulting in false discoveries and erroneous inferences. The phenomenon depicted in Figure 10.2 is an illustration of this. Many more examples can be found on a website and in a book devoted to the topic (Vigen, 2015). Fan et al. (2014) explain this phenomenon using simulated populations and relatively small sample sizes. They illustrate how, with 800 independent (i.e., uncorrelated) variables, the analyst has a 50% chance of observing an absolute correlation that exceeds 0.4. Their results suggest that there are considerable risks of false inference associated with a purely empirical approach to predictive analytics using high-dimensional data.

▶ http://www.tylervigen. com/spurious-correlations

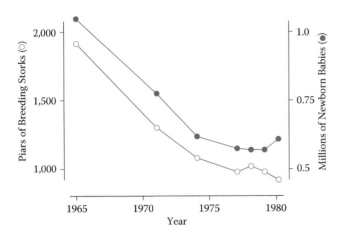

Figure 10.2. An illustration of coincidental correlation between two variables: stork die-off linked to human birth decline (Sies, 1988)

10.4.4   Incidental endogeneity

Finally, turning to incidental endogeneity, a key assumption in regression analysis is that the model covariates are uncorrelated with the residual error; endogeneity refers to a violation of this assumption. For high-dimensional models, this can occur purely by chance—a phenomenon Fan and Liao (2014) call *incidental endogeneity*. Incidental endogeneity leads to the modeling of spurious variation in the outcome variables resulting in errors in the model selection process and biases in the model predictions. The risks of incidental endogeneity increase as the number of variables in the model selection process grows large. Thus, it is a particularly important concern for big data analytics.

Fan et al. (2014) as well as a number of other authors (e.g., Stock and Watson, 2002; Fan et al., 2009; Hall and Miller, 2009; Fan and Liao, 2012) suggest robust statistical methods aimed at mitigating the risks of noise accumulation, spurious correlations, and incidental endogeneity. However, as previously noted, these issues and others are further compounded when data errors are present in a data set. Biemer and Trewin (1997) show that data errors will bias the results of traditional data analysis and inflate the variance of estimates in ways that are difficult to evaluate or mitigate in the analysis process.

10.4.5 Analysis errors resulting from inaccurate data

The previous sections have examined some of the issues social scientists face as either N or p in Figure 10.1 becomes extremely large. When row, column, and cell errors are added into the mix, these problems can be further exacerbated. For example, noise accumulation can be expected to accelerate when random noise (i.e., content errors) afflicts the data. Spurious correlations that give rise to both incidental endogeneity and coincidental correlations can render correlation analysis meaningless if the error levels in big data are high. In this section, we consider some of the issues that arise in classification, correlation, and regression analysis as a result of content errors that may be either variable or systematic.

There are various important findings in this section. First, for rare classes, even small levels of error can impart considerable biases in classification analysis. Second, variable errors will attenuate correlations and regression slope coefficients; however, these effects can be mitigated by forming meaningful aggregates of the data and substituting these aggregates for the individual units in these analyses. Third, unlike random noise, systematic errors can bias correlation and regression analysis is unpredictable ways, and these biases cannot be effectively mitigated by aggregating the data. Finally, multilevel modeling can—under certain circumstances—be an important mitigation strategy for addressing systematic errors emanating from multiple data sources. These issues will be examined in some detail in the remainder of this section.

We will start by focusing on two types of errors: variable (uncorrelated) errors and correlated errors. We first will describe these errors for continuous data and then extend it to categorical variables in the next section.

10.4.5.1 Variable (uncorrelated) and correlated error in continuous variables

Error models are essential for understanding the effects of error on data sets and the estimates that may be derived from them. They allow us to concisely and precisely communicate the nature of the errors that are being considered, the general conditions that give rise to them, how they affect the data, how they may affect the analysis of these data, and how their effects can be evaluated and mitigated. In the remainder of this chapter, we focus primarily on content errors and consider two types of error, variable errors and correlated errors, the latter a subcategory of systematic errors.

Variable errors are sometimes referred to as *random noise* or *uncorrelated* errors. For example, administrative databases often contain errors from a myriad of random causes, including mistakes in keying or other forms of data capture, errors on the part of the persons providing the data due to confusion about the information requested, difficulties in recalling information, the vagaries of the terms used to request the inputs, and other system deficiencies.

Correlated errors, on the other hand, carry a systematic effect that results in a nonzero covariance between the errors of two distinct units. For example, quite often, an analysis data set may combine multiple data sets from different sources and each source may impart errors that follow a somewhat different distribution. As we will show, these differences in error distributions can induce correlated errors in the merged data set. It is also possible that correlated errors are induced from a single source as a result of different operators (e.g., computer programmers, data collection personnel, data editors, coders, data capture mechanisms) handling the data. Differences in the way these operators perform their tasks have the potential to alter the error distributions so that data elements handled by the same operator have errors that are correlated (Biemer and Lyberg, 2003).

These concepts may be best expressed by a simple error model. Let y_{rc} denote the cell value for variable c on the rth unit in the data set, and let ε_{rc} denote the error associated with this value. Suppose it can be assumed that there is a true value underlying y_{rc}, which is denoted by μ_{rc}. Then, we can write

$$y_{rc} = \mu_{rc} + \varepsilon_{rc} \, . \qquad (10.1)$$

At this point, ε_{rc} is not stochastic in nature because a statistical process for generating the data has not yet been assumed. Therefore, it is not clear what *correlated error* really means. To remedy this problem, we can consider the hypothetical situation where the processes generating the data set can be repeated under the same general conditions (i.e., at the same point in time with the same external and internal factors operating). Each time the processes are repeated, a different set of errors may be realized. Thus, it is assumed that, although the true values μ_{rc} are fixed, the errors ε_{rc} can vary across the hypothetical infinite repetitions of the data set generating process. Let E($\cdot$) denote the expected value over all of these hypothetical repetitions, and define the variance Var($\cdot$) and covariance Cov($\cdot$) analogously.

For the present, error correlations between variables are not considered, and thus the subscript c is dropped to simplify the notation.

For the uncorrelated data model, we assume that $E(y_r|r) = \mu_r$, $Var(y_r|r) = \sigma_\varepsilon^2$, and $Cov(y_r, y_s|r, s) = 0$, for $r \neq s$. For the correlated data model, the latter assumption is relaxed. To add a bit more structure to the model, suppose the data set is the product of combining data from multiple sources (or operators) denoted by $j = 1, 2, \ldots, J$, and let b_j denote the systematic effect of the jth source. Here, we also assume that, with each hypothetical repetition of the data set generating process, these systematic effects can vary stochastically. (It is also possible to assume the systematic effects are fixed. See, for example, Biemer and Stokes (1991) for more details on this model.) Thus, we assume that $E(b_j) = 0$, $Var(b_j) = \sigma_b^2$, and $Cov(b_j, b_k) = 0$ for $j \neq k$.

Finally, for the rth unit within the jth source, let $\varepsilon_{rj} = b_j + e_{rj}$. Then, it follows that

$$Cov(\varepsilon_{rj}, \varepsilon_{sk}) = \begin{cases} \sigma_b^2 + \sigma_\varepsilon^2 & \text{for } r = s, j = k \\ \sigma_\varepsilon^2 & \text{for } r = s, j \neq k \\ 0 & \text{for } r \neq s, j \neq k \end{cases}.$$

The case where $\sigma_b^2 = 0$ corresponds to the uncorrelated error model (i.e., $b_j = 0$) and thus ε_{rj} is purely random noise.

Example: Speed sensor

Suppose that, due to calibration error, the jth speed sensor in a traffic pattern study underestimates the speed of vehicle traffic on a highway by an average of 4 miles per hour. Thus, the model for this sensor is that the speed for the rth vehicle recorded by this sensor (y_{rj}) is the vehicle's true speed (μ_{rj}) minus 4 mph (b_j) plus a random departure from -4 for the rth vehicle (ε_{rj}). Note that, to the extent that b_j varies across sensors $j = 1, \ldots, J$ in the study, σ_b^2 will be large. Further, to the extent that ambient noise in the readings for jth sensor causes variation around the values $\mu_{rc} + b_j$, then σ_ε^2 will be large. Both sources of variation will reduce the reliability of the measurements. However, as will be shown in Section 10.4.5.4, the systematic error component is particularly problematic for many types of analysis.

10.4.5.2   Extending variable and correlated error to categorical data

For variables that are categorical, the model of the previous section is not appropriate because the assumptions it makes about the error structure do not hold. For example, consider the case of a binary (0/1) variable. Because both y_r and μ_r should be either 1 or 0, the error in equation (10.1) must assume the values of -1, 0, or 1.

A more appropriate model is the misclassification model described by Biemer (2011), which we summarize here.

Let ϕ_r denote the probability of a false positive error (i.e., $\phi_r = \Pr(y_r = 1|\mu_r = 0)$), and let ∂_r denote the probability of a false negative error (i.e., $\partial_r = \Pr(y_r = 0|\mu_r = 1)$). Thus, the probability that the value for row r is correct is $1 - \partial_r$ if the true value is 1, and $1 - \phi_r$ if the true value is 0.

As an example, suppose an analyst wishes to compute the proportion, $P = \sum_r y_r/N$, of the units in the file that are classified as 1, and let $\pi = \sum_r \mu_r/N$ denote the true proportion. Then, under the assumption of uncorrelated error, Biemer (2011) shows that

$$P = \pi(1 - \partial) + (1 - \pi)\phi,$$

where $\partial = \sum_r \partial_r/N$ and $\phi = \sum_r \phi_r/N$.

In the classification error literature, the sensitivity of a classifier is defined as $1 - \partial$, that is, the probability that a true positive is correctly classified. Correspondingly, $1 - \phi$ is referred to as the specificity of the classifier, that is, the probability that a true negative is correctly classified. Two other quantities that will be useful in our study of misclassification error are the positive predictive value (PPV) and negative predictive value (NPV) given by

$$\text{PPV} = \Pr(\mu_r = 1|y_r = 1), \quad \text{NPV} = \Pr(\mu_r = 0|y_r = 0).$$

The PPV (NPV) is the probability that a positive (negative) classification is correct.

10.4.5.3 Errors when analyzing rare population groups

One of the attractions of newer sources of data, such as social media, is the ability to study rare population groups that seldom appear in large enough numbers in designed studies, such as surveys and clinical trials. Although this is true in theory, in practice content errors can affect the inferences that can be drawn from this data. We illustrate this using the following contrived and somewhat amusing example. The results in this section are particularly relevant to the approaches considered in Chapter 7.

Example: Thinking about probabilities

Suppose, using big data and other resources, we construct a terrorist detector and boast that the detector is 99.9% accurate. In other words, both the probability of a false negative (i.e., classifying a terrorist as a nonterrorist, ∂) and the probability of a false positive (i.e., classifying a nonterrorist as a terrorist, ϕ) are 0.001. Assume that about 1 person in 1 million in the population is a terrorist, that is, $\pi = 0.000001$ (hopefully, somewhat of an overestimate). Your friend, Terry, steps into the machine and, to Terry's chagrin (and your surprise), the detector declares that he is a terrorist! What are the odds that the machine is right? The surprising answer is only about 1 in 1,000. That is, 999 times out of 1,000 times the machine classifies a person as a terrorist, the machine will be wrong!

How could such an accurate machine be wrong so often in the terrorism example? Let us do the math.

The relevant probability is the PPV of the machine: given that the machine classifies an individual (Terry) as a terrorist, what is the probability the individual is truly a terrorist? Using the notation in Section 10.4.5.2 and Bayes' rule, we can derive the PPV as

$$
\begin{aligned}
\Pr(\mu_r = 1|y_r = 1) &= \frac{\Pr(y_r = 1|\mu_r = 1)\Pr(\mu_r = 1)}{\Pr(y_r = 1)} \\
&= \frac{(1 - \partial)\pi}{\pi(1 - \partial) + (1 - \pi)\varphi} \\
&= \frac{0.999 \times 0.000001}{0.000001 \times 0.999 + 0.99999 \times 0.001} \\
&\approx 0.001.
\end{aligned}
$$

This example calls into question whether security surveillance using emails, phone calls, etc. can ever be successful in finding rare threats such as terrorism, because to achieve a reasonably high PPV (e.g., 90%) would require a sensitivity and specificity of at least $1 - 10^{-7}$, or less than 1 chance in 10 million of an error.

To generalize this approach, note that any population can be regarded as a *mixture* of subpopulations. Mathematically, this can be written as

$$
f(y|\mathbf{x}; \partial) = \pi_1 f(y|\mathbf{x}; \partial_1) + \pi_2 f(y|\mathbf{x}; \partial_2) + \ldots + \pi_K f(y|\mathbf{x}; \partial_K),
$$

where $f(y|\mathbf{x}; \partial)$ denotes the population distribution of y given the vector of explanatory variables $\mathbf{x}$ and the parameter vector $\partial = (\partial_1, \partial_2, \ldots, \partial_K)$, π_k is the proportion of the population in the kth subgroup, and $f(y|\mathbf{x}; \partial_k)$ is the distribution of y in the kth subgroup. A rare subgroup is one where π_k is quite small (e.g., less than 0.01).

Table 10.1 shows the PPV for a range of rare subgroup sizes when the sensitivity is perfect (i.e., no misclassification of true positives) and specificity is not perfect but still high. This table reveals the

Table 10.1. Positive predictive value (%) for rare subgroups, high specificity, and perfect sensitivity

π_k	Specificity		
	99%	99.9%	99.99%
0.1	91.70	99.10	99.90
0.01	50.30	91.00	99.00
0.001	9.10	50.00	90.90
0.0001	1.00	9.10	50.00

fallacy of identifying rare population subgroups using fallible classifiers unless the accuracy of the classifier is appropriately matched to the rarity of the subgroup. As an example, for a 0.1% subgroup, the specificity should be at least 99.99%, even with perfect sensitivity, to attain a 90% PPV.

10.4.5.4 Errors in correlation analysis

In Section 10.4, we consider the problem of incidental correlation that occurs when an analyst correlates pairs of variables selected from big data stores containing thousands of variables. In this section, we discuss how errors in the data can exacerbate this problem or even lead to failure to recognize strong associations among the variables. We confine the discussion to the continuous variable model of Section 10.4.5.1 and begin with theoretical results that help explain what occurs in correlation analysis when the data are subject to variable and systematic errors.

For any two variables in the data set, c and d, define the covariance between y_{rc} and y_{rd} as

$$\sigma_{y|cd} = \frac{\sum_r \mathrm{E}(y_{rc} - \bar{y}_c)(y_{rd} - \bar{y}_d)}{N},$$

where the expectation is with respect to the error distributions and the sum extends over all rows in the data set. Let

$$\sigma_{\mu|cd} = \frac{\sum_r (\mu_{rc} - \bar{\mu}_c)(\mu_{rd} - \bar{\mu}_d)}{N}$$

denote the *population* covariance. (The population is defined as the set of all units corresponding to the rows of the data set.) For any variable c, define the variance components

$$\sigma_{y|c}^2 = \frac{\sum_r (y_{rc} - \bar{y}_c)^2}{N}, \quad \sigma_{\mu|c}^2 = \frac{\sum_r (\mu_{rc} - \bar{\mu}_c)^2}{N},$$

and let

$$R_c = \frac{\sigma_{\mu|c}^2}{\sigma_{\mu|c}^2 + \sigma_{b|c}^2 + \sigma_{\varepsilon|c}^2}, \quad \rho_c = \frac{\sigma_{b|c}^2}{\sigma_{\mu|c}^2 + \sigma_{b|c}^2 + \sigma_{\varepsilon|c}^2},$$

with analogous definitions for d. The ratio R_c is known as the *reliability ratio*, and ρ_c will be referred to as the *intra-source correlation*. Note that the reliability ratio is the proportion of total variance that is due to the variation of true values in the data set. If there were no errors, either variable or systematic, then this ratio would be 1. To the extent that errors exist in the data, R_c will be less than 1.

Similarly, ρ_c is also a ratio of variance components that reflects the proportion of total variance that is due to systematic errors with biases that vary by data source. A value of ρ_c that exceeds 0 indicates the presence of systematic error variation in the data. As we will show, even small values of ρ_c can cause significant problems in correlation analysis.

Using the results in Biemer and Trewin (1997), it can be shown that the correlation between y_{rc} and y_{rd}, defined as $\rho_{y|cd} = \sigma_{y|cd}/\sigma_{y|c}\sigma_{y|d}$, can be expressed as

$$\rho_{y|cd} = \sqrt{R_c R_d}\rho_{\mu|cd} + \sqrt{\rho_c \rho_d}. \tag{10.2}$$

Note that, if there are no errors (i.e., when $\sigma_{b|c}^2 = \sigma_{\varepsilon|c}^2 = 0$), then $R_c = 1$, $\rho_c = 0$, and the correlation between y_c and y_d is simply the population correlation.

Let us consider the implications of these results first without systematic errors (i.e., only variable errors) and then with the effects of systematic errors.

10.4.5.4.1 Variable errors only   If the only errors are due to random noise, then the additive term on the right in equation (10.2) is 0 and $\rho_{y|cd} = \sqrt{R_c R_d}\rho_{\mu|cd}$, which says that the correlation is attenuated by the product of the root reliability ratios. For example, suppose $R_c = R_d = 0.8$, which is considered excellent reliability. Then, the observed correlation in the data will be about 80% of the true correlation; that is, correlation is attenuated by random noise. Thus, $\sqrt{R_c R_d}$ will be referred to as the *attenuation factor* for the correlation between two variables.

Quite often in the analysis of big data, the correlations being explored are for aggregate measures, as in Figure 10.2. Therefore, suppose that, rather than being a single element, y_{rc} and y_{rd} are the means of n_{rc} and n_{rd} independent elements, respectively. For example, y_{rc} and y_{rd} may be the average rate of inflation and the average price of oil, respectively, for the rth year, for $r = 1, \ldots, N$ years. Aggregated data are less affected by variable errors because, as we sum the values in a data set, the positive and negative values of the random noise components combine and cancel each other under our assumption that $E(\varepsilon_{rc}) = 0$. In addition, the variance of the mean of the errors is of order $O(n_{rc}^{-1})$.

To simplify the result for the purposes of our discussion, suppose $n_{rc} = n_c$, that is, each aggregate is based upon the same sample size. It can be shown that equation (10.2) still applies if we replace R_c by its aggregated data counterpart denoted by $R_c^A = \sigma_{\mu|c}^2 / (\sigma_{\mu|c}^2 + \sigma_{\varepsilon|c}^2 / n_c)$. Note that R_c^A converges to 1 as n_c increases, which means that $\rho_{y|cd}$ will converge to $\rho_{\mu|cd}$. Figure 10.3 illustrates the speed at which this convergence occurs.

In this figure, we assume $n_c = n_d = n$ and vary n from 0 to 60. We set the reliability ratios for both variables to 0.5 (which is considered to be a "fair" reliability) and assume a population correlation of $\rho_{\mu|cd} = 0.5$. For n in the range $[2, 10]$, the attenuation is pronounced. However, above 10 the correlation is quite close to the population value. Attenuation is negligible when $n > 30$. These results suggest that variable error can be mitigated by aggregating like elements that can be assumed to have independent errors.

10.4.5.4.2 Both variable and systematic errors   If both systematic and variable errors contaminate the data, the additive term on the right in equation (10.2) is positive. For aggregate data, the reliability

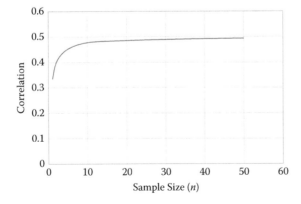

Figure 10.3. Correlation as a function of sample size (I)

ratio takes the form

$$R_c^A = \frac{\sigma_{\mu|c}^2}{\sigma_{\mu|c}^2 + \sigma_{b|c}^2 + n_c^{-1}\sigma_{\varepsilon|c}^2},$$

which converges not to 1 as in the case of variable error only, but to $\sigma_{\mu|c}^2/(\sigma_{\mu|c}^2 + \sigma_{b|c}^2)$, which will be less than 1. Thus, some attenuation is possible regardless of the number of elements in the aggregate. In addition, the intra-source correlation takes the form

$$\rho_c^A = \frac{\sigma_{b|c}^2}{\sigma_{\mu|c}^2 + \sigma_{b|c}^2 + n_c^{-1}\sigma_{\varepsilon|c}^2},$$

which converges to $\rho_c^A = \sigma_{b|c}^2/(\sigma_{\mu|c}^2 + \sigma_{b|c}^2)$, or approximately to $1 - R_c^A$ for large n_c. Thus, the systematic effects may still adversely affect correlation analysis without regard to the number of elements comprising the aggregates.

For example, consider the illustration in Figure 10.3 with $n_c = n_d = n$, reliability ratios (excluding systematic effects) set at 0.5 and population correlation at $\rho_{\mu|cd} = 0.5$. In this scenario, let $\rho_c = \rho_d = 0.25$. Figure 10.4 shows the correlation as a function of the sample size with systematic errors compared to the correlation without systematic errors. Correlation with systematic errors is both inflated and attenuated. However, at the assumed level of intra-source variation, the inflation factor overwhelms the attenuation factors and the result is a much inflated value of the correlation across all aggregate sizes.

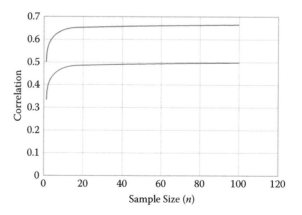

Figure 10.4. Correlation as a function of sample size (II)

To summarize these findings, correlation analysis is attenuated by variable errors, which can lead to null findings when conducting a correlation analysis and the failure to identify associations that exist in the data. Combined with systematic errors that may arise when data are extracted and combined from multiple sources, correlation analysis can be unpredictable because both attenuation and inflation of correlations can occur. Aggregating data mitigates the effects of variable error but may have little effect on systematic errors.

10.4.5.5   Errors in regression analysis

The effects of variable errors on regression coefficients are well known (Cochran, 1968; Fuller, 1991; Biemer and Trewin, 1997). The effects of systematic errors on regression have been less studied. We review some results for both types of errors in this section.

Consider the simple situation where we are interested in computing the population slope and intercept coefficients given by

$$b = \frac{\sum_r (y_r - \bar{y})(x_r - \bar{x})}{\sum_r (x_r - \bar{x})^2} \quad \text{and} \quad b_0 = \bar{y} - b\bar{x},$$

where, as before, the sum extends over all rows in the data set. When x is subject to variable errors, it can be shown that the observed regression coefficient will be attenuated from its error-free counterpart. Let R_x denote the reliability ratio for x. Then,

$$b = R_x B,$$

where $B = \sum_r (y_r - \bar{y})(\mu_{r|x} - \bar{\mu}_x) / \sum_r (\mu_{r|x} - \bar{\mu}_x)^2$ is the population slope coefficient, with $x_r = \mu_{r|x} + \varepsilon_{r|x}$, where $\varepsilon_{r|x}$ is the variable error with mean 0 and variance $\sigma^2_{\varepsilon|x}$. It also can be shown that $\text{Bias}(b_0) \approx B(1 - R_x)\bar{\mu}_x$.

As an illustration of these effects, consider the regressions displayed in Figure 10.5, which are based upon contrived data. The regression on the left is the population (true) regression with a slope of 1.05 and an intercept of −0.61. The regression on the left uses the same y- and x-values. The only difference is that, on the right, normal error has been added to the x-values, resulting in a reliability ratio of 0.73. As the theory predicts, the slope is attenuated toward 0 in direct proportion to the reliability R_x. As random error is added to the x-values, reliability is reduced and the fitted slope will approach 0.

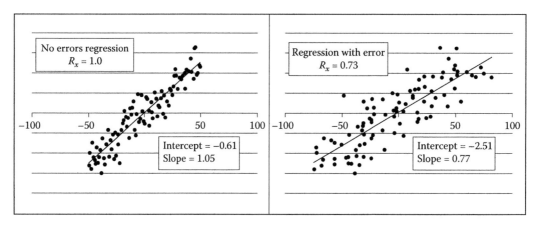

Figure 10.5. Regression of y on x with and without variable error. On the left is the population regression with no error in the x variable. On the right, variable error has been added to the x-values with a reliability ratio of 0.73. Note its attenuated slope, which is very near the theoretical value of 0.77

When only the dependent variable y is subject to variable error, the regression deteriorates, but the expected values of the slope and intercept coefficients are still equal to true to their population values. To see this, suppose $y_r = \mu_{y|r} + \varepsilon_{y|r}$, where $\mu_{r|y}$ denotes the error-free value of y_r and $\varepsilon_{r|y}$ is the associated variable error with variance $\sigma^2_{\varepsilon|y}$. The regression of y on x now can be rewritten as

$$\mu_{y|r} = b_0 + bx_r + e_r - \varepsilon_{r|y}, \qquad (10.3)$$

where e_r is the usual regression residual error with mean 0 and variance σ^2_e, which is assumed to be uncorrelated with $\varepsilon_{r|y}$. Letting $e' = e_r - \varepsilon_{r|y}$, it follows that the regression in equation (10.3) is equivalent to the previously considered regression of y on x where y is not subject to error, but now the residual variance is increased by the additive term, that is, $\sigma'^2_e = \sigma^2_{\varepsilon|y} + \sigma^2_e$.

Chai (1971) considers the case of systematic errors in the regression variables that may induce correlations both within and between variables in the regression. He shows that, in the presence of systematic errors in the independent variable, the bias in the slope coefficient may either attenuate the slope or increase its magnitude in ways that cannot be predicted without extensive knowledge of the error properties. Thus, similar to the results from correlation analysis, systematic errors greatly increase the complexity of the bias effects and their effects on inference can be quite severe.

One approach for managing systematic error at the source level in regression analysis is to model it using, for example, random effects (Hox, 2010). In brief, a random effects model specifies $y_{ijk} = \beta_{0i}^* + \beta x_{ijk} + \varepsilon_{ijk}$, where $\varepsilon_{ijk}' = b_i + \varepsilon_{ijk}$ and $\mathrm{Var}(\varepsilon_{ijk}') = \sigma_b^2 + \sigma_{\varepsilon|j}^2$. The next section considers other mitigation strategies that attempt to eliminate the error rather than model it.

10.5   Detecting and compensating for data errors

For survey data and other *designed* data collections, error mitigation begins at the data generation stage by incorporating design strategies that generate high-quality data that are at least adequate for the purposes of the data users. For example, missing data can be mitigated by repeated follow-up of nonrespondents, questionnaires can be perfected through pretesting and experimentation, interviewers can be trained in the effective methods for obtaining highly accurate responses, and computer-assisted interviewing instruments can be programmed to correct errors in the data as they are generated. For data where the data generation process is often outside the purview of the data collectors, as noted in Section 10.1, there is limited opportunity to address deficiencies in the data generation process. Instead, error mitigation must necessarily begin at the data processing stage. We illustrate this error mitigation process using two types of techniques—data editing and cleaning.

> ▶ Data errors further complicate analysis and exacerbate the analytical problems. There are essentially three solutions: prevention, remediation, and the choice of analysis methodology.

Data editing is a set of methodologies for identifying and correcting (or transforming) anomalies in the data. It often involves verifying that various relationships among related variables of the data set are plausible and, if they are not, attempting to make them so. Editing is typically a rule-based approach where rules can apply to a particular variable, a combination of variables, or an aggregate value that is the sum over all of the rows or a subset of the rows in a data set. Recently, data mining and machine learning techniques have been applied to data editing with excellent results (for a review, see Chandola et al., 2009). Tree-based methods, such as classification and regression trees and random forests, are particularly useful for creating editing rules for anomaly identification and resolution (Petrakos et al., 2004). However, some human review may be necessary to resolve the most complex situations.

For larger amounts of data, the identification of data anomalies could result in possibly billions of edit failures. Even if only a tiny proportion of these require some form of manual review for resolution, the task could still require the inspection of tens or hundreds of thousands of query edits, which would be infeasible for most applications. Thus, micro-editing must necessarily be a completely automated process unless it can be confined to a relatively small subset of the data. As an example, a representative (random) subset of the data set could be edited using manual editing for purposes of evaluating the error levels for the larger data set, or possibly to be used as a training data set, benchmark, or reference distribution for further processing, including recursive learning.

To complement fully automated micro-editing, data editing involving large amounts of data usually involves *top-down* or *macro-editing* approaches. For such approaches, analysts and systems inspect aggregated data for conformance to some benchmark values or data distributions that are known from either training data or prior experience. When unexpected or suspicious aggregates are identified, the analyst can "drill down" into the data to discover and, if possible, remove the discrepancy by either altering the value at the source (usually a microdata element) or deleting the edit-failed value.

There are a variety of methods that may be effective in macro-editing. Some of these are based upon data mining (Natarajan et al., 2010), machine learning (Clarke, 2014), cluster analysis (He et al., 2003; Duan et al., 2009), and various data visualization tools such as treemaps (Johnson and Shneiderman, 1991; Shneiderman, 1992; Tennekes and de Jonge, 2011) and tableplots (Tennekes et al., 2012, 2013; Puts et al., 2015). We further explore tableplots next.

10.5.1   TablePlots

Similar to other visualization techniques examined in Chapter 6, the tableplot has the ability to summarize a large multivariate data set in a single plot (Malik et al., 2010). In editing data, it can be used to detect outliers and unusual data patterns. Software for implementing this technique has been written in R and is available from the Comprehensive R Archive Network (https://cran.r-project.org/) (R Core Team, 2013). Figure 10.6 shows an example. The key idea is that micro-aggregates of two related variables should have similar data patterns. Inconsistent data patterns may signal errors in one of the aggregates that can be investigated and corrected in the edit-

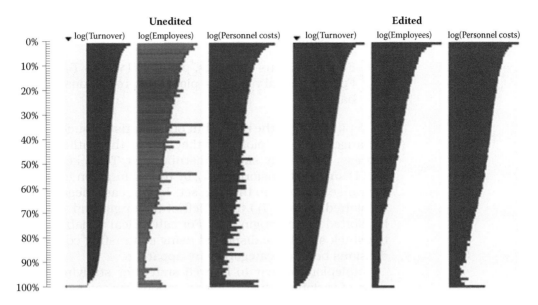

Figure 10.6. Comparison of tableplots for the Dutch Structural Business Statistics Survey for five variables before and after editing. Row bins with high missing and unknown numeric values are represented by lighter colored bars

ing process to improve data quality. The tableplot uses bar charts created for the micro-aggregates to identify these inconsistent data patterns.

Each column in the tableplot represents some variable in the data table, and each row is a "bin" containing a subset of the data. A statistic, such as the mean or total, is computed for the values in a bin and is displayed as a bar (for continuous variables) or as a stacked bar (for categorical variables).

The sequence of steps typically involved in producing a tableplot is as follows.

1. Sort the records in the data set by the key variable.

2. Divide the sorted data set into B bins containing the same number of rows.

3. For continuous variables, compute the statistic to be compared across variables for each row bin, e.g., T_b, for $b = 1, \ldots, B$, for each continuous variable V, ignoring missing values. The level of missingness for V may be represented by the color or brightness of the bar. For categorical variables with K categories, compute the proportion in the kth category, denoted by

P_{bk}. Missing values are assigned to a new $(K + 1)$th category ("missing").

4. For continuous variables, plot the B values T_b as a bar chart. For categorical variables, plot the B proportions P_{bk} as a stacked bar chart.

Typically, T_b is the mean, but other statistics such as the median or range could be plotted if they aid in the outlier identification process. For highly skewed distributions, Tennekes and de Jonge (2011) suggest transforming T_b by the log function to better capture the range of values in the data set. In that case, negative values can be plotted as $\log(-T_b)$ to the left of the origin and zero values can be plotted on the origin line. For categorical variables, each bar in the stack should be displayed using contrasting colors so that the divisions between categories are apparent.

Tableplots appear to be well suited for studying the distributions of variable values, the correlation between variables, and the occurrence and selectivity of missing values. Because they can help visualize massive multivariate data sets, they seem particularly well suited for big data. Currently, the R implementation of tableplot is limited to 2 billion records.

The tableplot in Figure 10.6 is taken from Tennekes and de Jonge (2011) for the annual Dutch Structural Business Statistics survey, a survey of approximately 58,000 business units annually. Topics covered in the questionnaire include turnover, number of employed persons, total purchases, and financial results. Figure 10.6 has been created by sorting on the first column, viz., log(turnover), and dividing the 57,621 observed units into 100 bins, so that each row bin contains approximately 576 records. To aid the comparisons between unedited and edited data, the two tableplots are displayed side by side, with the unedited graph on the left and the edited graph on the right. All variables have been transformed by the log function.

The unedited tableplot reveals that all four of the variables in the comparison with log(turnover) show some distortion by large values for some row bins. In particular, log(employees) has some relatively large nonconforming bins with considerable discrepancies. In addition, that variable suffers from a large number of missing values, as indicated by the brightness of the bar color. Overall, there are obvious data quality issues in the unprocessed data set for all four of these variables that should be addressed in the subsequent processing steps.

The edited tableplot reveals the effect of the data checking and editing strategy used in the editing process. Notice the much darker color for the number of employees for the graph on the left compared to same graph on the right. In addition, the lack of data in the lowest part of the turnover column has been somewhat improved. The distributions for the graph on the right appear smoother and are less jagged.

10.6 Summary

As social scientists, we are deeply concerned with ensuring that the inferences we make from our analysis are valid. Because many of the newer data sources we are using are not collected or generated from instruments and methods designed to produce valid and reliable data for scientific analysis and discovery, they can lead to inference errors. This chapter has described different types of errors that we encounter to make us aware of these limitations and take the necessary steps to understand and hopefully mitigate the effects of hidden errors on our results.

In addition to describing the types of errors, this chapter also provides an example of a solution to cleaning the data before analysis. Another option that has not been discussed is the possibility of using analytical techniques that attempt to model errors and compensate for them in the analysis. Such techniques include the use of latent class analysis for classification error (Biemer, 2011), multilevel modeling of systematic errors from multiple sources (Hox, 2010), and Bayesian statistics for partitioning massive data sets across multiple machines and then combining the results (Ibrahim and Chen, 2000; Scott et al., 2013).

While this chapter has focused on the accuracy of the data and the validity of the inference, other data quality dimensions, such as timeliness, comparability, coherence, and relevance, that we have not considered in this chapter are also important. For example, timeliness often competes with accuracy because achieving acceptable levels of the latter often requires greater expenditures of resources and time. In fact, some applications of data analysis prefer results that are less accurate for the sake of timeliness. Biemer and Lyberg (2003) discuss these and other issues in some detail.

It is important to understand that we will rarely, if ever, obtain perfect data for our analysis. Every data source will have some limitation—some will be inaccurate, some will become stale, and some will have sample bias. The key is to (1) be aware of the limitations of each data source, (2) incorporate that awareness into the analysis that is being conducted with it, and 3) understand what type of inference errors it can produce in order to appropriately communicate the results and make sound decisions.

10.7   Resources

▶ http://www.aapor.org

▶   https://www.degruyter.com/view/j/jos

The American Association of Public Opinion Research has a number of resources on its website. See, in particular, its report on big data (Japec et al., 2015).

The *Journal of Official Statistics* is a standard resource with many relevant articles. There is also an annual international conference (*International Total Survey Error Workshop*) on the total survey error framework, supported by major survey organizations.

▶   See   https://workbooks.coleridgeinitiative.org.

The *Errors and Inference* workbook in Chapter 13 provides an introduction to sensitivity analysis and imputation.

Chapter 11

Bias and Fairness

Kit T. Rodolfa, Pedro Saleiro, and Rayid Ghani

Interest in algorithmic fairness and bias has been growing recently (for good reason), but it is easy to become lost in the large number of definitions and metrics. There are many different, often competing, ways to measure whether a given model is statistically "fair," but it is important to remember to start from the social and policy goals for equity and fairness and map those to the statistical properties we want in our models to help achieve those goals. In this chapter, we will provide an overview of these statistical metrics along with some specific examples to help navigate these concepts and understand the tradeoffs involved in choosing to optimize to one metric over others, focusing on the metrics relevant to binary classification methods used frequently in risk-based models for policy settings.

11.1   Introduction

In Chapter 7, you learned about several of the concepts, tools, and approaches used in the field of machine learning and how they can be applied in the social sciences. In that chapter, we focus on evaluation metrics, such as precision (positive predictive value), recall (sensitivity), area-under-curve (AUC), and accuracy, that often are used to measure the performance of machine learning methods. In most (if not all) public policy problems, a key goal for the analytical systems being developed is to help achieve equitable outcomes for society, so we need to understand how to design systems that lead to equity.

When machine learning models are being used to make decisions, they cannot be separated from the social and ethical context in which they are applied, and those developing and deploying these models must take care to do so in a manner that accounts for both accuracy and fairness. In this chapter, we will discuss sources of potential bias in the modeling pipeline, as well as some of the ways that bias introduced by a model can be measured, with a particular focus on classification problems. Unfortunately, just as there is no single machine learning algorithm that is best suited to every application, no one fairness metric will fit every situation. However, we hope this chapter will provide you with a grounding in the available ways of measuring algorithmic fairness that will help you navigate the tradeoffs involved putting these into practice in your own applications.

11.2   Sources of bias

Bias may be introduced into a machine learning project at any step along the way, so it is important to carefully think through each potential source and how it may affect your results. In many cases, some sources may be difficult to measure precisely (or even at all), but this does not mean these potential biases can be readily ignored when developing interventions or performing analyses.

11.2.1   Sample bias

You probably are familiar with sampling issues as a potential source of bias in the contexts of causal inference and external validity in the social science literature. A biased sample can be as problematic for machine learning as it can be for inference, and predictions made on individuals or groups not represented in the training set are likely to be unreliable. As such, any application of machine learning should start with a careful understanding of the data-generating process for the training and test sets. What is the relevant population for the project and how might some individuals be incorrectly excluded or included from the data available for modeling or analysis?

If there is a mismatch between the available training data and the population to whom the model will be applied, you may want to consider whether it is possible to collect more representative data. A model to evaluate the risk of health violations at restaurants may be of limited applicability if the only training data available is based on inspections that resulted from reported complaints. In such a case, an initial trial of randomized inspections might provide a more representative dataset. However, this may not always be possible. For example, in the case of bail determinations, labeled data will only be available for individuals who are released under the existing system.

Even if the training data matches the population, are there underlying systemic biases involved in defining that population in general? For example, over-policing of minority neighborhoods might mean that the population of incarcerated individuals is unrepresentative of the population of individuals who have committed a given crime, and even a representative sample of the jail population might not be the appropriate universe for a given policy or social science question.

For data with a time component or models that will be deployed to aid future decisions, are there relevant policy changes in the past that may make data from certain periods of time less relevant? Are there pending policy changes going forward that may affect the modeling population?

Measurement here might be difficult, but it is nevertheless helpful to think through each of these questions in detail. Often, other sources of data (even in aggregate form) can provide some insight on how representative your data may be, including census data, surveys, and academic studies in the relevant area.

11.2.2   Label (outcome) bias

Regardless of whether your dataset reflects a representative sample of the relevant population for your intervention or analysis, there also may be bias inherent in the labels (i.e., the measured outcomes) associated with individuals in that data.

One mechanism by which bias may be introduced is in how the label/outcome itself is defined. For example, an analysis of recidivism might use a new arrest as an outcome variable when the appropriate outcome to look at may be the committing of a new crime. Thus, if some groups are policed more heavily than others, using arrests to define the outcome variable may introduce bias into the system's decisions. Similarly, a label that relies on the number of days an individual has been incarcerated would reflect known biases in sentence lengths based on race.

A related mechanism is measurement error. Even when the outcome of interest is well defined and can be measured directly, bias may be introduced through differential measurement accuracy across groups. For example, data collected through survey research might suffer from language barriers or cultural differences in social desirability that introduce measurement errors across groups.

11.2.3   Machine learning pipeline bias

Biases can be introduced by the handling and transformation of data throughout the machine learning pipeline as well, requiring careful consideration as you ingest data, create features, and model outcomes of interest. Below are a few examples at each stage of the process, but these are far from exhaustive and intended only to help motivate thinking about how bias might be introduced in your own projects.

Ingesting data   The process of loading, cleaning, and reconciling data from a variety of data sources (often referred to as ETL) can introduce a number of errors that might have differential downstream impacts on different populations.

- Are your processes for matching individuals across data sources equally accurate across different populations? For example, married vs. maiden names may bias match rates against women, while inconsistencies in handling of multi-part last names may make matching less reliable for Hispanic individuals.

- Nickname dictionaries used in record reconciliation might be derived from populations different than your population of interest.

- A data loading process that drops records with "special characters" might inadvertently exclude names with accents or tildes.

Feature engineering   Biases are easy to introduce during the process of constructing features, both in the handling of features that relate directly to protected classes as well as information that correlates with these populations (such as geolocation). A few examples include the following.

- Dictionaries to infer age or gender from name might be derived from a population that is not relevant to your problem.

- Handling of missing values and combining "other" categories can become problematic, especially for multi-racial individuals or people with non-binary gender.

- Thought should be given to how race and ethnicity indicators are collected—are these self-reported, recorded by a third party, or inferred from other data? The data collection process may inform the accuracy of the data and how errors differ across populations.

- Features that rely on geocoding to incorporate information based on distances or geographic aggregates may miss homeless individuals or provide less predictive power for more mobile populations.

Modeling   The model itself may introduce bias into decisions made from its scores by performing worse on some groups relative to others (many examples have been highlighted in the popular press recently, such as racial biases in facial recognition algorithms and gender biases in targeting algorithms for job advertisement on social media). Because of the complex correlation structure of the data, it generally is not sufficient to simply omit the protected attributes and assume this will result in fair outcomes. Instead, model performance across groups needs to be measured directly in order to understand and address any biases. However, there are many (often incompatible) ways to define fairness and Section 11.3 will take a closer look at these options in much more detail.

Much of the remainder of this chapter will focus on how we might define and measure fairness at the level of the machine learning pipeline itself. In Section 11.3, we will introduce several of the metrics used to measure algorithmic fairness and, in Section 11.4, we will discuss how these can be used in the process of evaluating and selecting machine learning models.

11.2.4   Application bias

A final potential source of bias worth considering is how the model or analysis might be put into use in practice. One way this might occur is through heterogeneity in the effectiveness of an intervention across groups. For example, imagine using a machine learning model that identifies individuals most at risk for developing diabetes in the next three years to determine who should receive a particular preventative treatment. If the treatment is much more effective for individuals with a certain genetic background relative to others, the overall outcome of the effort might be to exacerbate disparities in diabetes rates even if the model itself is modeling risk in an unbiased way.

Similarly, it is important to be aware of the risk of discriminatory applications of a machine learning model. Perhaps a model developed to screen out unqualified job candidates is only "trusted" by a hiring manager for female candidates but often ignored or overridden for men. In a perverse way, applying an unbiased model in such a context might serve to increase inequities by giving bad actors more information with which to (wrongly) justify their discriminatory practices.

Although there may be relatively little you can do to detect or mitigate these types of bias at the modeling stage, performing a trial to compare current practice with a deployed model can be instructive where doing so is feasible. Keep in mind, of course, that the potential for machine learning systems to be applied in biased ways should not be construed as an argument against developing these systems at all, nor would it be reasonable to suggest that current practices are likely to be free of bias. Rather, it is an argument for thinking carefully about both the status quo and how it may change in the presence of such a system, putting in place legal and technical safeguards to help ensure that these methods are applied in socially responsible ways.

11.2.5   Considering bias when deploying your model

Ultimately, what we care about is some global idea of how putting a model into practice will affect some overall concept of social welfare and fairness influenced by all of these possible sources of bias. While this is generally impossible to measure in a quantitative way, it can provide a valuable framework for qualitatively evaluating the potential impact of your model. For most of the remainder of this chapter, we consider a set of more quantitative metrics that can be applied to the predictions of a machine learning pipeline specifically. Even so, it is important to keep in mind that these metrics only apply to the sample and labels you have; ignoring other sources of bias that may be at play in the underlying data-generating process could result in unfair outcomes, even when applying a model that appears to be "fair" by your chosen metric.

11.3   Dealing with bias

11.3.1   Define bias

Section 11.2.3 provides some examples for how bias might be introduced in the process of using machine learning to work with a dataset. While far from exhaustive as a source of potential bias in an overall application, these biases can be more readily measured and addressed through choices made during data preparation, modeling, and model selection. This section focuses on detecting and understanding biases introduced at this stage of the process.

One key challenge, however, is that there is no universally-accepted definition of what it means for a model to be fair. Take the example of a model being used to make bail determinations. Different people might consider it "fair" if it results in the following.

- It makes mistakes about denying bail to an equal number of individuals of each racial category.

- The chances that a given person will be wrongly denied bail is equal, regardless of race.

- Among the jailed population, the probability of having been wrongly denied bail is independent of race.

- For people who should be released, the chances that a given person will be denied bail is equal regardless of race.

In different contexts, reasonable arguments can be made for each of these potential definitions, but unfortunately, not all of them can hold at the same time. The remainder of this section explores these competing options and how to approach them in more detail.

11.3.2 Definitions

Most of the metrics used to assess model fairness relate either to the types of errors a model might make or how predictive the model is across different groups. For binary classification models (on which we focus here), these are generally derived from values in the *confusion matrix* (see Figure 7.9 and Chapter 7 for more details).

- True positives (*TP*) are individuals for whom both the model prediction and actual outcome are positive labels.

- False positives (*FP*) are individuals for whom the model predicts a positive label, but the actual outcome is a negative label.

- True negatives (*TN*) are individuals for whom both the model prediction and actual outcome are negative labels.

- False negatives (*FN*) are individuals for whom the model predicts a negative label, but the actual outcome is a positive label.

Based on these four categories, we can calculate several ratios that are instructive for thinking about the equity of a model's predictions in different situations (Sections 11.3.4 and 11.3.5 provide detailed examples here).

★ $FPR = FP/(FP + TN)$.

- False positive rate (*FPR*) is the fraction of individuals with negative actual labels that the model misclassifies with a positive predicted label.[*]

★ $FNR = FN/(FN + TP)$.

- False negative rate (*FNR*) is the fraction of individuals with positive actual labels that the model misclassifies with a negative predicted label.[*]

★ $FDR = FP/(FP + TP)$.

- False discovery rate (*FDR*) is the fraction of individuals that the model predicts to have a positive label but for which the actual label is negative.[*]

★ $FOR = FN/(FN + TN)$.

- False omission rate (*FOR*) is the fraction of individuals that the model predicts to have a negative label but for which the actual label is positive.[*]

★ $Precision = TP/(FP + TP)$.

- Precision is the fraction of individuals that the model predicts to have a positive label for which this prediction is correct.[*]

★ $Recall = TP/(FN + TP)$.

- Recall is the fraction of individuals with positive actual labels that the model has correctly classified as such.[*]

For the first two metrics (*FPR* and *FNR*), notice that the denominator is based on actual outcomes (rather than model predictions), while in the next two (*FDR* and *FOR*) the denominator is based on model predictions (whether an individual falls above or below the threshold used to turn model scores into 0/1 predicted classes). The final two metrics relate to correct predictions rather than errors, but are directly related to error measurements (i.e., recall = $1 - FNR$ and precision = $1 - FDR$) and may sometimes have better properties for calculating model bias.

Notice that the metrics defined here require the use of a threshold to turn modeled scores into 0/1 predicted classes and are therefore most useful when either a threshold is well defined for the problem (e.g., when available resources mean a program can only serve a given number of individuals) or where calculating these metrics at different threshold levels might be used (along with model performance metrics) to choose a threshold for application. In some cases, it also may be of interest to think about equity across the full distribution of the modeled score (Chouldechova, 2017; Kleinberg et al., 2017). Common practices in these situations are to look at how model performance metrics, such as the area under the receiver-operator curve (AUC-ROC) or model calibration compared across subgroups (e.g., by race, gender, age). Or, in cases where the underlying causal relationships are well known, counterfactual methods (Kilbertus et al., 2017; Kusner et al., 2017) may be used to assess a model's bias (these methods also may be useful when you suspect label bias is an issue in your data). We do not explore these topics deeply here but refer you to the relevant references if you would like to learn more.

11.3.3 Choosing bias metrics

Any of the metrics defined above can be used to calculate disparities across groups in your data and (unless you have a perfect model) many of them cannot be balanced across subgroups at the same time. As a result, one of the most important—and frequently most challenging—aspects of measuring bias in your machine learning pipeline is simply understanding how "fairness" should be defined for your particular case.

In general, this requires consideration of the project's goals and a detailed discussion between the data scientists, decision makers, and those who will be affected by the application of the model. Each perspective may have a different concept of fairness and a different understanding of harm involved in making different types of errors, both at individual and societal levels. Importantly, data scientists have a critical role in this conversation, both as the experts in understanding how different concepts of fairness might translate into metrics and measurement and as individuals with experience deploying similar models. Although there is no universally correct definition of fairness, nor one that can be learned from the data, this does not excuse the data scientists from responsibility for taking part in the conversation around fairness and equity in their models and helping decision makers understand the options and tradeoffs involved.

Practically speaking, coming to an agreement on how fairness should be measured in a purely abstract manner is likely to be difficult. Often, it can be instructive instead to explore different options and metrics based on preliminary results, providing tangible context for potential tradeoffs between overall performance and different definitions of equity and helping guide stakeholders through the process of deciding what to optimize. The remainder of this section will discuss some of the metrics that may be of particular interest in different types of applications such as these.

- If your intervention is punitive in nature (e.g., determining to whom to deny bail), individuals may be harmed by intervening on them in error, so you may care more about metrics that focus on false positives. Section 11.3.4 provides an example to guide you through what some of these metrics mean in this case.

- If your intervention is assistive in nature (e.g., determining who should receive a food subsidy), individuals may be harmed by failing to intervene on them when they have need, so you may care more about metrics that focus on false negatives. Section 11.3.5 provides an example to guide you through metrics that may be applicable in this case.

- If your resources are significantly constrained such that you can only intervene on a small fraction of the population at need, some of the metrics described here may be of limited use. Section 11.3.6 describes this case in more detail.

Navigating the many options for defining bias in a given context is a difficult and nuanced process, even for those familiar with the underlying statistical concepts. In order to help facilitate these conversations between data scientists and stakeholders, we have developed the fairness tree depicted in Figure 11.1. Although it certainly cannot provide a single "right" answer for a given context, our

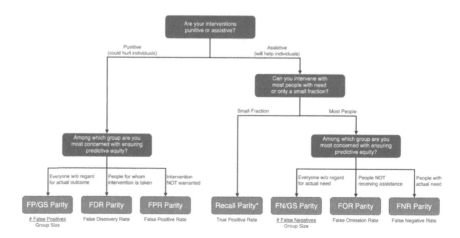

Figure 11.1. Fairness tree

hope is that the fairness tree can act as a tool to help structure the process of arriving at an appropriate metric (or set of metrics) on which to focus.

11.3.4 Punitive example

When the application of a risk model is punitive in nature, individuals may be harmed by being incorrectly included in the "high risk" population that receives an intervention. In an extreme case, we can think of this as incorrectly detaining an innocent person in jail. Hence, with punitive interventions, we focus on bias and fairness metrics based on false positives.

11.3.4.1 Count of false positives

We might naturally think about the number of people wrongly jailed from each group as a reasonable place to start for assessing whether our model is biased. Here, we are concerned with statements such as, "twice as many people from Group A were wrongly convicted as from Group B."*

However, it is unclear whether differences in the number of false positives across groups reflect unfairness in the model. For instance, if there are twice as many people in Group A as there are in Group B, some might deem the situation described above as fair from the standpoint that the composition of the false positives reflects the composition of the groups. This brings us to our second metric.

> ★ In probabilistic terms, we could express this as
>
> $$P(\text{wrongly jailed, group } i) = C \quad \forall i,$$
>
> where C is a constant value. Or, alternatively,
>
> $$\frac{FP_i}{FP_j} = 1 \quad \forall i, j,$$
>
> where FP_i is the number of false positives in group i.

11.3.4.2 Group size-adjusted false positives

By accounting for differently sized groups, we ask the question, "just by virtue of the fact that an individual is a member of a given group, what are the chances they will be wrongly convicted?"*

Although this metric might appear to meet a reasonable criteria of avoiding treating groups differently in terms of classification errors, there are other sources of disparities we also might need to consider. For example, suppose there are 10,000 individuals in Group A and 30,000 in Group B. Suppose further that 100 individuals from each group are in jail, with 10 Group A people wrongly convicted and 30 Group B people wrongly convicted. We have balanced the number of false positives by group size (0.1% for both groups) so there are no disparities with regard to this metric, but note that 10% of the jailed Group A individuals are innocent compared to 30%

> ★ In terms of probability,
>
> $$P(\text{wrongly jailed} \mid \text{group } i) = C \quad \forall i,$$
>
> where C is a constant value. Or, alternatively,
>
> $$\frac{FP_i}{FP_j} = \frac{n_i}{n_j} \quad \forall i, j,$$
>
> where FP_i is the number of false positives and n_i the total number of individuals in group i.

of the jailed Group B individuals. The next metric is concerned with unfairness in this way.

11.3.4.3 False discovery rate

The false discovery rate (*FDR*) focuses specifically on the people who are affected by the intervention—in the example above, among the 200 people in jail, what are the group-level disparities in rates of wrong convictions. The jail example is particularly instructive because we could imagine the social cost of disparities manifesting directly through inmates observing how frequently different groups are wrongly convicted.[*]

11.3.4.4 False positive rate

The false positive rate (*FPR*) focuses on a different subset, specifically, the individuals who should **not** be subject to the intervention. Here, this would ask, "for an *innocent* person, what are the chances they will be wrongly convicted by virtue of the fact that they are a member of a given group?"[*]

The difference between choosing to focus on the *FPR* and group size-adjusted false positives is somewhat nuanced and warrants highlighting.

- Having no disparities in group size-adjusted false positives implies that, if I were to choose a random person from a given group (regardless of group-level crime rates or their individual guilt or innocence), I would have the same chance of picking a wrongly convicted person across groups.

- Having no disparities in *FPR* implies that, if I were to choose a random *innocent* person from a given group, I would have the same chance of picking a wrongly convicted person across groups.

11.3.4.5 Tradeoffs in metric choice

By way of example, imagine you have a society with two groups (A and B) and a criminal process with equal *FDR* and group size-adjusted false positives such as the following.

- Group A has 1,000 total individuals, of whom 100 have been jailed with 10 wrongfully convicted. Suppose the other 900 are all guilty.

★ In probabilistic terms,

$$P(\text{wrongly jailed} \mid \text{jailed, group } i) = C \quad \forall\, i,$$

where C is a constant value. Or, alternatively,

$$\frac{FP_i}{FP_j} = \frac{k_i}{k_j} \quad \forall\, i, j,$$

where FP_i is the number of false positives and k_i the total number of *jailed* individuals in group i.

★ In probabilistic terms,

$$P(\text{wrongly jailed} \mid \text{innocent, group } i) = C \quad \forall\, i,$$

where C is a constant value. Or, alternatively,

$$\frac{FP_i}{FP_j} = \frac{n_i \times (1 - p_i)}{n_j \times (1 - p_j)} \quad \forall\, i, j,$$

where FP_i is the number of false positives, n_i is the total number of individuals, and p_i is the prevalence (here, rate of being truly guilty) in group i.

- Group B has 3,000 total individuals, of whom 300 have been jailed with 30 wrongfully convicted. Suppose the other 2,700 are all innocent.

In this case,

$$\frac{FP_A}{n_A} = \frac{10}{1000} = 1.0\%$$

$$FDR_A = \frac{10}{100} = 10.0\%$$

$$FPR_A = \frac{10}{10} = 100.0\%$$

while

$$\frac{FP_B}{n_B} = \frac{30}{3000} = 1.0\%$$

$$FDR_B = \frac{30}{300} = 10.0\%$$

$$FPR_B = \frac{30}{2730} = 1.1\%$$

that is,

- A randomly chosen individual has the same chance (1.0%) of being wrongly convicted regardless of the group to which they belong;

- In both groups, a randomly chosen person who is in jail has the same chance (10.0%) of actually being innocent;

- HOWEVER, an innocent person in Group A is certain to be wrongly convicted, nearly 100 times the rate of an innocent person in Group B.

While this is an exaggerated case for illustrative purposes, there is a more general principle at play here: when prevalences differ across groups, disparities cannot be eliminated from both the *FPR* and group size-adjusted false positives at the same time (in the absence of perfect prediction).

While there is no universal rule for choosing a bias metric (or set of metrics) to prioritize, it is important to keep in mind that there are both theoretical and practical limits on the degree to which these metrics can be jointly optimized.

Balancing these tradeoffs generally will require some degree of subjective judgment on the part of policymakers and should reflect both societal values established with the input of those impacted by model-assisted decisions as well as practical constraints. For example, if there is uncertainty in the quality of the labels (e.g., how well can we truly measure the size of the innocent population?), it may make more sense in practical terms to focus on the group size-adjusted false positives than *FPR*.

11.3.5 Assistive example

By contrast to the punitive case, when the application of a risk model is assistive in nature, individuals may be harmed by being incorrectly excluded from the "high risk" population that receives an intervention. Here, we use identifying families to receive a food assistance benefit as a motivating example. Where the punitive case focused on errors of inclusion through false positives, most of the metrics of interest in the assistive case focus on analogues that measure errors of omission through false negatives.

11.3.5.1 Count of false negatives

A natural starting point for understanding whether a program is being applied fairly is to count how many people it is missing from each group, focusing on statements such as, "twice as many families with need for food assistance from Group A were missed by the benefit as from Group B."*

Differences in the number of false negatives by group, however, may be relatively limited in measuring equity when the groups are very different in size. If there are twice as many families in Group A as in Group B in this example, the larger number of false negatives might not be seen as inequitable, which motivates our next metric.

11.3.5.2 Group size-adjusted false negatives

To account for differently sized groups, one way of phrasing the question of fairness is to ask, "just by virtue of the fact that an individual is a member of a given group, what are the chances they will be missed by the food subsidy?"*

While avoiding disparities on this metric focuses on the reasonable goal of treating different groups similarly in terms of classification errors, we also may want to directly consider two subsets within each group: (1) the set of families not receiving the subsidy and (2) the set of families who would benefit from receiving the subsidy. We take a closer look at each of these cases next.

★ In probabilistic terms, we could express this as

P(missed by benefit, group i) = C $\forall i$,

where C is a constant value. Or, alternatively,

$$\frac{FN_i}{FN_j} = 1 \quad \forall i,j,$$

where FN_i is the number of false negatives in group i.

★ That is, in terms of probability,

P(missed by benefit | group i) = C $\forall i$,

where C is a constant value. Or, alternatively,

$$\frac{FN_i}{FN_j} = \frac{n_i}{n_j} \quad \forall i,j,$$

where FN_i is the number of false negatives and n_i the total number of families in group i.

11.3.5.3 False omission rate

The false omission rate (*FOR*) focuses specifically on people on whom the program does not intervene—in our example, the set of families not receiving the food subsidy. Such families will either be true negatives (i.e., those not in need of the assistance) or false negatives (i.e., those who did need assistance but were missed by the program), and the *FOR* asks what fraction of this set fall into the latter category.[*]

In practice, the *FOR* can be a useful metric in many situations, particularly because often need can be more easily measured among individuals not receiving a benefit than among those who do (e.g., when the benefit affects the outcome on which need is measured). However, when resources are constrained such that a program can only reach a relatively small fraction of the population, its utility is more limited. See Section 11.3.6 for more details on this case.

> ★ In probabilistic terms,
>
> P(missed by program | no subsidy, group i)
> $$= C \quad \forall \, i,$$
>
> where C is a constant value. Or, alternatively,
> $$\frac{FN_i}{FN_j} = \frac{n_i - k_i}{n_j - k_j} \quad \forall \, i, j,$$
>
> where FN_i is the number of false negatives, k_i is the number of families receiving the subsidy, and n_i is the total number of families in group i.

11.3.5.4 False negative rate

The false negative rate (*FNR*) focuses instead on the set of people with need for the intervention. In our example, this asks the question, "for a family that needs food assistance, what are the chances they will be missed by the subsidy by virtue of the fact they are a member of a given group?"[*]

As with the punitive case, there is some nuance in the difference between choosing to focus on group size-adjusted false negatives and the *FNR* that are worth noting.

- Having no disparities in group size-adjusted false negatives implies that, if I were to choose a random family from a given group (regardless of group-level nutritional outcomes or their individual need), I would have the same chance of picking a family missed by the program across groups.

- Having no disparities in *FNR* implies that, if I were to choose a random family *with need for assistance* from a given group, I would have the same chance of picking one missed by the subsidy across groups.

- Unfortunately, disparities in both of these metrics cannot be eliminated at the same time, except where the level of need is identical across groups, or in the generally unrealistic case of perfect prediction.

> ★ In probabilistic terms,
>
> P(missed by subsidy | need assistance, group i)
> $$= C \quad \forall \, i,$$
>
> where C is a constant value. Or, alternatively,
> $$\frac{FN_i}{FN_j} = \frac{n_i \times p_i}{n_j \times p_j} \quad \forall \, i, j,$$
>
> where FN_i is the number of false negatives, n_i is the total number of individuals, and p_i is the prevalence (here, rate of need for food assistance) in group i.

> ★ In probabilistic terms,
>
> P(received subsidy |
> need assistance, group i)
> $= C \quad \forall\, i,$
>
> where C is a constant value.
> Or, alternatively,
>
> $$\frac{TP_i}{TP_j} = \frac{n_i \times p_i}{n_j \times p_j} \quad \forall\, i,\,,j$$
>
> where TP_i is the number of true positives, n_i is the total number of individuals, and p_i is the prevalence (here, rate of need for food assistance) in group i.

> ★ Note that $recall$ = 1 − FNR.

> ★ $\dfrac{recall_i}{recall_j} = \dfrac{p_i}{p_j} \quad \forall\, i, j$

11.3.6 Special case: Resource-constrained programs

In many real-world applications, programs may only have sufficient resources to serve a small fraction of individuals who might benefit. In these cases, some of the metrics described here may prove less useful. For instance, where the number of individuals served is much smaller than the number of individuals with need, the false omission rate will converge on the overall prevalence, and it will prove impossible to balance *FOR* across groups.

In such cases, group-level recall may provide a useful metric for thinking about equity, asking the question, "given that the program cannot serve everyone with need, is it at least serving different populations in a manner that reflects their level of need?"[*]

Note that, unlike the metrics described above, using recall as an equity metric does not explicitly focus on the mistakes being made by the program, but rather on how it is addressing need within each group. Nevertheless, balancing recall is equivalent to balancing the false negative rate across groups,[*] but it may be a more well-behaved metric for resource-constrained programs in practical terms. When the number of individuals served is small relative to need, *FNR* will approach 1 and ratios between group-level *FNR* values will not be particularly instructive, while ratios between group-level recall values will be meaningful.

Additionally, a focus on recall can provide a lever that a program can use to consider options for achieving programmatic or social goals. For example, if underlying differences in prevalence across groups is believed to be a result of social or historical inequities, a program may want to go further than balancing recall across groups, focusing even more heavily on historically under-served groups. One rule of thumb we have used in these cases is to balance recall relative to prevalence (however, there is no theoretically "right" choice here and generally it is best to consider a range of options).[*]

Here, the idea is that (assuming the program is equally effective across groups) balancing recall will seek to improve outcomes at an equal rate across groups without impacting underlying disparities while a heavier focus on previously under-served groups might seek to improve outcomes across groups while also attempting to close these gaps.

11.4 Mitigating bias

The metrics described in this chapter can be put to use in a variety of ways: auditing existing models and methods for achieving more equitable results, in the process of choosing a model to deploy, or in making choices about how a chosen model is put into use. This section provides some details about how you might approach each of these tasks.

11.4.1 Auditing model results

Because the metrics described here rely only on the predicted and actual labels, no specific knowledge of the process by which the predicted labels are determined is needed to make use of them to assess bias and fairness in the results. Given this type of labeled outcome data for any existing or proposed process (and our knowledge of how trustworthy the outcomes data may be), bias audit tools, such as Aequitas, can be applied to help understand whether that process is yielding equitable results (for the various possible definitions of "equitable" we have described).

▶ https://github.com/dssg/aequitas

Note that the existing process need not be a machine learning model: these equity metrics can be calculated for any set of decisions and outcomes, regardless of whether the decisions are derived from a model, judge, case worker, heuristic rule, or other process. And, in fact, it generally will be useful to make measures of equity in any existing processes which a model might augment or replace to help understand whether application of the model might improve, degrade, or leave unchanged the fairness of the existing system.

11.4.2 Model selection

As described in Chapter 7, many different types of models (each, in turn, with many tunable hyperparameters) can be utilized for a given machine learning problem, making the task of selecting a specific model an important step in the process of model development. Chapter 7 describes how this might be done by considering a model's performance on various evaluation metrics, as well as how consistent that performance is across time or random splits of the data. This framework for model selection can be naturally extended to incorporate equity metrics; however, doing so introduces a layer of complexity in determining how to evaluate tradeoffs between overall performance and predictive equity.

Just as there is no one-size-fits-all metric for measuring equity that works in all contexts, you might choose to incorporate fairness in the model selection process in a variety of different ways. The following are a few options we have considered (although certainly not an exhaustive list).

- If many models perform similarly on overall evaluation metrics of interest (e.g., above some reasonable threshold), how do they vary in terms of equitability?

- How much "cost" in terms of performance do you have to pay to reach various levels of fairness? Think of this as creating a menu of options to explicitly show the tradeoffs involved. For example, imagine your best-performing model has a precision of 0.75 but *FDR* ratio of 1.3, but you can reach an *FDR* ratio of 1.2 by selecting a model with a precision of 0.73, or a ratio of 1.1 at a precision of 0.70, or an *FDR* parity at a precision of 0.64.

- You may want to consider several of the equity metrics we have described, determine which model performs best on each metric of interest (perhaps above some overall performance threshold), and then consider choosing between these options.

- If you are concerned about fairness across several subgroups (e.g., multiple categories of race/ethnicity, different age groups, etc), you might consider exploring the models that perform best for each subgroup in addition to those that perform similarly across groups.

- Another option might be to develop a single model selection parameter that penalizes performance by how far a model is from equity and explore how model choice changes based on how heavily you weight the equity parameter. Note, however, that when you are comparing equity across more than two groups, you will need to find a means of aggregating these to a single value (e.g., you might look at the average disparity, largest disparity, or use some weighting scheme to reflect different costs of disparities favoring different groups).

In most cases, this process will yield a number of options for a final model to deploy: some with better overall performance, some with better overall equity measures, and some with better performance for specific subgroups. Unlike model selection based on performance metrics alone, the final choice between these generally will involve a judgment call that reflects the project's dual goals of balancing accuracy and equity. As such, the final choice of model from this narrowed menu of options is best treated as a discussion between the data scientists and stakeholders in the same manner as choosing how to define fairness in the first place.

11.4.3   Other options for mitigating bias

Beyond incorporating measurements of equity into your model selection process, they also can inform how you put the model you choose into action. In general, disparities will vary as you vary the threshold used for turning continuous scores into 0/1 predicted classes. While many applications will dictate the total number of individuals who can be selected for intervention, it still may be useful to consider lower thresholds. For example, one of our projects has shown large *FDR* disparities across age and race in our models when selecting the top 150 individuals for an intervention (a number dictated by programmatic capacity), but these disparities are mitigated by considering the top 1,000 with relatively little cost in precision. This result has suggested a strategy for deployment: use the model to select the 1,000 highest risk individuals and randomly select 150 individuals from this set to stay within the program's capacity while balancing equity and performance.

Another approach that can work well in some situations is to consider using different thresholds across groups to achieve more equitable results, which we have explored in detail through a recent case study (Rodolfa et al., 2020). This is perhaps most robust where the metric of interest in monotonically increasing or decreasing with the number of individuals chosen for intervention (such as recall). This can be formulated in two ways.

- For programs that have a target scale but may have some flexibility in budgeting, you can consider to what extent the overall size of the program would need to increase to achieve equitable results (or other fairness goals, such as those described in Section 11.3.6). In this case, interventions do not need to be denied to any individuals in the interest of fairness, but the program would incur some additional cost in order to achieve a more equitable result.

- If the program's scale is a hard constraint, you still may be able to use subgroup-specific thresholds to achieve more equitable results by selecting fewer of some groups and more of others relative to the single threshold. In this case, the program would not incur additional costs of expansion, but some individuals who might have received the intervention based only on their score would need to be substituted for individuals with somewhat lower scores from under-represented subgroups.

As you are thinking about equity in the application of your machine learning models, it also is particularly important to keep in mind that measuring fairness in a model's predictions is only a proxy for what you fundamentally care about: fairness in outcomes in the presence of the model. As a model is put into practice, you may find that the program itself is more effective for some groups than others, motivating either additional changes in your model selection process or customizing interventions to the specific needs of different populations (or individuals). Incorporating fairness into decisions about who is chosen to receive an intervention is an important first step, but it should not be mistaken for a comprehensive solution to disparities in a program's application and outcomes.

In addition, some work is being done investigating means for incorporating bias and fairness more directly in the process of model development itself. For example, in many cases different numbers of examples across groups or unmeasured variables may contribute to a model having higher error rates on some populations than others, so additional data collection (either more examples or new features) may help mitigate these biases where doing so is feasible (Chen et al., 2018). Other work is being done to explore the results of incorporating equity metrics directly into the loss functions used to train some classes of machine learning models, making balancing accuracy and equity an aspect of model training itself (Celis et al., 2019; Zafar et al., 2017). Although we do not explore these more advanced topics in depth here, we refer you to the cited articles for more detail.

11.5 Further considerations

11.5.1 Compared to what?

Although building machine learning models that are completely free of bias is an admirable goal, it may not always be an achievable one. Nevertheless, even an imperfect model may provide an improvement over current practices depending on the degree of bias involved in existing processes. It is important to be cognizant of the existing context and make measurements of equity for current practices as well as new algorithms that might replace or augment them. The status quo should not be assumed to be free of bias because it is familiar any more than algorithms should be assumed capable of achieving perfection simply because they are complex. In practice, a more nuanced view is likely to yield better results: new models should be rigorously compared with current results and implemented when they are found to yield improvements but continually refined to improve on their outcomes over time as well.

11.5.2   Costs to both errors

In the examples in Section 11.3, we focus on programs that could be considered purely assistive or purely punitive to illustrate some of the relevant metrics for such programs. While this classification may work for some real-world applications, in many others there will be costs associated with both errors of inclusion and errors of exclusion that need to be considered together in deciding both on how to think about fairness and how to put those definitions into practice through model selection and deployment. For the bail example, there are, of course, real costs to society both of jailing innocent people and releasing someone who does, in fact, commit a subsequent crime. In many programs where individuals may be harmed by being omitted, errors of inclusion may mean wasted resources or even political backlash about excessive budgets.

In theory, you might imagine being able to assign some cost to each type of error—as well as to disparities in these errors across groups—and make a single unified cost-benefit calculation of the net result of putting a given model into application in a given way. In practice, of course, making an even reasonable quantitative estimate of the individual and social costs of these different types of errors is likely infeasible in most cases. Instead, a more practical approach generally involves exploring a number of different options through different choices of models and parameters and using these options to motivate a conversation about the program's goals, philosophy, and constraints.

11.5.3   What is the relevant population?

Related to the sample bias discussed in bias sources, understanding the relevant population for your machine learning problem is important both to the modeling itself and to your measures of equity. Calculation of metrics, such as the group size-adjusted false positive rate or false negative rate will vary depending on who is included in the denominator.

For example, imagine modeling who should be selected to receive a given benefit using data from previous applicants and looking at racial equity based on these metrics. What population is actually relevant to thinking about equity in this case? It might be the pool of applicants available in your data, or it might be the set of people who may apply if they have knowledge of the program (regardless of whether or not they actually apply), or perhaps even the population at large (e.g., as measured by the census). Each of those choices could potentially lead to different conclusions about the fairness of the program's decisions (either in the presence or absence of a machine learning model), highlighting the importance of understanding the relevant population and who might potentially be missing from your data as an element of how fairness is defined in your context. Keep in mind that determining (or at least making a reasonable estimate of) the correct population sometimes may require collecting additional data.

11.5.4 Continuous outcomes

For the sake of simplicity, we have been focusing on binary classification problems to help illustrate the types of considerations you might encounter when thinking about fairness in the application of machine learning techniques. However, these considerations, of course, extend to other types of problems, such as regression models of continuous outcomes.

In these cases, bias metrics can be formulated around aggregate functions of the errors a model makes on different types of individuals (e.g., the mean squared error and mean absolute error metrics familiar to you from regression) or tests for similarity of the distributions of these errors across subgroups. Working with continuous outcomes adds an additional layer of complexity in terms of defining fairness to account for the magnitude of the errors being made (e.g., how do you choose between a model that makes very large errors on a small number of individuals vs. one that makes relatively small errors on a large number of individuals?).

Unfortunately, the literature on bias and fairness in machine learning problems in other contexts (such as regression with continuous outcomes) is less rich than the work focused on classification; if you would like to learn more about what has been done in this regard, however, we suggest consulting Chouldechova and Roth (2018) for a good starting point (see, in particular, Section 3.5 of their discussion).

11.5.5 Considerations for ongoing measurement

The role of a data scientist is far from over when their machine learning model is put into production. Making use of these models requires ongoing curation, both to guard against degradation in terms of performance or fairness as well as to constantly improve outcomes. The vast majority of models you put into production will make mistakes, and a responsible data scientist will seek to look closely at these mistakes and understand—on both individual and population levels—how to learn from them to improve the model. Ensuring errors are balanced across groups is a good starting point, but seeking to reduce these errors over time is an important aspect of fairness as well.

One challenge you may face in making these ongoing improvements to your model is with measuring outcomes in the presence of a program that seeks to change them. In particular, the measurement of true positives and false positives in the absence of knowledge of a counterfactual (i.e., what would have happened in the absence of intervention) may be difficult or impossible. For example, among families who have improved nutritional outcomes after receiving a food subsidy, you may not be able to ascertain which families' outcomes actually were helped by the program versus which would have improved on their own, obfuscating any measure of recall you might use to judge performance or equity. Similarly, for individuals denied bail, you cannot know if they actually would have fled or committed a crime had they been released, making metrics such as false discovery rate impossible to calculate.

During a model's pilot phase, initially making predictions without taking action or using the model in parallel with existing processes can help mitigate some of these measurement problems over the short term. Similarly, when resources are limited such that only a fraction of individuals can receive an intervention, using some degree of randomness in the decision-making process can help establish the necessary counterfactual. However, in many contexts, this may not be practical or ethical, and you will need to consider other means for ongoing monitoring of the model's performance. Even in these contexts, however, it is important to continually review the performance of the model and seek to improve its performance in terms of both equity and efficiency. In practice, this may include some combination of considering proxies for the counterfactual, quasi-experimental inference methods, and expert/stakeholder review of the model's results (both in aggregate and of selected individual cases).

11.5.6   Equity in practice

Much of our discussion has been about fairness in the machine learning pipeline, focusing on the ways in which a model may be correct or incorrect in its predictions and how these might vary across groups. As a responsible practitioner of data science, undoubtedly these issues are important for you to understand and strive to handle correctly in your models, but fundamentally they can only serve as a proxy for a larger concept of fairness—ultimately, we care about equity in terms of differences in outcomes across groups. Ensuring fairness in decisions (whether made by machines, humans, or some combination) is an element of achieving this goal, but it is not the only element nor, in many cases, is it likely to be the largest one. In the face of other potential sources of bias—sample, label, application, historical, and societal—even fair decisions at the machine learning level may not lead to equitable results in society and the decision-making process may need to compensate for these other inequities. Some of these other sources of bias may be more challenging to quantify or incorporate into models directly, but data scientists have a shared responsibility to understand the broader context in which their models will be applied and seek equitable outcomes in these applications.

11.5.7 Additional terms for metrics

The metrics described here have been given a variety of names in the literature. While we have attempted to use language focused on the underlying statistics in this chapter, here are additional terms you may see in the literature for several of these metrics.

- Equalizing false discovery rates (*FDR*) sometimes is referred to as predictive parity or the "outcome test." Note that this is mathematically equivalent to having equal precision (also called positive predictive value) across groups.

- Equalizing false omission rates (*FOR*) is mathematically equivalent to equalizing the negative predictive value (*NPV*).

- When both *FOR* and *FDR* are equal across groups at the same time, sometimes this is referred to as sufficiency or conditional use accuracy equality.

- Equalizing the false negative rate (*FNR*), which is equivalent to equalizing recall (also called the true positive rate or sensitivity), sometimes is called equal opportunity.

- Equalizing the false positive rates (*FPR*), which is equivalent to equalizing the true negative rate (also known as specificity), sometimes is called predictive equality.

- When both *FNR* and *FPR* are equal across groups (i.e., when both equal opportunity and predictive equality are satisfied), various authors have referred to this as error rate balance, separation, equalized odds, conditional procedure accuracy equality, or the absence of disparate mistreatment.

- When members of every group have an equal probability of being assigned to the positive predictive class, this condition is referred to as group fairness, statistical parity, equal acceptance rate, demographic parity, or benchmarking. When this is true up to the contributions of only a set of "legitimate" attributes allowed to affect the prediction, this is called conditional statistical parity or conditional demographic parity.

- One definition, termed treatment equality, suggests considering disparities in the ratio of false negatives to false positives across groups.

- Metrics that look at the entirety of the score distribution across groups include AUC parity and calibration (also called test fairness, matching conditional frequencies, or under certain conditions well-calibration). Similarly, balance for the positive class and balance for the negative class look at average scores across groups among individuals with positive or negative labels, respectively.

- Additional work is being done looking at fairness through the lens of similarity between individuals (Dwork et al., 2012; Zemel et al., 2013) and causal reasoning (Kilbertus et al., 2017; Kusner et al., 2017).

As a field, we have yet to settle on a single widely-accepted terminology for thinking about bias and fairness; but, rather than becoming lost in competing naming conventions, we would encourage you to focus on what disparities in the different underlying metrics actually mean for how models you build might actually affect different populations in your particular context.

11.6   Case studies

The active conversation about algorithmic bias and fairness in both the academic and popular press has contributed to a more well-rounded evaluation of many of the models and technologies that are already in everyday use. This section highlights several recent cases, discussing them through the context of the metrics we have described as well as providing some resources for you to read further about each one.

11.6.1 Recidivism predictions with COMPAS

Over the course of the last two decades, models of recidivism risk have been put into use in many jurisdictions across the country. These models show a wide variation in methods (from heuristic rule-based scores to machine learning models) and have been developed by a variety of academic researchers, government employees, and private companies, many built with little or no evaluation of potential biases (Desmarais and Singh, 2013). Different jurisdictions put these models to use in a variety of ways, including identifying individuals for diversion and treatment programs, making bail determinations, and even in the course of sentencing decisions.

In May 2016, journalists at ProPublica undertook an exploration of accuracy and racial bias in these algorithms, focusing on one of the commercial solutions, Correctional Offender Management Profiling for Alternative Sanctions (COMPAS), built by Northpointe Inc. (Angwin et al., 2016; Larson et al., 2016). Their analysis focused on some of the errors made by the model, finding dramatic disparities between black and white defendants: among black defendants who did not have another arrest in the subsequent two years, 45% had been labeled by the system as high risk, almost twice the rate as that for white defendants (23%). Similarly, among individuals who did recidivate within two years, 48% of white defendants had been labeled as low risk, compared with 28% of black defendants.

In this evaluation, ProPublica is focusing on *FPR* and *FNR* metrics for their definition of fairness: e.g., if you are a person who, in fact, will not recidivate, do your chances of being mislabeled as high risk by the model differ depending on your race? In their response (The Northpointe Suite, 2016), Northpointe argues that this is the wrong fairness metric in this context—instead, they claim *FDR* should be the focus: e.g., if the model labels you as high risk, do the chances that it was wrong in doing so depend on your race? By this definition, COMPAS is fair: 37% of black defendants labeled as high risk did not recidivate, compared to 41% of white defendants. Table 11.1 summarizes these metrics for both racial groups.

In a follow-up article in December 2016 (Angwin and Larson, 2016), the ProPublica authors remark on the surprising result that a model could be "simultaneously fair and unfair." The public debate around COMPAS also has inspired a number of academic

Table 11.1. COMPAS fairness metrics

Metric	Caucasian (%)	African American (%)
False positive rate (*FPR*)	23	45
False negative rate (*FNR*)	48	28
False discovery rate (*FDR*)	41	37

researchers to look closer at these definitions of fairness, showing that, in the presence of different base rates, it would be impossible for a model to satisfy both definitions of fairness at the same time. The case of COMPAS demonstrates the potentially dramatic impact of decisions about how equity is defined and measured in real applications with considerable implications for individual's lives. Indeed, it is incumbent upon the researchers developing such a model, the policymakers deciding to put it into practice, and the users making decisions based upon its scores to understand and explore the options for measuring fairness as well as the tradeoffs involved in making that choice.

11.6.2 Facial recognition

A growing number of applications for facial recognition software are arising, from tagging friends in photos on social media to recognizing suspects by police departments, and off-the-shelf software is available from several large technology firms, including Microsoft, IBM, and Amazon. However, growth in the use of these technologies has seen a number of embarrassing stumbles related to how well they work across race in practice, including an automated image annotation app released by Google in 2015 that mistakenly tagged several African American users as gorillas (Conor Dougherty, 2015) and a number of early applications deployed on digital cameras that would erroneously tell Asian users to open their eyes or fail to detect users with darker skin tones entirely (Rose, 2010).

Despite the broad uses of these technologies, even in policing, relatively little work had been done to quantify their racial bias prior to 2018 when a researcher at MIT's Media Lab undertook a study to assess racial bias in the ability to correctly detect gender of three commercial facial recognition applications (developed by Microsoft, Face++, and IBM) (Buolamwini and Gebru, 2018). She developed a benchmark dataset reasonably well-balanced across race and gender by collecting 1,270 photos of members of parliament in several African and European nations, scoring each photo for skin tone using the Fitzpatrick skin type classification system commonly used in dermatology.

The results of this analysis show stark differences across gender and skin tone, focusing on false discovery rates for predictions of each gender. Overall, *FDR* is very low for individuals predicted to be male in the dataset, ranging from 0.7% to 5.6% between systems, while it is much higher among individuals predicted to be female (ranging from 10.7% to 21.3%). Notice that the models here are making a binary classification of gender, so individuals with a score on one side of a threshold are predicted as male and on the other side are predicted as female. Thus, the overall gender disparities seen here indicate that, at least relative to this dataset, all three thresholds have been chosen in such a way that the models are more certain when making a prediction that an individual is male than making a prediction that they are female. In theory, these thresholds could be readily tuned to produce a better balance in errors, but Buolamwini and Gebru (2018) note that all three APIs provide only predicted classes rather than the underlying scores, precluding users from choosing a different balance of error rates by predicted gender.

The disparities are even more stark when considering skin tone and gender jointly. In general, model performance is much worse for individuals with darker skin tones than those with lighter skin. Most dramatically, the *FDR* for individuals with darker skin who are predicted to be female ranges from 20.8% to 34.7%. At the other extreme, the largest *FDR* for lighter-skinned individuals predicted to be male is less than 1%. Table 11.2 shows these results in more detail.

One factor contributing to these disparities is likely sample bias. Although the training data used for these particular commercial models is not available, many of the widely available public data sets for developing similar facial recognition algorithms have been heavily skewed, with as many as 80% of training images being of white individuals and 75% being of men. Improving the representativeness of these data sets may be helpful, but it will not eliminate the need for ongoing studies of disparate performance of facial recognition across populations that might arise from other characteristics of the underlying models as well.

Table 11.2. The *FDR* values by skin tone and predicted gender (F = female, M = male, D = dark skin, L = light skin)

System	All **F** (%)	All **M** (%)	**DF** (%)	**DM** (%)	**LF** (%)	**LM** (%)
Microsoft	10.7	2.6	20.8	6.0	1.7	0.0
Face++	21.3	0.7	34.5	0.7	6.0	0.8
IBM	20.3	5.6	34.7	12.0	7.1	0.3

These technologies also can provide a case study for when policymakers might decide against putting a given model to use for certain applications entirely. In 2019, the City of San Francisco, CA announced that it would become the first city in the country to ban the use of facial recognition technologies entirely from city services, including its police department (Harwell, 2019). There, city officials reached the conclusion that any potential benefits of these technologies were outweighed by the combination of potential biases and overall privacy concerns, with the city's Board of Supervisors voting 8–1 to ban the technology. While the debate around appropriate uses for facial recognition likely will continue for some time across jurisdictions, San Francisco's decision highlighted the role of legal and policy constraints around how models are used in addition to ensuring that the models are fair when and where they are applied.

11.6.3   Facebook advertisement targeting

Social media has created new opportunities for advertisers to quickly and easily target their advertisements to particular subsets of the population. Additionally, regardless of this user-specified targeting, these advertising platforms will make automated decisions about who is shown a given advertisement, generally optimizing to some metric of cost efficiency. Recently, however, these tools have begun to come under scrutiny surrounding the possibility that they might violate US Civil Rights laws that make it illegal for individuals to be excluded from job or housing opportunities on the basis of protected characteristics such as age, race, or sex.

Public awareness that Facebook's tools allowed advertisers to target content based on these protected characteristics began to form in 2016 with news reports highlighting the feature (Angwin and Parris, 2016). While the company initially responded that their policies forbid advertisers from targeting ads in discriminatory ways, there were legitimate use cases for these technologies as well, suggesting that the responsibility fell to the people placing the ads. By 2018, however, it was clear that the platform was allowing some advertisers to do just that and the American Civil Liberties Union filed a complaint of gender discrimination with the US Equal Employment Opportunity Commission (Campbell, 2018). The complaint pointed to 10 employers that had posted job ads targeted exclusively to men, including positions such as truck drivers, tire salesmen, mechanics, and security engineers. Similar concerns were cited by the US Department of Housing and Urban Development in 2019 when it filed charges against the social media company alleging it had served ads that violate the Fair Housing Act (Brandom, 2019). Responding to the growing criticism, Facebook began to limit the attributes advertisers could use to target their content.

However, these limitations might not be sufficient in light of the platform's machine learning algorithms that are determining who is shown a given ad regardless of the specific targeting parameters. Research performed by Ali et al. (2019) confirms that the content of an advertisement could dramatically impact to whom it is served despite broad targeting parameters. Users who have been shown a job posting for a position as a lumberjack were more than 90% men and more than 70% white, while those seeing a posting for a janitorial position were more than 75% women and 65% black. Similarly wide variety has been observed for housing advertisements, ranging from an audience nearly 65% black in some conditions to 85% white in others. A separate study of placement of STEM career ads with broad targeting has found similar gender biases in actual impressions, with content shown to more men than women (Lambrecht and Tucker, 2019).

Unlike the other case studies we have described, the concept of fairness for housing and job advertisements is provided by existing legislation, focusing not on errors of inclusion or exclusion but rather on representativeness itself. As such, the metric of interest here is disparity in the probability of being assigned to the predicted positive class (e.g., being shown the ad) across groups, unrelated to potentially differential propensities of each group to respond. To address these disparities, Facebook (as well as other ad servers) may need to modify their targeting algorithms to directly ensure job and housing ads are shown to members of protected groups at similar rates. This, in turn, would require a reliable mechanism for determining whether a given ad is subject to these requirements, which poses its own technical challenges. As of this writing, understanding how best to combat discrimination in ad targeting is an ongoing area of research as well as an active public conversation.

11.7 Aequitas: A toolkit for auditing bias and fairness in machine learning models

▶ https://github.com/dssg/aequitas

To help data scientists and policymakers make informed decisions about bias and fairness in their applications, we have developed Aequitas, an open source bias and fairness audit toolkit that was released in May 2018. It is an intuitive and easy to use addition to the machine learning workflow, enabling users to seamlessly audit models for several bias and fairness metrics in relation to multiple population sub-groups. Aequitas can be used directly as a Python library, via command line interface or a web application, making it accessible to a wide range of users (from data scientists to policy-makers).

▶ http://www.datascience publicpolicy.org/aequitas/

Because the concept of fairness is highly dependent on the particular context and application, Aequitas provides comprehensive information on how its results should be used in a public policy context, taking the resulting interventions and its implications into consideration. It is intended to be used not only by data scientists but also policymakers, through both seamless integration in the machine learning workflow as well as a web app tailored for non-technical users auditing these models' outputs.

In Aequitas, bias assessments can be made prior to model selection, evaluating the disparities of the various models developed based on whatever training data has been used to tune it for its task. The audits can be performed prior to a model being operationalized, based on operational data of how biased the model has proven to be in holdout data. Or, they can involve a bit of both, auditing bias in an A/B testing environment in which limited trials of revised algorithms are evaluating whatever biases have been observed in those same systems in prior production deployments.

Aequitas has been designed to be used by two types of users.

1. Data scientists and AI researchers: As they build systems for use in risk assessment tools, data scientists and artificial intelligence researchers will use Aequitas to compare bias measures and check for disparities in different models they are building during the process of model building and selection.

2. Policymakers: Before "accepting" an AI system to use in a policy decision, policymakers will run Aequitas to understand what biases exist in the system and what (if anything) they need to do in order to mitigate those biases. This process must be repeated periodically to assess the fairness degradation through time of a model in production.

11.7.1 Aequitas in the larger context of the machine learning pipeline

Audits must be conducted internally by data scientists before evaluation and model selection. Policymakers (or clients) must audit

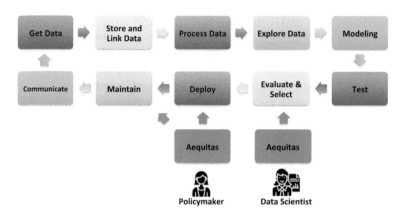

Figure 11.2. The machine learning pipeline

externally before accepting a model in production as well as perform periodic audits to detect any fairness degradation over time.

Figure 11.2 puts Aequitas in the context of the machine learning workflow and shows each type of user and when the audits must be made. The main goal of Aequitas is to standardize the process of understanding model biases. By providing a toolkit for auditing by both data scientists and decision makers, it makes it possible for these different actors to take bias and fairness into consideration at all stages of decision making in the modeling process: model selection, whether or not to deploy a model, when to retrain, the need to collect more and better data, and so on.

For a more hands-on tutorial using Aequitas, see Section 13.2.10 (Bias and Fairness workbook).

Chapter 12
Privacy and Confidentiality

Stefan Bender, Ron S. Jarmin, Frauke Kreuter, and Julia Lane

This chapter addresses the issue that sits at the core of any study of human beings—ensuring that the privacy and confidentiality of the people and organizations being studied is protected appropriately.

The challenge faced by social science researchers relative to data users in other contexts (e.g., agencies administering public programs or within private enterprises) is the need to compute accurate statistics from sensitive databases, share their results broadly and facilitate scientific review and replication. However, the quantity and accuracy of statistics computed from databases containing sensitive private information increase the risk that private information is released in ways not originally intended. Thus, the stewards of sensitive databases and the researchers accessing them must find ways to conduct their work that protects the confidentiality of the data and balances potential harms from releasing too much private information and the benefits from the statistics they publish. Historically, social scientists have used two methods of minimizing that risk: anonymizing the data so that an individual's information cannot be reidentified and asking human subjects for the consent to the use of their data (National Academies, 2014). Those twin approaches have become obsolete, for reasons we will discuss in this chapter, and have not been replaced by an alternative framework. This concluding chapter identifies the issues that must be addressed for responsible research.

12.1 Introduction

Much social science research uses data on individuals, households, different types of businesses, and organizations, such as educational and government institutions. Indeed, the example running throughout this book involves data on individuals (such as faculty and students) and organizations (such as universities and firms). In circumstances such as these, researchers must ensure that sensitive data are used responsibly—that there is a clear understanding of possible harms and that appropriate measures are taken to mitigate them. In practical terms, the study of human subjects requires that the interests of individual privacy and data confidentiality be balanced against the social benefits of research access and use.

We begin by defining terms.

Utility: Data utility is the value resulting from data use. That utility has soared in the private sector—the largest companies in the US are the data companies Amazon, Google, Microsoft, Facebook, and Apple (Galloway, 2017). They make their money by providing utility to their customers. The goal of social science research, as stated in Chapter 1, is to use new data and tools to answer questions such as:

- "What are the earnings and employment outcomes of individuals graduating from two and four year colleges?"

- "How does placement in different types of firms change the likelihood of recidivism of formerly incarcerated workers?"

- "How do regulatory agencies move from reactive, complaint-based, health and safety inspections for workplaces and housing to a more proactive approach that focuses on prevention?"

Privacy: Privacy is a right of the subjects in a database. It "encompasses not only the famous 'right to be left alone,' or keeping one's personal matters and relationships secret, but also the ability to share information selectively but not publicly" (President's Council of Advisors on Science and Technology, 2014). A useful way of thinking about privacy is the notion of preserving the appropriate flow of information (Nissenbaum, 2009). There is no specific data type or piece of information that is too sensitive to be shared in all circumstances. In some context, providing detailed information about one's health is very appropriate, e.g., if it helps in finding the correct treatment for a disease. Generally, it is important to understand the context and the contextual integrity of the data flow when deciding which data to collect and how to analyze and share them.

Confidentiality: Confidentiality refers to the duty that the steward and authorized users of a database have to ensure the privacy rights of the subjects are protected. While confidentiality can be defined as "preserving authorized restrictions on information access and disclosure, including means for protecting personal privacy and proprietary information" (McCallister et al., 2010), doing so is not easy—the challenge to the research community is how to balance the *risk* of providing access with the associated utility (Duncan et al., 2011). To give a simple example, if means and percentages are presented for a *large* number of people, it will be difficult to infer an individual's value from such output, even if one knows that a certain individual or unit has contributed to the formation of that mean or percentage. However, if those means and percentages are presented for subgroups or in multivariate tables with small cell sizes, the risk for disclosure increases (Doyle et al., 2001).

Risk: In our context, generally risk is thought of as the risk of an intruder reidentifying an individual or a business in a research dataset (Duncan and Stokes, 2004). That is, risk refers to the ways that confidentiality can be broken and privacy rights violated. Often, it is argued that those risks increase every year as increasing volumes of data are available on individuals on the internet or in the databases of large corporations and as there are more and better tools available to make such linkages (Herzog et al., 2007; Shlomo, 2014). However, it also could be argued that the proliferation of data and tools reduces risk because it is so much easier for an intruder to discover information on an individual through a Google search (Lane, 2020). Regardless, it is generally accepted that greater research access to data and their original values increases the risk of reidentification for individual units.

Harm: Harm refers to the costs (e.g., financial or reputational) that incur to the subjects of a sensitive database due to the release of too much private information. Although much of the discussion of privacy and confidentiality has been driven by the imperatives of the legislation governing statistical agencies, which imposes civil and criminal penalties for any reidentification, statistical agencies no longer have a monopoly on data access and use. As a result, there is more attention being paid to the potential for harm based on the type of information being shared, rather than the fact that a piece of information is shared (Nissenbaum, 2019). Intuitively, if an intruder discovers that an individual in a dataset is a woman, or is married, that may cause less harm than if information about income, sexual history, or criminal records is recovered.

There is an explicit tradeoff between privacy and data utility. The greater the number and accuracy of statistics released from a sensitive private database, the more utility we obtain from its use. But increasing the number and/or accuracy of the statistics also increased the risk that too much private information is

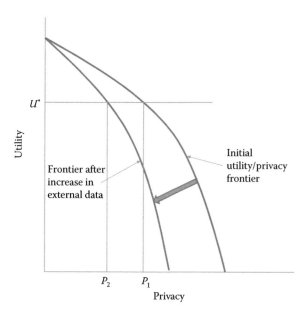

Figure 12.1. The privacy–utility tradeoff

► https://dl.acm.org/doi/
10.1145/773153.773173

released (Dinur and Nissim, 2003). We depict this tradeoff graphically in Figure 12.1. The concave curves in this hypothetical example depict the technological relationship between data utility and privacy for an organization such as a business firm or a statistical agency. At one extreme, all information is available to anyone about all units, and therefore high analytic utility is associated with the data that are not at all protected. At the other extreme, no one has access to any data and no utility is achieved. Initially, assume the organization is on the outer frontier. Increased external data resources (those not used by the organization) increase the risk of reidentification. This is represented by an inward shift of the utility/privacy frontier in Figure 12.1. Before the increase in external data, the organization could achieve a level of data utility U^* and privacy P_1. The increase in externally available data now means, that in order to maintain utility at U^*, privacy is reduced to P_2. This simple example represents the challenge to all organizations that release statistical or analytical products obtained from underlying identifiable data. As more data from external sources becomes available, it becomes more difficult to maintain privacy.

Previously, national statistical agencies had the capacity and the mandate to make dissemination decisions: they assessed the risk, they understood the data user community and the associated utility from data releases. And, they had the wherewithal to address the legal, technical, and statistical issues associated with protecting confidentiality (Trewin et al., 2007).

But in a world of massive amounts of data, many once-settled issues have new complications, and wholly new issues arise that need to be addressed, albeit under the same rubrics. The new types of data have much greater potential utility, often because it is possible to study small cells or the tails of a distribution in ways not possible with small data. In fact, in many social science applications, the tails of the distribution are often the most interesting and most difficult-to-reach parts of the population being studied: consider health care costs for a small number of ill people (Stanton and Rutherford, 2006) or economic activity such as rapid employment growth by a small number of firms (Decker et al., 2016).

Example: The importance of activity in the tails

Spending on health care services in the US is highly concentrated among a small proportion of people with extremely high use. For the overall civilian population living in the community, the latest data indicate that more than 20% of all personal health care spending in 2009 ($275 billion) was on behalf of only 1% of the population (Schoenman, 2012).

It is important to understand the source of the risk of privacy breaches. Let us assume for a moment that we have conducted a traditional small-scale survey with 1,000 respondents. The survey contains information on political attitudes, spending and saving in a given year, and income, as well as background variables on income and education. If name and address are saved together with this data and someone obtains access to the data, obviously it is easy to identify individuals and gain access to information that is otherwise not public. If the personal identifiable information (name and address) are removed from this data file, the risk is much reduced. If someone has access to the survey data and sees all of the individual values, it might be difficult to assess with certainty which of the more than 330 million inhabitants in the US is associated with an individual data record. However, the risk is higher if one knows some of this information (e.g., income) for a person, and knows that this person is in the survey. With these two pieces of information, it is likely possible to uniquely identify the person in the survey data.

Larger amounts of data will increase the risk precisely for this reason. Much data has been made available for reidentification purposes (Ohm, 2010). Most obviously, the risk of reidentification becomes much greater because the new types of data have much richer detail and a much larger public community has access to ways to reidentify individuals. There have been many famous examples of reidentification occurring even when obvious personal information, such as name and social security number, has been removed and the data provider has believed that the data were consequently deidentified. In the 1990s, Massachusetts Group Insurance released "deidentified" data on the hospital visits of state employees; researcher Latanya Sweeney quickly reidentified the hospital records of the then Governor William Weld using nothing more than state voter records about residence and date of birth (Sweeney, 2001). In 2006, the release of supposedly deidentified web search data by AOL resulted in two *New York Times* reports being able to reidentify a customer simply from her browsing habits (Barbaro et al., 2006). And in 2012, statisticians at the department store, Target, used a young teenager's shopping patterns to determine that she was pregnant before her father did (Hill, 2012).

But, there are also less obvious problems. What is the legal framework when the ownership of data is unclear? In the past, when data have been more likely to be collected and used within the same entity—for example, within an agency that collects administrative data or within a university that collects data for research purposes—organization-specific procedures (usually) have been in place and sufficient to regulate the usage of these data. Today, legal ownership is less clear (Lane et al., 2014). There are many unresolved issues, such as who has the legal authority to make decisions about permission, access, and dissemination and under what circumstances. The challenge today is that data sources often are combined, collected for one purpose, and used for another. Data providers often have a poor understanding of whether or how their data will be used. Think, for example, about cell phone calls. *The New York Times* has produced a series of thought-provoking articles about the access to and use of cell phone data, such as the one entitled, "Your Apps Know Where You Were Last Night, and They're Not Keeping It Secret" (Valentino-DeVries et al., 2018). Who owns your cell phone calls? Should it be you, as the initiator of the call, your friend as the recipient, your cell phone company, your friend's cell phone company, the cloud server on which the data are stored for billing purposes, or the satellite company that connects the two of you? And, what laws should regulate access and use? The state (or country) where you are located when you make the call? Or, your friend's state (or country)? The state (or country) of your cell phone provider? And so on. The legal framework is, at best, murky.

Example: Knowledge is power

In a discussion of legal approaches to privacy in the context of big data, Strandburg (2014) says: " 'Big data' has great potential to benefit society. At the same time, its availability creates significant potential for mistaken, misguided, or malevolent uses of personal information. The conundrum for the law is to provide space for big data to fulfill its potential for societal benefit, while protecting citizens adequately from related individual and social harms. Current privacy law evolved to address different concerns and must be adapted to confront big data's challenges."

12.2 Why is access important?

Previous chapters in this book have provided detailed examples of the potential of data to provide insights into a variety of social science questions—particularly the relationship between investments in R&D and innovation. But, that potential is only realized if researchers have access to the data (Lane, 2007): not only to perform primary analyses but also to validate the data generation process (in particular, data linkage), replicate analyses, and build a knowledge infrastructure around complex data sets.

12.2.1 Validating the data-generating process

Research designs requiring a combination of data sources and/or analysis of the tails of populations challenge the traditional paradigm of conducting statistical analysis on deidentified or aggregated data. In order to combine data sets, someone in the chain that transforms raw data into research outputs needs access to link keys contained in the data sets to be combined. High-quality link keys uniquely identify the subjects under study and typically are derived from items such as individual names, birth dates, social security numbers, and business names, addresses, and tax ID numbers. From a privacy and confidentiality perspective, link keys are among the most sensitive information in many data sets of interest to social scientists. This is why many organizations replace link

★ The PII is "any information about an individual maintained by an agency, including (1) any information that can be used to distinguish or trace an individual's identity, such as name, social security number, date and place of birth, mother's maiden name, or biometric records; and (2) any other information that is linked or linkable to an individual, such as medical, educational, financial, and employment information" (McCallister et al., 2010).

keys containing personal identifiable information (PII)* with privacy-protecting identifiers (Schnell et al., 2009). Regardless, at some point in the process, those must be generated out of the original information, thus access to the latter is important.

12.2.2   Replication

John Ioannidis has claimed that most published research findings are false (Ioannidis, 2005); for example, the unsuccessful replication of genome-wide association studies, at less than 1%, is staggering (Bastian, 2013). Inadequate understanding of coverage, incentive, and quality issues, together with the lack of a comparison group, can result in biased analysis—famously in the case of using administrative records on crime to make inference about the role of death penalty policy in crime reduction (Donohue and Wolfers, 2006; Levitt and Miles, 2006). Similarly, overreliance on, e.g., Twitter data, in targeting resources after hurricanes can lead to the misallocation of resources towards young internet-savvy people with cell phones and away from elderly or impoverished neighborhoods (Shelton et al., 2014), or the bad survey methodology leading to the *Literary Digest* incorrectly calling the 1936 election (Squire, 1988). The first step to replication is data access; such access can enable other researchers to ascertain whether the assumptions of a particular statistical model are met, what relevant information is included or excluded, and whether valid inferences can be drawn from the data (Kreuter and Peng, 2014).

12.2.3   Building knowledge infrastructure

Creating a community of practice around a data infrastructure can result in tremendous new insights, as the Sloan Digital Sky Survey and the Polymath project have shown (Nielsen, 2012). In the social science arena, the Census Bureau has developed a productive ecosystem that is predicated on access to approved external experts to build, conduct research using, and improve key data assets, such as the Longitudinal Business Database (Jarmin and Miranda, 2002) and Longitudinal Employer Household Dynamics (Abowd et al., 2004), which have yielded a host of new data products and critical policy-relevant insights on business dynamics (Haltiwanger et al., 2013) and labor market volatility (Brown et al., 2008), respectively. Without providing robust, but secure, access to confidential data, researchers at the Census Bureau would have been unable to undertake the innovations that have made these new products and insights possible.

12.3 Providing access

The approaches to providing access have evolved over time. Statistical agencies often employ a range of approaches depending on the needs of heterogeneous data users (Doyle et al., 2001; Foster et al., 2009). Dissemination of data to the public usually occurs in three steps: an evaluation of disclosure risks, followed by the application of an anonymization technique, and finally an evaluation of disclosure risks and analytical quality of the candidate data release(s). The two main approaches have been *statistical disclosure* control techniques to produce anonymized public use data sets and controlled access through a *research data center* (Shlomo, 2018).

12.3.1 Statistical disclosure control techniques

Statistical agencies have made data available in a number of ways: through tabular data, public use files, licensing agreements, and, more recently, through synthetic data (Reiter, 2012). Hundepool et al. (2010) define statistical disclosure control as:

> concepts and methods that ensure the confidentiality of micro and aggregated data that are to be published. It is methodology used to design statistical outputs in a way that someone with access to that output cannot relate a known individual (or other responding unit) to an element in the output.

Traditionally, confidentiality protection has been accomplished by releasing only *aggregated tabular data*. This practice has worked well in settings where the primary purpose is enumeration, such as census taking. However, tabular data are poorly suited to describing the underlying distributions and covariance across variables that are often the focus of applied social science research (Duncan et al., 2011).

To provide researchers access to data that permits analysis of the underlying variance–covariance structure of the data, some agencies have constructed public use microdata samples. To protect confidentiality in such *public use files*, typically a number of statistical disclosure control procedures are applied. These include stripping all identifying (e.g., PII) fields from the data, topcoding highly skewed variables (e.g., income), and swapping records (Doyle et al., 2001; Zayatz, 2007). However, the mosaic effect—where disparate pieces of information can be combined to reidentify individuals—dramatically increases the risk of releasing public use files (Czajka et al., 2014). In addition, there is growing evidence that the statistical disclosure procedure applied to produce them decreases their utility across many applications (Burkhauser et al., 2010).

Some agencies provide access to confidential microdata through *licensing* arrangements. A contract specifies the conditions of use and what safeguards must be in place. In some cases, the agency has the authority to conduct random inspections. However, this approach has led to a number of operational challenges, including version control, identifying and managing risky researcher behavior, and management costs (Doyle et al., 2001).

Another approach to providing access to confidential data that has been proposed by a group of theoretical computer scientists: Cynthia Dwork, Frank McSherry, Kobbi Nissim, and Adam Smith (Dwork and Roth, 2014). Here, statistics or other reported outputs are injected with noise, and they are called "differentially private" if the inclusion or exclusion of the most at-risk person in the population does not change the probability of any output by more than a given factor. The parameter driving this factor (usually referred to as epsilon) quantifies how sensitive the aggregate output is to any one person's data. If it is low, the output is highly "private" in the sense that it will be very difficult to reconstruct anything based on it. If it is high, reconstruction is easy. For a discussion of the applications to census data, see Abowd (2018) and Ruggles et al. (2019).

Although the research agenda is an interesting and important one, there are a number of concerns about the practical implications. The Census Bureau, for example, has spent many millions of dollars to implement differential privacy techniques for the 2020 Decennial Census, and researchers who have studied the potential impact on small towns worry that small towns will "disappear" from official statistics—a major issue when data are used for decision making (Wezerek and Van Riper, 2020).

Another approach that has had some resurgence is the use of *synthetic data* where certain properties of the original data are preserved but the original data are replaced by "synthetic data" so that no individual or business entity can be found in the released data (Drechsler, 2011). One of the earlier examples of such work is the IBM Quest system (Agrawal and Srikant, 1994) that generated synthetic transaction data. Two more recent examples of synthetic data sets are the SIPP Synthetic-Beta (Abowd et al., 2006) of linked Survey of Income and Program Participation (SIPP) and Social Security Administration earnings data, and the Synthetic Longitudinal Business Database (SynLBD) (Kinney et al., 2011). Jarmin et al. (2014) discuss how synthetic data sets lack utility in many research settings but are useful for generating flexible data sets underlying data tools and apps, such as the Census Bureau's OnTheMap. It is important to keep in mind that the utility of synthetic data sets as a general purpose "anonymization" tool is relatively limited. Synthetic data generation typically requires explicitly defining which properties of the original data need to be preserved (such as univariate or bivariate distributions of certain variables), and as such can be of limited use in most social science research.

12.3.2 Research data centers

The second approach to data access is establishing research data centers (RDC). The RDC present an established operational approach to facilitate access to confidential microdata for research and/or statistical purposes. This approach is based on the theoretical framework of the "Five Safes," which initially was developed by Felix Ritchie at the UK Office of National Statistics in 2003 (Desai et al., 2016). The first dimension refers to safe projects. This dimension mainly refers to whether or not the intended use of the data conforms with the use specified in legislation or regulations. For example, a legislation may specifically allow users to use the data only for independent scientific research. Safe people, the second dimension of the Five Safes framework, requires data users to be able to use the data in an appropriate way. A certain amount of technical skill or minimum educational level may be required to access the data. In contrast, safe data, the third aspect, refers to the potential for deidentifying individuals or entities in the data. The fourth, safe settings, relates to the practical controls on how the data are accessed. Different channels may exist which, in turn, may depend on the deidentification risk. In practice, the lower the deidentification risk, the more restrictive the setting will be. Lastly, safe output refers to the risk of deidentification in publications from confidential microdata. Strong input and output controls are in place to ensure that published findings comply with the privacy and confidentiality regulations (Hayden, 2012).

12.4 Non-tabular data

In addition to tabular data, many new sources of data consist of text, audio, image, and video content. The above approaches primarily address maintaining the privacy and confidentiality of entities in tabular data, but it is equally important to do the same in

Box 12.1: Federal statistical research data centers

It is not easy to use the federal statistical research data centers (FSRDCs). Every stage of the research process is significantly more time-consuming than using public use data, and only the most persistent researchers are successful. In addition, most of the branches charge high fees for anyone unaffiliated with an institution sponsoring an FSRDC. Projects are approved only if they benefit the Census Bureau, which by itself makes most research topics ineligible. Prospective users must prepare detailed proposals, including the precise models they intend to run and the research outputs they hope to remove from the center, which generally are restricted to model coefficients and supporting statistics. Most descriptive statistics are prohibited. Researchers are not allowed to "browse" the data or change the outputs based on their results. Under census law, researchers must become (unpaid) Census Bureau employees to gain access to non-public data. To meet this requirement, once a project is approved, researchers must obtain Special Sworn Status, which involves a level 2 security clearance and fingerprint search. Applicants must be US citizens or US residents for three years, so most international scholars are excluded. Researchers then undergo data stewardship training. If researchers wish to modify their original model specifications or outputs, they must submit a written request and wait for approval. When the research is complete, the results must be cleared before publication by the Center for Disclosure Avoidance Research at the Census Bureau. Any deviations from the original proposal must be documented, justified, and approved. The FSRDCs have never been intended as a substitute for public use microdata, and they cannot fulfill that role. Even if the number of seats in the centers could be multiplied several hundred-fold to accommodate the current number of users of public use data, substantial hurdles remain. Applying for access and gaining approval to use the FSRDC takes at least six months and usually more. Eligibility for using FSRDCs is limited to investigators (1) affiliated with an FSRDC (or with significant financial resources), (2) with sufficient time to wait for review and approvals, and (3) conducting work deemed valuable by the Bureau (UnivTask Force on Differential Privacy for Census Data, 2019).

Box 12.2: The administrative data research facility

Additional approaches are becoming available. The Commission on Evidence-based Policy has identified new technologies, such as remote access cloud-based virtual data facilities, as a promising approach to providing scalable secure access to microdata without the disadvantages of the "brick and mortar" approach used by the FSRDC system. One new approach, the Administrative Data Research Facility, has incorporated the "five safes" principles—safe projects, safe people, safe settings, safe data, and safe outputs (https://en.wikipedia.org/wiki/Five_safes)—into its design. In addition to winning a Government Innovation Award (Government Computer News Staff, 2018), it has been used to provide secure access to confidential data to more than 450 government analysts and researchers in the past three years (Kreuter et al., 2019).

---

non-tabular data. Medical records, sensitive crime records, notes and comments in administrative records, camera footage (e.g., from security cameras or police body-cams) are all examples of data that is being used for analysis and that requires robust techniques to maintain the privacy and confidentiality of individuals. Although the techniques for these are not as mature, some work is being accomplished here.

Text anonymization: Typical approaches for anonymizing text range from simply removing personally identifiable information (PII) through regular expressions and dictionaries (Neamatullah et al., 2008) to machine learning-based approaches that balance the confidentiality of the entities in the data and the utility of the text (Cumby and Ghani, 2011).

Image and video anonymization: The most common use of anonymization techniques in image and video data is to redact, blur, or remove faces of individuals in order to protect their identity. This can be extended to other attributes of the person, such as clothing or the rest of the body, but the primary focus to date has been on detecting and then blurring or modifying the faces of individuals in the data. Sah et al. (2017) provide a survey of video redaction methods. Hukkelas et al. (2019) present a method to automatically anonymize faces in images while retaining the original data distribution.

12.5   The new challenges

While there are well-established policies and protocols surrounding access to and use of survey and administrative data, a major new challenge is the lack of clear guidelines governing the collection of data about human activity in a world in which all public, and some private, actions generate data that can be harvested (Ohm, 2010; President's Council of Advisors on Science and Technology, 2014; Strandburg, 2014). The twin pillars on which so much of social science have rested—informed consent and anonymization—are virtually useless in a big data setting where multiple data sets can be and are linked together using individual identifiers by a variety of players beyond social scientists with formal training and whose work is overseen by institutional review boards. This rapid expansion in data and their use is very much driven by the increased utility of the linked information to businesses, policymakers, and ultimately the taxpayer. In addition, there are no obvious data stewards and custodians who can be entrusted with preserving the privacy and confidentiality with regard to both the source data collected from sensors, social media, and many other sources, and the related analyses (Lane and Stodden, 2013).

It is clear that informed consent as historically construed is no longer feasible. As Nissenbaum (2011) notes, notification is either comprehensive or comprehensible but not both. While ideally human subjects are offered true freedom of choice based on a sound and sufficient understanding of what the choice entails, in reality the flow of data is so complex and the interest in the data usage so diverse that simplicity and clarity in the consent statement unavoidably result in losses of fidelity, as anyone who has accepted a Google Maps agreement is likely to understand (Hayden, 2015). In addition, informed consent requires a greater understanding of the breadth of type of privacy breaches, the nature of harm as diffused over time, and an improved valuation of privacy in the big data context. Consumers may value their own privacy in variously flawed ways. They may, for example, have incomplete information, or an overabundance of information rendering processing impossible, or use heuristics that establish and routinize deviations from rational decision making (Acquisti, 2014).

It is also nearly impossible to truly anonymize data. Big data often are structured in such a way that essentially everyone in the file is unique, either because so many variables exist or because they are so frequent or geographically detailed that they make it easy to reidentify individual patterns (Narayanan and Shmatikov, 2008).

It is also no longer possible to rely on sampling or measurement error in external files as a buffer for data protection, because most data are not in the hands of statistical agencies.

There are no data stewards controlling access to individual data. Data are often so interconnected (e.g., social media network data) that one person's action can disclose information about another person without that person even knowing that their data are being accessed. The group of students posting pictures about a beer party is an obvious example, but, in a research context, if the principal investigator grants access to the proposal, information could be divulged about colleagues and students. In other words, volunteered information of a minority of individuals can unlock the same information about many—a type of "tyranny of the minority" (Barocas and Nissenbaum, 2014b).

There are particular issues raised by the new potential to link information based on a variety of attributes that do not include PII. Barocas and Nissenbaum (2014a) write as follows:

> Rather than attempt to deanonymize medical records, for instance, an attacker (or commercial actor) might instead infer a rule that relates a string of more easily observable or accessible indicators to a specific medical condition, rendering large populations vulnerable to such inferences even in the absence of PII. Ironically, this is often the very thing about big data that generate the most excitement: the capacity to detect subtle correlations and draw actionable inferences. But it is this same feature that renders the traditional protections afforded by anonymity (again, more accurately, pseudonymity) much less effective.

In light of these challenges, Barocas and Nissenbaum continue:

> the value of anonymity inheres not in namelessness, and not even in the extension of the previous value of namelessness to all uniquely identifying information, but instead to something we called "reachability," the possibility of knocking on your door, hauling you out of bed, calling your phone number, threatening you with sanction, holding you accountable—with or without access to identifying information (Barocas and Nissenbaum, 2014a).

It is clear that the concepts used in the larger discussion of privacy and big data require updating. How we understand and assess harms from privacy violations needs updating. And, we must rethink established approaches to managing privacy in the big data context. The next section discusses the framework for doing so.

12.6 Legal and ethical framework

The Fourth Amendment to the US Constitution, which constrains the government's power to "search" the citizenry's "persons, houses, papers, and effects" usually is cited as the legal framework for privacy and confidentiality issues. Even so, the US has a "sectoral" approach to privacy regulation, for example, the Family Education Rights and Privacy Act and the Health Insurance Portability and Accountability Act both are used in situations where different economic areas have separate privacy laws (Ohm, 2014). In addition, current legal restrictions and guidance on data collection in the industrial setting include the Fair Information Practice Principles dating from 1973 and underlying the Fair Credit Reporting Act from 1970 and the Privacy Act from 1974 (Strandburg, 2014). Federal agencies often have statutory oversight, such as Title 13 of the US Code for the Census Bureau, the Confidential Information Protection and Statistical Efficiency Act for federal statistical agencies, and Title 26 of the US Code for the Internal Revenue Service.

The generation of data resulting from the commercial transactions of businesses, or the social interactions of consumers, is not governed by a similar legal framework. There are major questions as to what is reasonably private and what constitutes unwarranted intrusion (Strandburg, 2014). There is a lack of clarity on who owns the new types of data—whether it is the person who is the subject of the information; the person or organization who collects these data (the data custodian); the person who compiles, analyzes, or otherwise adds value to the information; the person who purchases an interest in the data; or society at large. The lack of clarity is exacerbated because some laws treat data as property and some treat it as information (Cecil and Eden, 2003).

The ethics of the use of big data are also not clear, because analysis may result in being discriminated against unfairly, being limited in one's life choices, being trapped inside stereotypes, being unable to delineate personal boundaries, or being wrongly judged, embarrassed, or harassed. There is an entire research agenda to be pursued that examines the ways in which big data may threaten interests and values, distinguishes the origins and nature of threats to individual and social integrity, and identifies different solutions (Boyd and Crawford, 2012). The approach should be to describe what norms and expectations are likely to be violated if a person agrees to provide data, rather than to describe what will be done during the research.

What is clear is that most data no longer are housed in statistical agencies, with well-defined rules of conduct, but in businesses or administrative agencies. In addition, because digital data can be alive forever, ownership could be claimed by yet-to-be-born relatives whose personal privacy could be threatened by release of information about blood relations.

The new European Data Protection Regulation (GDPR), which is in effect since May 2018, has been designed to address some of these challenges. In addition to ensuring lawful data collection practices, GDPR pushes for purpose limitation and data minimization. This principle requires organisations to clearly state for what purpose personal data is collected, to collect the data only for the time needed to complete the purpose, and to collect only those personal data that are needed to achieve the specified processing purposes. In the US, the California Consumer Privacy Act (CCPA) has been in effect since January 2020, so here too companies now have time limits to process customer data.

However, GDPR and other regulations of this type still rely on traditional regulatory tools for managing privacy, which are notice and consent. Both have failed to provide a viable market mechanism allowing a form of self-regulation governing industry data collection. Going forward, a more nuanced assessment of tradeoffs in the big data context, moving away from individualized assessments of the costs of privacy violations, is needed (Strandburg, 2014).

Ohm advocates for a new conceptualization of legal policy regarding privacy in the big data context that uses five guiding principles for reform: first, that rules take into account the varying levels of inherent risk to individuals across different data sets; second, that traditional definitions of PII need to be rethought; third, that regulation has a role in creating and policing walls between data sets; fourth, that those analyzing big data must be reminded, with a frequency in proportion to the sensitivity of the data, that they are dealing with people; and finally, that the ethics of big data research must be an open topic for continual reassessment (Ohm, 2014).

12.7 Summary

The excitement about how big data can change the social science research paradigm should be tempered by a recognition that existing ways of protecting privacy confidentiality are no longer viable (Karr and Reiter, 2014). There is a great deal of research that can be used to inform the development of such a structure, but it has been siloed into disconnected research areas, such as statistics, cyber-security, and cryptography, as well as a variety of different practical applications, including the successful development of remote access secure data enclaves. We must piece together the knowledge from these various fields to develop ways in which vast new sets of data on human beings can be collected, integrated, and analyzed while protecting them (Lane et al., 2014).

It is possible that the confidentiality risks of disseminating data may be so high that traditional access models will no longer hold; that the data access model of the future will be to take the analysis to the data rather than the data to the analyst or the analyst to the data. One potential approach is to create an integrated system including (1) unrestricted access to highly redacted data, most likely some version of synthetic data, followed by (2) methods for approved researchers to access the confidential data via remote access solutions, combined with (3) verification servers that allows users to assess the quality of their inferences with the redacted data so as to be more efficient with their use (if necessary) of the remote data access. Such verification servers might be a web-accessible system based on a confidential database with an associated public micro-data release, which helps to analyze the confidential database (Karr and Reiter, 2014). Such approaches are starting to be developed, both in the US and Europe (Jones and Elias, 2006; Elias, 2014).

There is also some evidence that people do not require com-plete protection, and they will gladly share even private informa-tion provided that certain social norms are met (Pentland et al., 2014; Wilbanks, 2014). There is a research agenda around identi-fying those norms as well, characterizing the interests and wishes of actors (the information senders and recipients or providers and users), the nature of the attributes (especially types of informa-tion about the providers, including how these might be transformed or linked), and identifying transmission principles (the constraints underlying the information flows).

However, it is likely that it is no longer possible for a lone social scientist to address these challenges. One-off access agreements to individuals are conducive to neither the production of high-quality science nor the high-quality protection of data (Schermann et al., 2014). The curation, protection, and dissemination of data on human subjects cannot be an artisan activity but should be viewed as a major research infrastructure investment, similar to investments in the physical and life sciences (Abazajian et al., 2009; Human Microbiome Jumpstart Reference Strains Consortium et al., 2010; Bird, 2011). In practice, this means that linkages become professionalized and replicable, research is fostered within research data centers that protect privacy in a systematic manner, knowledge is shared about the process of privacy protections disseminated in a professional fashion, and there is ongoing documentation about the value of evidence-based research. It is thus that the risk–utility tradeoff depicted in Figure 12.1 can be shifted in a manner that serves the public good.

12.8   Resources

The American Statistical Association's Privacy and Confidentiality website provides a useful source of information.

> ► http://community.amstat.org/cpc/home

An overview of federal activities is provided by the Confidentiality and Data Access Committee of the Federal Committee on Statistics and Methodology.

> ► https://nces.ed.gov/FCSM/cdac_resources.asp

The World Bank and International Household Survey Network provide a good overview of data dissemination "best practices."

> ► http://www.ihsn.org/home/projects/dissemination

There is a *Journal of Privacy and Confidentiality* based at Carnegie Mellon University, and the journal *Transactions in Data Privacy*.

> ► http://repository.cmu.edu/jpc/

> ► http://www.tdp.cat/

The United Nations Economic Commission on Europe hosts workshops and conferences and produces occasional reports.

> ► http://www.unece.org/stats/mos/meth/confidentiality.html

A collection of lectures is available from the semester on privacy at the Simons Institute for the Theory of Computing.

> ► https://simons.berkeley.edu/programs/privacy2019, available on youtube: https://www.youtube.com/user/SimonsInstitute/.

Chapter 13

Workbooks

Brian Kim, Christoph Kern, Jonathan Scott Morgan,
Clayton Hunter, and Avishek Kumar

This final chapter provides an overview of the Python workbooks that accompany this book. The workbooks combine text explanation and code you can run, implemented in *Jupyter notebooks*, to explain techniques and approaches selected from each chapter and to provide thorough implementation details, enabling students and interested practitioners to quickly "get up to speed" and start using the technologies covered in this book. We hope you have a lot of fun with them.

▶ https://jupyter.org/

13.1 Introduction

We provide accompanying Juptyer workbooks for most chapters in this book. The workbooks provide a thorough overview of the work needed to implement the selected technologies. They combine explanation, basic exercises, and substantial additional Python and SQL code to provide a conceptual understanding of each technology, give insight into how key parts of the process are implemented through exercises, and then lay out an end-to-end pattern for implementing each in your own work. The workbooks are implemented using Jupyter notebooks, interactive documents that mix formatted text and Python code samples that can be edited and run in real time in a Jupyter notebook server, allowing you to run and explore the code for each technology as you read about it.

The workbooks are centered around two main substantive examples. The first example is used in the *Databases*, *Dataset Exploration and Visualization*, *Machine Learning*, *Bias and Fairness*, and *Errors and Inference* workbooks, which focus on corrections data. The second example includes the *APIs*, *Record Linkage*, *Text Analysis*, and *Network Analysis* workbooks, which primarily use patent

▶ https://www.patentsview.org/

▶ https://federalreporter.nih.gov/

▶ https://mybinder.org/

▶ https://www.anaconda.com/distribution/

data from PatentsView and grant data from Federal RePORTER to investigate innovation and funding.

The Jupyter notebooks are designed to be run online in a cloud environment using Binder and do not need additional software installed locally. Individual workbooks can be opened by following the corresponding Binder link, and everything can be done in your browser. The full set of workbooks is available at https://workbooks.coleridgeinitiative.org, and additional workbooks may be added over time and made available in this repository.

The workbooks also can be run locally. In that case, you will need to install Python on your system, along with some additional Python packages needed by the workbooks. The easiest way to make all of this work is to install the free Anaconda Python distribution. Anaconda includes a Jupyter server and precompiled versions of many packages used in the workbooks. It includes multiple tools for installing and updating both Python and installed packages. It is separate from any OS-level version of Python, and it is easy to completely uninstall.

13.2   Notebooks

Below is a list of the workbooks, along with a short summary of the content that each covers.

13.2.1   Databases

The *Databases* notebook builds the foundation of using SQL to query data. Much of the subsequent notebooks will involve using these tools. This workbook also introduces you to the main data source that is used in the online workbooks, the North Carolina Department of Corrections Data (https://webapps.doc.state.nc.us/opi/downloads.do?method=view). In this notebook, you will:

- Build basic queries using SQL,

- Understand and perform various joins.

13.2.2   Dataset Exploration and Visualization

The *Dataset Exploration and Visualization* notebook further explores the North Carolina Department of Corrections data, demonstrating how to work with missing values and date variables and join tables by using SQL in Python. Although some of the SQL from the *Databases* notebook is revisited here, the focus is on practicing Python code and using Python for data analysis. The workbook also explains how to pull data from a database into a dataframe in Python and continues by exploring the imported data using the `numpy` and `pandas` packages, as well as `matplotlib` and `seaborn` for visualizations. In this workbook, you will learn how to:

- Connect to and query a database through Python,

- Explore aggregate statistics in Python,

- Create basic visualizations in Python.

13.2.3 APIs

The *APIs* notebook introduces you to the use of Internet-based web service APIs for retrieving data from online data stores. This notebook walks through the process of retrieving data about patents from the PatentsView API from the US Patent and Trademark Office. The data consist of information about patents, inventors, companies, and geographic locations since 1976. In this workbook, you will learn how to:

- Construct a URL query,

- Obtain a response from the URL,

- Retrieve the data in JSON form.

13.2.4 Record Linkage

In the *Record Linkage* workbook, you will use Python to implement the basic concepts behind record linkage using data from PatentsView and Federal RePORTER. This workbook will cover how to pre-process data for linkage before demonstrating multiple methods of record linkage, including probabilistic record linkage, in which different types of string comparators are used to compare multiple pieces of information between two records to produce a score that indicates how likely it is that the records are data about the same underlying entity. In this workbook, you will learn how to:

- Prepare data for record linkage,

- Use and evaluate the results of common computational string comparison algorithms, including Levenshtein distance, Levenshtein–Damerau distance, and Jaro–Winkler distance,

- Understand the Fellegi–Sunter probabilistic record linkage method, with step-by-step implementation guide.

13.2.5   Text Analysis

In the *Text Analysis* notebook, you will use the data that you have pulled from the PatentsView API in the *API* notebook to find topics from patent abstracts. This will involve going through every step of the process, from extracting the data, to cleaning and preparing it, to using topic modeling algorithms. In this workbook, you will learn how to:

- Clean and prepare text data,

- Apply Latent Dirichlet allocation for topic modeling,

- Improve and iterate models to improve their focus on identified topics.

13.2.6   Networks

In the *Networks* workbook, you will create network data where the nodes are researchers who have been awarded grants, and ties are created between each researcher on a given grant. You will use Python to read the grant data and translate them into network data, calculate node- and graph-level network statistics, and create network visualizations. In this workbook, you will learn how to:

- Calculate node- and graph-level network statistics,

- Create graph visualizations.

13.2.7   Machine Learning—Creating Labels

The *Machine Learning—Creating Labels* workbook is the first of a three-part Machine Learning workbook sequence, starting with how to create an outcome variable (label) for a machine learning task by using SQL in Python. It uses the North Carolina Department of Corrections Data to build an outcome that measures recidivism, i.e., whether a former inmate returns to jail in a given period of time. It also shows how to define a Python function to automate programming tasks. In this workbook, you will learn how to:

- Define and compute a prediction target in the machine learning framework,

- Use SQL with data that has a temporal structure (multiple records per observation).

13.2.8 Machine Learning—Creating Features

The *Machine Learning–Creating Features* workbook prepares predictors (features) for the machine learning task that has been introduced in the *Machine Learning–Creating Labels* workbook. It shows how to use SQL in Python for generating features that are expected to predict recidivism, such as the number of times someone has been admitted to prison prior to a given date. In this workbook, you will learn how to:

- Generate features with SQL for a given prediction problem,

- Automate SQL tasks by defining Python functions.

13.2.9 Machine Learning—Model Training and Evaluation

The *Machine Learning–Model Training and Evaluation* workbook uses the label and features that have been created in the previous workbooks to construct a training and test set for model building and evaluation. It walks through examples on how to train machine learning models using `scikit-learn` in Python and how to evaluate prediction performance for classification tasks. In addition, it demonstrates how to construct and compare many different machine learning models in Python. In this workbook, you will learn how to:

- Pre-process data to provide valid inputs for machine learning models,

- Properly divide data with a temporal structure into training and test sets,

- Train and evaluate machine learning models for classification using Python.

13.2.10 Bias and Fairness

The *Bias and Fairness* workbook demonstrates an example of using the bias and fairness audit toolkit Aequitas in Python. This workbook uses an example from criminal justice and demonstrates how Aequitas can be used to detect and evaluate biases of a machine learning system in relation to multiple (protected) subgroups. You will learn how to:

- Calculate confusion matrices for subgroups and visualize performance metrics by groups,

- Measure disparities by comparing, e.g., false positive rates between groups,

- Assess model fairness based on various disparity metrics.

13.2.11 Errors and Inference

The *Errors and Inference* workbook walks through how one might think critically about issues that may arise in their analysis. In this notebook, you will evaluate the machine learning models from previous notebooks and learn about ways to improve the data to use as much information as possible to make conclusions. Specifically, you will learn how to:

- Perform sensitivity analysis with machine learning models,

- Use imputation to fill in missing values.

13.2.12 Additional workbooks

An additional set of workbooks that accompanied the first edition of this book is available at https://github.com/BigDataSocialScience/Big-Data-Workbooks. This repository provides two different types of workbooks, each needing a different Python setup to run. The first type of workbook is intended to be downloaded and run locally by individual users. The second type is designed to be hosted, assigned, worked on, and graded on a single server, using `jupyterhub` to host and run the notebooks and `nbgrader` to assign, collect, and grade.

▶ https://github.com/jupyter/
jupyterhub

▶ https://github.com/jupyter/
nbgrader

13.3 Resources

We note in Section 1.8 the importance of Python, SQL, and Git/GitHub for the social scientist who intends to work with large data. See that section for pointers to useful online resources, and also see https://github.com/BigDataSocialScience, where we have collected many useful web links, including the following.

For more on getting started with Anaconda, see the Anaconda documentation.

► http://docs.continuum.io/anaconda

For more information on IPython and the Jupyter notebook server, see the IPython site (http://ipython.org/), IPython documentation (http://ipython.readthedocs.org/), Jupyter Project site (http://jupyter.org/), and Jupyter Project documentation (http://jupyter.readthedocs.org/).

For more information on using `jupyterhub` and `nbgrader` to host, distribute, and grade workbooks using a central server, see the `jupyterhub` GitHub repository (https://github.com/jupyter/jupyterhub/), `jupyterhub` documentation (http://jupyterhub.readthedocs.org/), `nbgrader` GitHub repository (https://github.com/jupyter/nbgrader/), and `nbgrader` documentation (http://nbgrader.readthedocs.org/).

Bibliography

Abadi, D., Ailamaki, A., Andersen, D., Bailis, P., Balazinska, M., Bernstein, P. A., Boncz, P. et al. (2010). The Seattle Report on database research. *ACM SIGMOD Record*, **48**(4):44–53. https://sigmodrecord.org/publications/sigmodRecord/1912/pdfs/07_Reports_Abadi.pdf.

Abazajian, K. N., Adelman-McCarthy, J. K., Agüeros, M. A., Allam, S. S., Prieto, C. A., An, D., Anderson, K. S. J. et al. (2009). The seventh data release of the Sloan Digital Sky Survey. *Astrophysical Journal Supplement Series*, **182**(2).

Abowd, J. M. (2018). The US Census Bureau adopts differential privacy. In *Proceedings of the 24th ACM SIGKDD International Conference on Knowledge Discovery & Data Mining.*, pages 2867–2867. ACM.

Abowd, J. M., Haltiwanger, J., and Lane, J. (2004). Integrated longitudinal employer-employee data for the United States. *American Economic Review*, **94**(2):224–229.

Abowd, J. M., Stinson, M., and Benedetto, G. (2006). Final report to the Social Security Administration on the SIPP/SSA/IRS Public Use File Project. Technical report, Census Bureau, Longitudinal Employer-Household Dynamics Program.

Acquisti, A. (2014). The economics and behavioral economics of privacy. In Lane, J., Stodden, V., Bender, S. and Nissenbaum, H., editors, *Privacy, Big Data, and the Public Good: Frameworks for Engagement*, pages 98–112. Cambridge University Press.

Agrawal, R., and Srikant, R. (1994). Fast algorithms for mining association rules in large databases. In *Proceedings of the 20th International Conference on Very Large Data Bases*.

Ahlberg, C., Williamson, C., and Shneiderman, B. (1992). Dynamic queries for information exploration: An implementation and evaluation. In *Proceedings of the SIGCHI Conference on Human Factors in Computing Systems*, pages 619–626. ACM.

Al Aghbari, Z., Bahutair, M., and Kamel, I. (2019). GeoSimMR: A mapreduce algorithm for detecting communities based on distance and interest in social networks. *Data Science Journal*, **18**(1):13.

Ali, M., Sapiezynski, P., Bogen, M., Korolova, A., Mislove, A., and Rieke, A. (2019). Discrimination through optimization: How facebooks ad delivery can lead to biased outcomes. In *Proceedings of the ACM on Human-Computer Interaction*, pages 3.

Allison, P. D. (2001). *Missing Data*. Sage Publications.

Amaya, A., Biemer, P.P., and Kinyon, D. (2020). Total error in a big data world: Adapting the TSE framework to big data. *Journal of Survey Statistics and Methodology*, **8**(1):89–119. https://doi.org/10.1093/jssam/smz056

Angwin, J., and Larson, J. (2016). Bias in Criminal Risk Scores Is Mathematically Inevitable, Researchers Say. https://www.propublica.org/article/bias-in-criminal-risk-scores-is-mathematically-inevitable-researchers-say. Accessed February 12, 2020.

Angwin, J., Larson, J., Mattu, S., and Kirchner, L. (2016). Machine Bias. https://www.propublica.org/article/machine-bias-risk-assessments-in-criminal-sentencing. Accessed February 12, 2020.

Angwin, J., and Parris, T Jr. (2016). Facebook Lets Advertisers Exclude Users by Race. https://www.propublica.org/article/facebook-lets-advertisers-exclude-users-by-race. Accessed February 12, 2020.

Anscombe, F. J. (1973). Graphs in statistical analysis. *American Statistician*, **27**(1):17–21.

Apache Hadoop (n.d.). HDFS architecture. http://hadoop.apache.org/docs/stable/hadoop-project-dist/hadoop-hdfs/HdfsDesign.html. Accessed February 1, 2020.

Apache Software Foundation (n.d.). Apache ambari. http://ambari.apache.org. Accessed February 1, 2020.

Armbrust, M., Fox, A., Griffith, R., Joseph, A. D., Katz, R., Konwinski, A., Lee, G. et al. (2010). A view of cloud computing. *Communications of the ACM*, **53**(4):50–58.

Athey, S., and Imbens, G. W. (2017). The state of applied econometrics: Causality and policy evaluation. *Journal of Economic Perspectives*, **31**(2):3–32.

Athey, S., and Wager, S. (2019). Estimating treatment effects with causal forests: An application. Technical report, https://arxiv.org/abs/1902.07409

Barabási, A.-L., and Albert, R. (1999). Emergence of scaling in random networks. *Science*, **286**(5439):509–512.

Barbaro, M., Zeller, T., and Hansell, S. (August 9, 2006). A face is exposed for AOL searcher no. 4417749. *New York Times*. https://www.nytimes.com/2006/08/09/technology/09aol.html

Barocas, S., and Nissenbaum, H. (2014a). Big data's end run around procedural privacy protections. *Communications of the ACM*, **57**(11):31–33.

Barocas, S., and Nissenbaum, H. (2014b). The limits of anonymity and consent in the big data age. In Lane, J., Stodden, V., Bender, S. and Nissenbaum, H., editors, *Privacy, Big Data, and the Public Good: Frameworks for Engagement*. Cambridge University Press.

Bastian, H. (2013). Bad research rising: The 7th Olympiad of research on biomedical publication. https://blogs.scientificamerican.com/absolutely-maybe/bad-research-rising-the-7th-olympiad-of-research-on-biomedical-publication/ Accessed February 1, 2020.

Batagelj, V., and Mrvar, A. (1998). Pajek—program for large network analysis. *Connections*, **21**(2):47–57.

Bell, A. (2012). Python for economists. https://scholar.harvard.edu/files/ambell/files/python_for_economists.pdf Accessed February 1, 2020.

Bengio, Y., Courville, A., and Vincent, P. (2013). Representation learning: A review and new perspectives. *IEEE Transactions on Pattern Analysis and Machine Intelligence*, **35**(8):1798–1828.

Bhuvaneshwar, K., Sulakhe, D., Gauba, R., Rodriguez, A., Madduri, R., Dave, U., Lacinski, L., Foster, I., Gusev, Y., and Madhavan, S. (2015). A case study for cloud based high throughput analysis of NGS data using the Globus Genomics system. *Computational and Structural Biotechnology Journal*, **13**:64-74.

Biemer, P. P. (2010). Total survey error: Design, implementation, and evaluation. *Public Opinion Quarterly*, **74**(5):817-848.

Biemer, P. P. (2011). *Latent Class Analysis of Survey Error*. John Wiley & Sons.

Biemer, P. P., and Lyberg, L. E. (2003). *Introduction to Survey Quality*. John Wiley & Sons.

Biemer, P. P., and Stokes, S. L. (1991). Approaches to modeling measurement error. In Biemer, P. P., Groves, R. M., Lyberg, L. E., Mathiowetz, N. A. and Sudman, S., editors, *Measurement Errors in Surveys*, pages 54-68. John Wiley.

Biemer, P. P., and Trewin, D. (1997). A review of measurement error effects on the analysis of survey data. In Lyberg, L., Biemer, P. P., Collins, M., Leeuw, E. D., Dippo, C., Schwarz, N. and Trewin, D., editors, *Survey Measurement and Process Quality*, pages 601-632. John Wiley & Sons.

Bird, I. (2011). Computing for the Large Hadron Collider. *Annual Review of Nuclear and Particle Science*, **61**:99-118.

Bird, S., Klein, E., and Loper, E. (2009). Natural Language Processing with Python: Analyzing Text with the Natural Language Toolkit. In *O'Reilly Media*. Available online at http://www.nltk.org/book/

Blei, D. M., and Lafferty, J. (2009). Topic models. In Srivastava, A. and Sahami, M., editors, *Text Mining: Theory and Applications*. Taylor & Francis.

Blei, D. M., and McAuliffe, J. D. (2007). Supervised topic models. In *In Advances in Neural Information Processing Systems*. MIT Press.

Blei, D. M., Ng, A., and Jordan, M. (2003). Latent Dirichlet allocation. *Journal of Machine Learning Research*, **3**:993-1022.

Blitzer, J., Dredze, M., and Pereira, F. (2007). Biographies, Bollywood, boom-boxes and blenders: Domain adaptation for sentiment classification. In *ACL*, pages 187-205.

Börner, K. (2010). *Atlas of Science: Visualizing What We Know*. MIT Press.

Boy, J., Rensink, R., Bertini, E., Fekete, J.-D. et al. (2014). A principled way of assessing visualization literacy. *IEEE Transactions on Visualization and Computer Graphics*, **20**(12):1963–1972.

Boyd, D., and Crawford, K. (2012). Critical questions for big data: Provocations for a cultural, technological, and scholarly phenomenon. *Information, Communication & Society*, **15**(5):662–679.

Boyd-Graber, J., Hu, Y., and Mimno, D. (2017). *Applications of Topic Models*, volume 11 of *Foundations and Trends in Information Retrieval*. NOW Publishers.

Brady, H. E. (2019). The challenge of big data and data science. *Annual Review of Political Science*, **22**:297–323.

Brandom, R. (2019). Facebook has been charged with housing discrimination by the US government. https://www.theverge.com/2019/3/28/18285178/facebook-hud-lawsuit-fair-housing-discrimination. Accessed February 12, 2020.

Brewer, E. (2012). CAP twelve years later: How the "rules" have changed. *Computer*, **45**(2):23–29.

Broekstra, J., Kampman, A., and Van Harmelen, F. (2002). Sesame: A generic architecture for storing and querying RDF and RDF schema. In *The Semantic Web—ISWC 2002*, pages 54–68. Springer.

Brown, C., Haltiwanger, J., and Lane, J. (2008). *Economic Turbulence: Is a Volatile Economy Good for America?* University of Chicago Press.

Brynjolfsson, E., Hitt, L. M., and Kim, H. H. (2011). Strength in numbers: How does data-driven decisionmaking affect firm performance? Available at SSRN: https://ssrn.com/abstract=1819486.

Buolamwini, J., and Gebru, T. (2018). Gender shades: Intersectional accuracy disparities in commercial gender classification. In Friedler, S. A. and Wilson, C., editors, *Proceedings of the 1st Conference on Fairness, Accountability and Transparency*, volume 81 of *Proceedings of Machine Learning Research*, pages 77–91, New York, NY, USA: PMLR.

Burkhauser, R. V., Feng, S., and Larrimore, J. (2010). Improving imputations of top incomes in the public-use current population survey by using both cell-means and variances. *Economics Letters*, **108**(1):69–72.

Burt, R. S. (1993). The social structure of competition. In *Explorations in Economic Sociology*. New York: Russel Sage Foundation.

Burt, R. S. (2004). Structural holes and good ideas. *American Journal of Sociology*, **110**(2):349–399.

Butler, D. (2013). When Google got flu wrong. *Nature*, **494**(7436): 155.

Campbell, A. F. (2018). Women accuse Facebook of illegally posting job ads that only men can see. https://www.vox.com/business-and-finance/2018/9/18/17874506/facebook-job-ads-discrimination. Accessed February 12, 2020.

Card, S. K., and Nation, D. (2002). Degree-of-interest trees: A component of an attention-reactive user interface. In *Proceedings of the Working Conference on Advanced Visual Interfaces*, pages 231–245. ACM.

Carton, S., Helsby, J., Joseph, K., Mahmud, A., Park, Y., Walsh, J., Cody, C., Patterson, C. E., Haynes, L., and Ghani, R. (2016). Identifying police officers at risk of adverse events. In *Proceedings of the 22nd ACM SIGKDD International Conference on Knowledge Discovery and Data Mining, KDD 16*, pages 67–76. New York, NY, USA. Association for Computing Machinery.

Caruana, R., Lou, Y., Gehrke, J., Koch, P., Sturm, M., and Elhadad, N. (2015). Intelligible models for healthcare: Predicting pneumonia risk and hospital 30-day readmission. In *Technical report, Proceedings of the 21th ACM SIGKDD International Conference on Knowledge Discovery and Data Mining (KDD 15)*, pages 1721–1730. New York, NY, USA. Association for Computing Machinery.

Catlett, C., Malik, T., Goldstein, B., Giuffrida, J., Shao, Y., Panella, A., Eder, D., Zanten, E. V., Mitchum, R., Thaler, S., and Foster, I. (2014). Plenario: An open data discovery and exploration platform for urban science. *Bulletin of the IEEE Computer Society Technical Committee on Data Engineering*, **37**:27–42.

Cavallo, A., and Rigobon, R. (2016). The billion prices project: Using online prices for measurement and research. *Journal of Economic Perspectives*, **30**(2):151–178.

Cecil, J., and Eden, D. (2003). The legal foundations of confidentiality. In *In Key Issues in Confidentiality Research: Results of an NSF workshop*. National Science Foundation.

Celis, L. E., Huang, L., Keswani, V., and Vishnoi, N. K. (2019). Classification with fairness constraints: A meta-algorithm with provable guarantees. In *Proceedings of the Conference on Fairness, Accountability, and Transparency, FAT* 19*, pages 319–328. New York, NY, USA. Association for Computing Machinery.

Centers for Disease Control, Prevention (2014). United States cancer statistic: An interactive cancer atlas. http://nccd.cdc.gov/DCPC_INCA. Accessed February 1, 2016.

Chai, J. J. (1971). Correlated measurement errors and the least squares estimator of the regression coefficient. *Journal of the American Statistical Association*, **66**(335):478–483.

Chandola, V., Banerjee, A., and Kumar, V. (2009). Anomaly detection: A survey. *ACM Computing Surveys*, **41**(3).

Chapelle, O., and Keerthi, S. S. (2010). Efficient algorithms for ranking with SVMs. *Information Retrieval*, **13**(3):201–215.

Chapelle, O., Schoelkopf, B., and Zien, A., editors, (2006). *Semi-Supervised Learning*. London, U.K.: MIT Press.

Chawla, N. V. (2005). Data mining for imbalanced datasets: An overview. In Maimon, O. and Rokach, L., editors, *The Data Mining and Knowledge Discovery Handbook*, pages 853–867. Springer.

Chen, I., Johansson, F. D., and Sontag, D. (2018). Why is my classifier discriminatory?. In *Proceedings of the 32nd International Conference on Neural Information Processing Systems, NIPS 18*, pages 3543–3554. Inc., Red Hook, NY, USA. Curran Associates.

Cheng, J., Adamic, L. A., Kleinberg, J. M., and Leskovec, J. (2016). Do cascades recur?. In *Proceedings of the 25th International Conference on World Wide Web*, pages 671–681.

Chetty, R. (2012). The transformative potential of administrative data for microeconometric research. In http://conference.nber.org/confer/2012/SI2012/LS/ChettySlides.pdf. Accessed February 1, 2016

Ching, A., Murthy, R., Dmytro Molkov, D., Vadali, R., and Yang, P. (2012). Under the hood: Scheduling MapReduce jobs more efficiently with Corona.

Chouldechova, A. (2017). Fair prediction with disparate impact: A study of bias in recidivism prediction instruments. *Big Data*, **5**(2):153–163.

Chouldechova, A., and Roth, A. (2018). The frontiers of fairness in machine learning. arXiv preprint arXiv:1810.08810.

Christen, P. (2012a). *Data matching: concepts and techniques for record linkage, entity resolution, and duplicate detection*. Springer Science & Business Media.

Christen, P. (2012b). A survey of indexing techniques for scalable record linkage and deduplication. *IEEE Transactions on Knowledge and Data Engineering*, **24**(9):1537–1555.

Clarke, C. (2014). Editing big data with machine learning methods. In *Paper presented at the Australian Bureau of Statistics Symposium, Canberra*.

Cleveland, W. S., and McGill, R. (1984). Graphical perception: Theory, experimentation, and application to the development of graphical methods. *Journal of the American Statistical Association*, **79**(387):531–554.

Clifton, C., Kantarcioglu, M., Doan, A., Schadow, G., Vaidya, J., Elmagarmid, A., and Suciu, D. (2006). Privacy-preserving data integration and sharing. In *9th ACM SIGMOD Workshop on Research Issues in Data Mining and Knowledge Discovery*, pages 19–26. ACM.

Clouder (n.d.). Cloudera Manager. https://www.cloudera.com/content/www/en-us/products/cloudera-manager.html. Accessed February 1, 2020.

Cochran, W. G. (1968). Errors of measurement in statistics. *Technometrics*, **10**(4):637–666.

Conor Dougherty (2015). Google Photos Mistakenly Labels Black People 'Gorillas'. https://bits.blogs.nytimes.com/2015/07/01/google-photos-mistakenly-labels-black-people-gorillas/. Accessed February 12, 2020.

Corti, P., Kraft, T. J., Mather, S. V., and Park, B. (2014). *PostGIS Cookbook*. Packt Publishing.

Crammer, K., and Singer, Y. (2002). On the algorithmic implementation of multiclass kernel-based vector machines. *Journal of Machine Learning Research*, **2**:265–292.

Crossno, P. J., Cline, D. D., and Jortner, J. N. (1993). A heterogeneous graphics procedure for visualization of massively parallel solutions. *ASME FED*, **156**:65–65.

Cumby, C., and Ghani, R. (2011). A machine learning based system for semi-automatically redacting documents. In *Technical report, AAAI Publications, Twenty-Third IAAI Conference*.

Czajka, J., Schneider, C., Sukasih, A., and Collins, K. (2014). Minimizing disclosure risk in HHS open data initiatives. Technical report, US Department of Health & Human Services.

Dean, J., and Ghemawat, S. (2004). MapReduce: Simplified data processing on large clusters. In *In Proceedings of the 6th Conference on Symposium on Opearting Systems Design & Implementation—Volume 6, OSDI'04*. USENIX Association.

DeBelius, D. (2015). Let's tesselate: Hexagons for tile grid maps. *NPR Visuals Team Blog*, http://blog.apps.npr.org/2015/05/11/hex-tile-maps.html.

Decker, R. A., Haltiwanger, J., Jarmin, R. S., and Miranda, J. (2016). Where has all the skewness gone? The decline in high-growth (young) firms in the US. *European Economic Review*, **86**:4–23.

Desai, T., Ritchie, F., and Welpton, R. (2016). Five safes: designing data access for research. Technical report, Working Papers 20161601, Department of Accounting, Economics and Finance, Bristol Business School, University of the West of England, Bristol.

Desmarais, S. L., and Singh, J. P. (2013). Risk assessment instruments validated and implemented in correctional settings in the united states. Technical report, Lexington, KY: Council of State Governments.

Devlin, J., Chang, M.-W., Lee, K., and Toutanova, K. (2019). BERT: Pre-training of deep bidirectional transformers for language understanding. In *Conference of the North American Chapter of the Association for Computational Linguistics*.

DeWitt, D. J., and Stonebraker, M. (2008). MapReduce: A major step backwards. http://www.dcs.bbk.ac.uk/dell/teaching/cc/paper/dbc08/dewitt_mr_db.pdf.

Donohue, III J. J., and Wolfers, J. (2006). Uses and abuses of empirical evidence in the death penalty debate. Technical report, National Bureau of Economic Research.

Doyle, P., Lane, J. I., Theeuwes, J. J. M., and Zayatz, L. V. (2001). Confidentiality, Disclosure, and Data Access: Theory and Practical Applications for Statistical Agencies. In *Elsevier Science*.

Drechsler, J. (2011). *Synthetic Datasets for Statistical Disclosure Control: Theory and Implementation*. Springer.

Duan, L., Xu, L., Liu, Y., and Lee, J. (2009). Cluster-based outlier detection. *Annals of Operations Research*, **168**(1):151–168.

DuGoff, E. H., Schuler, M., and Stuart, E. A. (2014). Generalizing observational study results: Applying propensity score methods to complex surveys. *Health Services Research*, **49**(1):284–303.

Duncan, G. T., Elliot, M., Juan Jose Salazar, G. (2011). Statistical Confidentiality: Principles and Practice. In *Springer*.

Duncan, G. T., and Stokes, S. L. (2004). Disclosure risk vs. data utility: The ru confidentiality map as applied to topcoding. *Chance*, **17**(3):16–20.

Dunne, C., and Shneiderman, B. (2013). Motif simplification: Improving network visualization readability with fan, connector, and clique glyphs. In *Proceedings of the SIGCHI Conference on Human Factors in Computing Systems*, pages 3247–3256. ACM.

Dunning, T. (1993). Accurate methods for the statistics of surprise and coincidence. *Computational Linguistics*, **19**(1):61–74.

Dutwin, D., and Buskirk, T. D. (2017). Reply. *Public Opinion Quarterly*, **81**(S1):246–249.

Dwork, C., Hardt, M., Pitassi, T., Reingold, O., and Zemel, R. (2012). Fairness through awareness. In *Proceedings of the 3rd Innovations in Theoretical Computer Science Conference, ITCS 12*, pages 214–226. New York, NY, USA. Association for Computing Machinery.

Dwork, C., and Roth, A. (2014). The algorithmic foundations of differential privacy. *Foundations and Trends in Theoretical Computer Science*, **9**(3–4):211–407.

Economic and Social Research Council (2016). Administrative Data Research Network.

Edelman, B., Luca, M., and Svirsky, D. (2017). Racial discrimination in the sharing economy: Evidence from a field experiment. *American Economic Journal: Applied Economics*, **9**(2):1–22.

Elias, P. (2014). A European perspective on research and big data access. In Lane, J., Stodden, V., Bender, S. and Nissenbaum, H., editors, *Privacy, Big Data, and the Public Good: Frameworks for Engagement*, pages 98–112. Cambridge University Press.

Elliott, J., Kelly, D., Chryssanthacopoulos, J., Glotter, M., Jhunjhnuwala, K., Best, N., Wilde, M., and Foster, I. (2014). The parallel system for integrating impact models and sectors (pSIMS). *Environmental Modelling & Software*, **62**:509–516.

Elmagarmid, A. K., Ipeirotis, P. G., and Verykios, V. S. (2007). Duplicate record detection: A survey. *IEEE Transactions on Knowledge and Data Engineering*, **19**(1):1–16.

Evans, D. S. (1987). Tests of alternative theories of firm growth. *Journal of Political Economy*, **95**:657–674.

Evans, J. A., and Foster, J. G. (2011). Metaknowledge. *Science*, **331**(6018):721–725.

Fan, J., Han, F., and Liu, H. (2014). Challenges of big data analysis. *National Science Review*, **1**(2):293–314.

Fan, J., and Liao, Y. (2012). Endogeneity in ultrahigh dimension. Technical report, Princeton University.

Fan, J., and Liao, Y. (2014). Endogeneity in high dimensions. *Annals of Statistics*, **42**(3):872.

Fan, J., Samworth, R., and Wu, Y. (2009). Ultrahigh dimensional feature selection: Beyond the linear model. *Journal of Machine Learning Research*, **10**:2013–2038.

Fekete, J.-D. (2015). ProgressiVis: A toolkit for steerable progressive analytics and visualization. Paper presented at 1st Workshop on Data Systems for Interactive Analysis, Chicago, IL, October 26.

Fekete, J.-D., and Plaisant, C. (2002). Interactive information visualization of a million items. In *IEEE Symposium on Information Visualization*, pages 117–124. IEEE.

Feldman, R., and Sanger, J. (2006). *Text Mining Handbook: Advanced Approaches in Analyzing Unstructured Data*. Cambridge University Press.

Fellegi, I. P., and Sunter, A. B. (1969). A theory for record linkage. *Journal of the American Statistical Association*, **64**(328):1183–1210.

Feng, S., Wallace, E., II, A. G., Rodriguez, P., Iyyer, M., and Boyd-Graber, J. (2018). Pathologies of neural models make interpretation difficult. In *Empirical Methods in Natural Language Processing*

Ferragina, P., and Scaiella, U. (2010). Tagme: On-the-fly annotation of short text fragments (by wikipedia entities). In *Proceedings of the 19th ACM International Conference on Information and Knowledge Management, CIKM 10*, pages 1625–1628. New York, NY, USA. Association for Computing Machinery.

Few, S. (2009). *Now You See It: Simple Visualization Techniques for Quantitative Analysis*. Analytics Press.

Few, S. (2013). *Information Dashboard Design: Displaying Data for At-a-Glance Monitoring*. Analytics Press.

Fielding, R. T., and Taylor, R. N. (2002). Principled design of the modern Web architecture. *ACM Transactions on Internet Technology*, **2**(2):115–150.

Fisher, D., Popov, I., Drucker, S., and Schraefel, M. (2012). Trust me, I'm partially right: Incremental visualization lets analysts explore large datasets faster. In *Proceedings of the SIGCHI Conference on Human Factors in Computing Systems*, pages 1673–1682. ACM.

Flach, P. (2012). *Machine Learning: The Art and Science of Algorithms That Make Sense of Data*. Cambridge University Press.

Fortuna, B., Grobelnik, M., and Mladenic, D. (2007). OntoGen: Semi-automatic ontology editor. In *Proceedings of the 2007 Conference on Human Interface: Part II*, pages 309–318. Springer.

Foster, L., Jarmin, R. S., and Riggs, T. L. (2009). Resolving the tension between access and confidentiality: Past experience and future plans at the US Census Bureau. In *Technical Report 09-33, US Census Bureau Center for Economic Studies*.

Fox, A., Gribble, S. D., Chawathe, Y., Brewer, E. A., and Gauthier, P. (1997). Cluster-based scalable network services. *ACM SIGOPS Operating Systems Review*, **31**(5).

Francis, W. N., and Kucera, H. (1979). Brown corpus manual. Technical report, Department of Linguistics, Brown University, Providence, Rhode Island, US.

Freeman, L. C. (1979). Centrality in social networks conceptual clarification. *Social Networks*, **1**(3):215–239.

Fuller, W. A. (1991). Regression estimation in the presence of measurement error. In Biemer, P. P., Groves, R. M., Lyberg, L. E., Mathiowetz, N. A. and Sudman, S., editors, *Measurement Errors in Surveys*, pages 617–635. John Wiley & Sons.

Galloway, S. (2017). *The four: the hidden DNA of Amazon, Apple, Facebook and Google*. Random House.

Geman, S., and Geman, D. (1990). Stochastic relaxation, Gibbs distributions, and the Bayesian restoration of images. In Shafer, G. and Pearl, J., editors, *Readings in Uncertain Reasoning*, pages 452–472. Morgan Kaufmann.

Gerrish, S. M., and Blei, D. M. (2012). The issue-adjusted ideal point model. Technical report, https://arxiv.org/abs/1209.6004.

Girone, M. (2008). CERN database services for the LHC computing grid. In *In Journal of Physics: Conference Series*, volume **119**. IOP Publishing.

Girvan, M., and Newman, M. E. J. (2002). Community structure in social and biological networks. *Proceedings of the National Academy of Sciences*, **99**(12):7821–7826.

Glaeser, E. (2019). Urban management in the 21st century: Ten insights from professor ed glaeser. Technical report, Centre for Development and Enterprise (CDE).

Glennon, B. (2019). How do restrictions on high-skilled immigration affect offshoring? evidence from the h-1b program. Technical report, http://brittaglennon.com/research/.

Glueck, M., Khan, A., and Wigdor, D. J. (2014). Dive in! Enabling progressive loading for real-time navigation of data visualizations. In *Proceedings of the SIGCHI Conference on Human Factors in Computing Systems*, pages 561–570. ACM.

Göbel, S., and Munzert, S. (2018). Political advertising on the wikipedia marketplace of information. *Social Science Computer Review*, **36**(2):157–175.

Goodfellow, I., Bengio, Y., and Courville, A. (2016). *Deep Learning*. MIT Press. http://www.deeplearningbook.org.

Government Computer News Staff (2018). Data mashups at government scale: The Census Bureau ADRF. *GCN Magazine*.

Gray, J. (1981). The transaction concept: Virtues and limitations. In *Proceedings of the Seventh International Conference on Very Large Data Bases, volume 7*, pages 144–154.

Green, D. P., and Kern, H. L. (2012). Modeling heterogeneous treatment effects in survey experiments with Bayesian additive regression trees. *Public Opinion Quarterly*, **76**:491–511.

Greenwood, D., Stopczynski, A., Sweatt, B., Hardjono, T., and Pentland, A. (2014). The new deal on data: A framework for institutional controls. In Lane, J., Stodden, V., Bender, S. and Nissenbaum, H., editors, *Privacy, Big Data, and the Public Good: Frameworks for Engagement*. Cambridge University Press.

Griffiths, T. L., and Steyvers, M. (2004). Finding scientific topics. *Proceedings of the National Academy of Sciences*, **101**(Suppl. 1):5228–5235.

Grimmer, J., and Stewart, B. M. (2013). Text as data: The promise and pitfalls of automatic content analysis methods for political texts. *Political Analysis*, **21**(3):267–297.

Gropp, W., Lusk, E., and Skjellum, A. (2014). *Using MPI: Portable Parallel Programming with the Message-Passing Interface*. MIT Press.

Groves, R. M. (2004). *Survey Errors and Survey Costs*. John Wiley & Sons.

Haak, L. L., Fenner, M., Paglione, L., Pentz, E., and Ratner, H. (2012). ORCID: A system to uniquely identify researchers. *Learned Publishing*, **25**(4):259–264.

Hadoop, A. (n.d.). HDFS architecture. http://spark.apache.org/docs/latest/programming-guide.html#transformations.

Hainmueller, J., and Hazlett, C. (2014). Kernel regularized least squares: Reducing misspecification bias with a flexible and interpretable machine learning approach. *Political Analysis*, **22**(2):143–168.

Halevy, A., Norvig, P., and Pereira, F. (2009). The unreasonable effectiveness of data. *IEEE Intelligent Systems*, **24**(2):8–12.

Hall, P., and Miller, H. (2009). Using generalized correlation to effect variable selection in very high dimensional problems. *Journal of Computational and Graphical Statistics*, **18**:533–550.

Haltiwanger, J., Jarmin, R. S., and Miranda, J. (2013). Who creates jobs? Small versus large versus young. *Review of Economics and Statistics*, **95**(2):347–361.

Han, H., Giles, L., Zha, H., Li, C., and Tsioutsiouliklis, K. (2004). Two supervised learning approaches for name disambiguation in author citations. In *Proceedings of the Joint ACM/IEEE Conference on Digital Libraries*, pages 296–305. IEEE.

Hansen, D., Shneiderman, B., and Smith, M. A. (2010). *Analyzing Social Media Networks with NodeXL: Insights from a Connected World*. Morgan Kaufmann.

Hansen, M. H., Hurwitz, W. N., and Madow, W. G. (1993). *Sample Survey Methods and Theory*. John Wiley & Sons.

Harford, T. (2014). Big data: A big mistake? *Significance*, **11**(5): 14–19.

Harrison, L., Reinecke, K., and Chang, R. (2015). Infographic aesthetics: Designing for the first impression. In *Proceedings of the 33rd Annual ACM Conference on Human Factors in Computing Systems*, pages 1187–1190. ACM.

Hart, N. (2019). Two Years of Progress on Evidence-Based Policymaking in the United States. *Data Coalition Blog*.

Harwell, D. (2019). San Francisco becomes first city in U.S. to ban facial-recognition software: https://www.washingtonpost.com/technology/2019/05/14/san-francisco-becomes-first-city-us-ban-facial-recognition-software/. Accessed February 12, 2020.

Hastie, T., Tibshirani, R., and Friedman, J. (2001). *The Elements of Statistical Learning*. Springer.

Hayden, E. C. (2012). A broken contract. *Nature*, **486**(7403):312–314.

Hayden, E. C. (2015). Researchers wrestle with a privacy problem. *Nature*, **525**(7570).

He, Z., Xu, X., and Deng, S. (2003). Discovering cluster-based local outliers. *Pattern Recognition Letters*, **24**(9):1641–1650.

Healy, K., and Moody, J. (2014). Data visualization in sociology. *Annual Review of Sociology*, **40**:105–128.

Henry, N., and Fekete, J.-D. (2006). MatrixExplorer: A dual-representation system to explore social networks. *IEEE Transactions on Visualization and Computer Graphics*, **12**(5):677–684.

Herzog, T. N., Scheuren, F. J., and Winkler, W. E. (2007). *Data Quality and Record Linkage Techniques*. Springer Science & Business Media.

Hill, K. (February 16, 2012). How Target figured out a teen girl was pregnant before her father did. *Forbes*, http://www.forbes.com/sites/kashmirhill/2012/02/16/how-target-figured-out-a-teen-girl-was-pregnant-before-her-father-did/#7280148734c6.

Hofmann, T. (1999). Probabilistic latent semantic analysis. In *Proceedings of Uncertainty in Artificial Intelligence*.

Holmberg, A., and Bycroft, C. (2017). Statistics new zealand's approach to making use of alternative data sources in a new era of integrated data. In Biemer, P. P., de Leeuw, E. D., Eckman, S., Edwards, B., Kreuter, F., Lyberg, L. E., Tucker, N. C. and West, B. T., editors, *Total survey error in practice*. Hoboken, NJ: John Wiley and Sons.

Hox, J. (2010). *Multilevel Analysis: Techniques and Applications*. Routledge.

Hsieh, Y. P., and Murphy, J. (2017). Total twitter error: Decomposing public opinion measurement on twitter from a total survey error perspective. In Biemer, P. P., de Leeuw, E. D., Eckman, S., Edwards, B., Kreuter, F., Lyberg, L. E., Tucker, N. C. and West, B. T., editors, *Total survey error in practice*. Hoboken, NJ: John Wiley and Sons.

Hu, Y., Zhai, K., Eidelman, V., and Boyd-Graber, J. (2014). Polylingual tree-based topic models for translation domain adaptation. In *Proceedings of the 52nd Annual Meeting of the Association for Computational Linguistics.*

Huang, A. (2008). Similarity measures for text document clustering. In *Paper presented at New Zealand Computer Science Research Student Conference, Christchurch, New Zealand, April 14–18*

Huang, J., Ertekin, S., and Giles, C. L. (2006). Efficient name disambiguation for large-scale databases. In *Knowledge Discovery in Databases: PKDD 2006*, pages 536–544. Springer.

Hukkelas, H., Mester, R., and Lindseth, F. (2019). Deepprivacy: A generative adversarial network for face anonymization. Technical report, https://arxiv.org/abs/1909.04538.

Human Microbiome Jumpstart Reference Strains Consortium, Nelson, K. E., Weinstock, G. M. et al. (2010). A catalog of reference genomes from the human microbiome. *Science*, **328**(5981):994–999.

Hundepool, A., Domingo-Ferrer, J., Franconi, L., Giessing, S., Lenz, R., Longhurst, J., Nordholt, E. S., Seri, G., and Wolf, P. (2010). Handbook on statistical disclosure control. Technical report, Network of Excellence in the European Statistical System in the Field of Statistical Disclosure Control.

Husbands Fealing, K., Lane, J. I., Marburger, J., and Shipp, S. (2011). *Science of Science Policy: The Handbook*. Stanford University Press.

Ibrahim, J. G., and Chen, M.-H. (2000). Power prior distributions for regression models. *Statistical Science*, **15**(1):46–60.

Imai, K., Ratkovic, M. et al. (2013). Estimating treatment effect heterogeneity in randomized program evaluation. *Annals of Applied Statistics*, **7**(1):443–470.

Imbens, G. W., and Rubin, D. B. (2015). *Causal Inference for Statistics, Social, and Biomedical Sciences: An Introduction*. Cambridge University Press.

Inselberg, A. (2009). *Parallel Coordinates*. Springer.

Institute For Research On Innovation, Science (IRIS) Research (2019). *Summary Documentation for the IRIS UMETRICS 2019 Data Release*. Institute for Research on Innovation and Science (IRIS).

Institute for Social Research (2013). PSID file structure and merging PSID data files. Technical report. http://psidonline.isr.umich.edu/Guide/FileStructure.pdf.

Ioannidis, J. P. A. (2005). Why most published research findings are false. *PLoS Medicine*, **2**(8):e124.

Iyyer, M., Manjunatha, V., Boyd-Graber, J., Daumé III, H. (2015). Deep unordered composition rivals syntactic methods for text classification. In *Association for Computational Linguistics*.

James, G., Witten, D., Hastie, T., and Tibshirani, R. (2013). *An Introduction to Statistical Learning*. Springer.

Japec, L., Kreuter, F., Berg, M., Biemer, P., Decker, P., Lampe, C., Lane, J., O'Neil, C., and Usher, A. (2015). Big data in survey research: AAPOR Task Force Report. *Public Opinion Quarterly*, **79**(4):839–880.

Jarmin, R. S., Louis, T. A., and Miranda, J. (2014). Expanding the role of synthetic data at the US Census Bureau. *Statistical Journal of the IAOS*, **30**(2):117–121.

Jarmin, R. S., and Miranda, J. (2002). The longitudinal business database. Available at SSRN: https://ssrn.com/abstract=2128793.

Johnson, B., and Shneiderman, B. (1991). Tree-maps: A space-filling approach to the visualization of hierarchical information structures. In *Proceedings of the IEEE Conference on Visualization*, pages 284–291. IEEE.

Jones, P., and Elias, P. (2006). Administrative data as a research resource: A selected audit. Technical report, ESRC National Centre for Research Methods.

Jovanovic, B. (1982). Selection and the evolution of industry. *Econometrica: Journal of the Econometric Society*, **50**(3):649–670.

Kabo, F., Hwang, Y., Levenstein, M., and Owen-Smith, J. (2015). Shared paths to the lab: A sociospatial network analysis of collaboration. *Environment and Behavior*, **47**(1):57–84.

Karr, A., and Reiter, J. P. (2014). Analytical frameworks for data release: A statistical view. In Lane, J., Stodden, V., Bender, S. and Nissenbaum, H., editors, *Privacy, Big Data, and the Public Good: Frameworks for Engagement*. Cambridge University Press.

Keshif (n.d.). Infographics aesthetics dataset browser. http://keshif. me/demo/infographics_aesthetics. Accessed February 1, 2020.

Kilbertus, N., Rojas Carulla, M., Parascandolo, G., Hardt, M., Janzing, D., and Schölkopf, B. (2017). Avoiding discrimination through causal reasoning. In *Advances in Neural Information Processing Systems 30*, pages 656–666. Curran Associates, Inc.

Kim, K., Khabsa, M., and Giles, C. L. (2016a). Inventor name disambiguation for a patent database using a random forest and dbscan. In *2016 IEEE/ACM joint conference on digital libraries (JCDL)*, pages 269–270. IEEE.

Kim, Y., Huang, J., and Emery, S. (2016b). Garbage in, garbage out: Data collection, quality assessment and reporting standards for social media data use in health research, infodemiology and digital disease detection. *Journal of medical Internet Research*, **18**(2).

King, G., Pan, J., and Roberts, M. E. (2013). How censorship in china allows government criticism but silences collective expression. *American Political Science Review*, **107**(2):1–18.

Kinney, S. K., Reiter, J. P., Reznek, A. P., Miranda, J., Jarmin, R. S., and Abowd, J. M. (2011). Towards unrestricted public use business microdata: The synthetic Longitudinal Business Database. *International Statistical Review*, **79**(3):362–384.

Kirk, A. (2012). Data Visualization: A Successful Design Process. In *Packt Publishing*

Kiss, T., and Strunk, J. (2006). Unsupervised multilingual sentence boundary detection. *Computational Linguistics*, **32**(4):485–525.

Kleinberg, J., Ludwig, J., Mullainathan, S., and Obermeyer, Z. (2015). Prediction policy problems. *American Economic Review*, **105**(5):491–495.

Kleinberg, J., Mullainathan, S., and Raghavan, M. (2017). Inherent Trade-Offs in the Fair Determination of Risk Scores. In Papadimitriou, C. H., editors, *8th Innovations in Theoretical Computer Science Conference (ITCS 2017)*, volume **67**, Dagstuhl, Germany: Schloss Dagstuhl–Leibniz-Zentrum fuer Informatik.

Kohler, U., and Kreuter, F. (2012). *Data Analysis Using Stata*. Stata Press, 3rd edition.

Kolb, L., Thor, A., and Rahm, E. (2012). Dedoop: Efficient deduplication with hadoop. *Proceedings of the VLDB Endowment*, **5**(12):1878–1881.

Kong, L., Schneider, N., Swayamdipta, S., Bhatia, A., Dyer, C., and Smith, N. A. (2014). A dependency parser for tweets. In *Proceedings of the 2014 Conference on Empirical Methods in Natural Language Processing (EMNLP)*, pages 1001–1012. Association for Computational Linguistics.

Köpcke, H., Thor, A., and Rahm, E. (2010). Evaluation of entity resolution approaches on real-world match problems. *Proceedings of the VLDB Endowment*, **3**(1–2):484–493.

Kopf, D. (2018). This year's Nobel Prize in economics was awarded to a Python convert. *Quartz*.

Kraak, M.-J. (2014). *Mapping Time: Illustrated by Minard's Map of Napoleon's Russian Campaign of 1812*. ESRI Press.

Kreuter, F., Ghani, R., and Lane, J. (2019). Change through data: A data analytics training program for government employees. *Harvard Data Science Review*, **1**(2).

Kreuter, F., and Peng, R. D. (2014). Extracting information from big data: Issues of measurement, inference, and linkage. In Lane, J., Stodden, V., Bender, S. and Nissenbaum, H., editors, *Privacy, Big Data, and the Public Good: Frameworks for Engagement*, pages 257–275. Cambridge University Press.

Kuhn, H. W. (2005). The Hungarian method for the assignment problem. *Naval Research Logistics*, **52**(1):7–21.

Kuhn, M., and Johnson, K. (2013). *Applied Predictive Modeling*. Springer Science & Business Media.

Kullback, S., and Leibler, R. A. (1951). On information and sufficiency. *Annals of Mathematical Statistics*, **22**(1):79–86.

Kumar, M., Ghani, R., and Mei, Z.-S. (2010). Data mining to predict and prevent errors in health insurance claims processing. In *Proceedings of the 16th ACM SIGKDD International Conference on Knowledge Discovery and Data Mining*, KDD '10, pages 65–74. ACM.

Kusner, M. J., Loftus, J., Russell, C., and Silva, R. (2017). Counterfactual fairness. In *Advances in Neural Information Processing Systems 30*, pages 4066–4076. Curran Associates, Inc.

Lafferty, J. D., McCallum, A., and Pereira, F. C. N. (2001). Conditional random fields: Probabilistic models for segmenting and labeling sequence data. In *Proceedings of the Eighteenth International Conference on Machine Learning*, pages 282–289. Morgan Kaufmann.

Lahiri, P., and Larsen, M. D. (2005). Regression analysis with linked data. *Journal of the American Statistical Association*, **100**(469):222–230.

Lakkaraju, H., Aguiar, E., Shan, C., Miller, D., Bhanpuri, N., Ghani, R., and Addison, K. L. (2015). A machine learning framework to identify students at risk of adverse academic outcomes. In *Proceedings of the 21th ACM SIGKDD International Conference on Knowledge Discovery and Data Mining, KDD '15*, pages 1909–1918. ACM.

Lam, H., Bertini, E., Isenberg, P., Plaisant, C., and Carpendale, S. (2012). Empirical studies in information visualization: Seven scenarios. *IEEE Transactions on Visualization and Computer Graphics*, **18**(9):1520–1536.

Lambrecht, A., and Tucker, C. (2019). Algorithmic bias? an empirical study of apparent gender-based discrimination in the display of stem career ads. *Management Science*, **65**(7):2966–2981.

Landauer, T., and Dumais, S. (1997). Solutions to Plato's problem: The latent semantic analysis theory of acquisition, induction and representation of knowledge. *Psychological Review*, **104**(2):211–240.

Lane, J. (2007). Optimizing access to micro data. *Journal of Official Statistics*, **23**:299–317.

Lane, J. (2020). Tiered Access: Risk and Utility.

Lane, J., Owen-Smith, J., Rosen, R. F., and Weinberg, B. A. (2015). New linked data on research investments: Scientific workforce, productivity, and public value. *Research Policy*, **44**:1659–1671.

Lane, J., Owen-Smith, J., Staudt, J., and Weinberg, B. A. (2018). New Measurement of Innovation. In U.S. Census Bureau, editors,

Center for Economic Studies and Research Data Centers Research Report: 2017. Washington DC: U.S. Census Bureau.

Lane, J., and Stodden, V. (2013). What? Me worry? what to do about privacy, big data, and statistical research. *AMSTAT News*, **438**:14.

Lane, J., Stodden, V., Bender, S. and Nissenbaum, H., editors (2014). *Privacy, Big Data, and the Public Good: Frameworks for Engagement.* Cambridge University Press.

Larson, J., Mattu, S., Kirchner, L., and Angwin, J. (2016). How We Analyzed the COMPAS Recidivism Algorithm. https://www.propublica.org/article/how-we-analyzed-the-compas-recidivism-algorithm. Accessed February 12, 2020.

Lazer, D., Kennedy, R., King, G., and Vespignani, A. (2014). The parable of Google Flu: Traps in big data analysis. *Science*, **343**.

Levitt, S. D., and Miles, T. J. (2006). Economic contributions to the understanding of crime. *Annual Review of Law Social Science*, **2**:147–164.

Lewis, D. D. (1998). Naive (Bayes) at forty: The independence assumption in information retrieval. In *Proceedings of European Conference of Machine Learning*, pages 4–15.

Li, G.-C., Lai, R., D'Amour, A., Doolin, D. M., Sun, Y., Torvik, V. I., Amy, Z. Y., and Fleming, L. (2014). Disambiguation and co-authorship networks of the us patent inventor database (1975–2010). *Research Policy*, **43**(6):941–955.

Lifka, D., Foster, I., Mehringer, S., Parashar, M., Redfern, P., Stewart, C., and Tuecke, S. (2013). XSEDE cloud survey report. Technical report, National Science Foundation, USA, http://hdl.handle.net/2142/45766.

Lin, J., and Dyer, C. (2010). *Data-Intensive Text Processing with MapReduce.* In *Morgan & Claypool Publishers*

Lins, L., Klosowski, J. T., and Scheidegger, C. (2013). Nanocubes for real-time exploration of spatiotemporal datasets. *IEEE Transactions on Visualization and Computer Graphics*, **19**(12):2456–2465.

Little, R. J. A., and Rubin, D. B. (2014). *Statistical Analysis with Missing Data.* John Wiley & Sons.

Liu, Z., and Heer, J. (2014). The effects of interactive latency on exploratory visual analysis. *IEEE Transactions on Visualization and Computer Graphics*, **20**(12):2122–2131.

Lockwood, G. K. (October 9, 2015). Conceptual overview of map-reduce and hadoop. http://www.glennklockwood.com/data-intensive/hadoop/overview.html.

Lohr, S. (2009). *Sampling: Design and Analysis*. Cengage Learning.

Lundberg, S. M., and Lee, S.-I. (2017). A unified approach to interpreting model predictions. In *Proceedings of the 31st International Conference on Neural Information Processing Systems, NIPS 17*, pages 4768–4777. Red Hook, NY, USA. Curran Associates Inc.

Lynch, J. (2018). Not even our own facts: Criminology in the era of big data. *Criminology*, **56**(3):437–454.

MacEachren, A. M., Crawford, S., Akella, M., and Lengerich, G. (2008). Design and implementation of a model, web-based, GIS-enabled cancer atlas. *Cartographic Journal*, **45**(4):246–260.

MacKinlay, J. (1986). Automating the design of graphical presentations of relational information. *ACM Transactions on Graphics*, **5**(2):110–141.

Malik, W. A., Unwin, A., and Gribov, A. (2010). An interactive graphical system for visualizing data quality–tableplot graphics. In *Classification as a Tool for Research*, pages 331–339. Springer.

Malmkjær, K. (2002). *The Linguistics Encyclopedia*. Routledge.

Manning, C. D., Surdeanu, M., Bauer, J., Finkel, J., Bethard, S. J., and McClosky, D. (2014). The Stanford CoreNLP natural language processing toolkit. In *Proceedings of 52nd Annual Meeting of the Association for Computational Linguistics: System Demonstrations*, pages 55–60.

Marburger, J. H. (2005). Wanted: Better benchmarks. *Science*, **308**(5725):1087.

Marcus, M. P., Santorini, B., and Marcinkiewicz, M. A. (1993). Building a large annotated corpus of English: The Penn treebank. *Computational Linguistics*, **19**(2):313–330.

Mas, A., and Moretti, E. (2009). Peers at work. *American Economic Review*, **99**(1):112–145.

Maskeri, G., Sarkar, S., and Heafield, K. (2008). Mining business topics in source code using latent Dirichlet allocation. In *Proceedings of the 1st India Software Engineering Conference*, pages 113–120. ACM.

McCallister, E., Grance, T., and Scarfone, K. A. (2010). SP 800-122. In *Guide to Protecting the Confidentiality of Personally Identifiable Information (PII)*. National Institute of Standards and Technology.

McCallum, A. K. (2002). Mallet: A machine learning for language toolkit. http://mallet.cs.umass.edu.

Meij, E., Bron, M., Hollink, L., Huurnink, B., and Rijke, M. (2009). Learning semantic query suggestions. In *Proceedings of the 8th International Semantic Web Conference, ISWC '09*, pages 424–440. Springer.

Meng, X.-L. (2018). Statistical paradises and paradoxes in big data (i): Law of large populations, big data paradox, and the 2016 US presidential election. *The Annals of Applied Statistics*, **12**(2):685–726.

Mikolov, T., Sutskever, I., Chen, K., Corrado, G. S., and Dean, J. (2013). Distributed representations of words and phrases and their compositionality. In *Advances in Neural Information Processing Systems*, pages 3111–3119. Morgan Kaufmann.

Mitchell, T. M. (1997). *Machine Learning*. McGraw-Hill.

Moffatt, C. L., (February 3, 1999). Visual representation of SQL joins. http://www.codeproject.com/Articles/33052/Visual-Representation-of-SQL-Joins.

Molina, G., AlGhamdi, F., Ghoneim, M., Hawwari, A., Rey-Villamizar, N., Diab, M., and Solorio, T. (2016). Overview for the second shared task on language identification in code-switched data. In *Proceedings of the Second Workshop on Computational Approaches to Code Switching*, pages 40–49. Austin, Texas. Association for Computational Linguistics.

Molinaro, A. (2005). *SQL Cookbook: Query Solutions and Techniques for Database Developers*. O'Reilly Media.

Moreno, J. L. (1934). *Who Shall Survive?: A New Approach to the Problem of Human Interrelations*. Nervous and Mental Disease Publishing Co.

Mortensen, P. S., and Bloch, C. W. (2005). *Oslo Manual: Guidelines for Collecting and Interpreting Innovation Data*. Organisation for Economic Co-operation and Development.

Munzner, T. (2014). *Visualization Analysis and Design*. CRC Press.

Narayanan, A., and Shmatikov, V. (2008). Robust de-anonymization of large sparse datasets. In *IEEE Symposium on Security and Privacy*, pages 111–125. IEEE.

Natarajan, K., Li, J., and Koronios, A. (2010). *Data Mining Techniques for Data Cleaning*. Springer.

National Academies (2014). Proposed Revisions to the Common Rule for the Protection of Human Subjects in the Behavioral and Social Sciences. Technical report, National Academies of Sciences, Washington DC.

National Academies of Sciences, Engineering, and Medicine and others (2018). *Federal Statistics, Multiple Data Sources, and Privacy Protection: Next Steps*. National Academies Press.

National Center for Health Statistics (2019). The linkage of national center for health statistics survey data to the national death index – 2015 linked mortality file (lmf): Methodology overview and analytic considerations. Technical report, https://www.cdc.gov/nchs/data-linkage/mortality-methods.htm.

Navigli, R., Faralli, S., Soroa, A., de Lacalle, O., and Agirre, E. (2011). Two birds with one stone: Learning semantic models for text categorization and word sense disambiguation. In *In Proceedings of the 20th ACM International Conference on Information and Knowledge Management*. ACM.

Neamatullah, I., Douglass, M. M., wei H. Lehman, L., Reisner, A., Villarroel, M., Long, W. J., Szolovits, P., Moody, G. B., Mark, R. G., and Clifford, G. D. (2008). Automated de-identification of free-text medical records. *BMC Medical Informatics and Decision Making*, **8**.

Nelson, R. K. (2010). Mining the dispatch. http://dsl.richmond.edu/dispatch/.

Newman, M. (2005). A measure of betweenness centrality based on random walks. *Social Networks*, **27**(1):39–54.

Newman, M. (2010). *Networks: An Introduction*. Oxford University Press.

Nguyen, V.-A., Boyd-Graber, J., and Resnik, P. (2012). SITS: A hierarchical nonparametric model using speaker identity for topic segmentation in multiparty conversations. In *Proceedings of the Association for Computational Linguistics*.

Nguyen, V.-A., Boyd-Graber, J., and Resnik, P. (2013). Lexical and hierarchical topic regression. In *Advances in Neural Information Processing Systems*.

Nguyen, V.-A., Boyd-Graber, J., Resnik, P., and Chang, J. (2014). Learning a concept hierarchy from multi-labeled documents. In *In Proceedings of the Annual Conference on Neural Information Processing Systems*. Morgan Kaufmann.

Nguyen, V.-A., Boyd-Graber, J., Resnik, P., and Miler, K. (2015). Tea Party in the House: A hierarchical ideal point topic model and its application to Republican legislators in the 112th Congress. In *Association for Computational Linguistics*.

Niculae, V., Kumar, S., Boyd-Graber, J., and Danescu-Niculescu-Mizil, C. (2015). Linguistic harbingers of betrayal: A case study on an online strategy game. In *Association for Computational Linguistics*.

Nielsen, M. (2012). *Reinventing Discovery: The New Era of Networked Science*. Princeton University Press.

Nissenbaum, H. (2009). *Privacy in context: Technology, policy, and the integrity of social life*. Stanford University Press.

Nissenbaum, H. (2011). A contextual approach to privacy online. *Daedalus*, **140**(4):32–48.

Nissenbaum, H. (2019). Contextual integrity up and down the data food chain. *Theoretical inquiries in law*, **20**(1):221–256.

Obe, R. O., and Hsu, L. S. (2015). *PostGIS in Action*, 2nd Edition. Manning Publications.

Obstfeld, D. (2005). Social networks, the tertius iungens orientation, and involvement in innovation. *Administrative Science Quarterly*, **50**(1):100–130.

Office of Management, Budget (2019). M-19-23: Phase 1 implementation of the foundations for evidence-based policymaking act of 2018: Learning agendas, personnel, and planning guidance. Technical report, https://www.whitehouse.gov/wp-content/uploads/2019/07/M-19-23.pdf, Washington DC.

Ohm, P. (2010). Broken promises of privacy: Responding to the surprising failure of anonymization. *UCLA Law Review*, **57**:1701.

Ohm, P. (2014). The legal and regulatory framework: what do the rules say about data analysis? In Lane, J., Stodden, V., Nissenbaum, H. and Bender, S., editors, *Privacy, Big Data, and the Public Good: Frameworks for Engagement*. Cambridge University Press.

Olson, J. M., and Brewer, C. A. (1997). An evaluation of color selections to accommodate map users with color-vision impairments. *Annals of the Association of American Geographers*, **87**(1):103–134.

Organisation of Economic Co-operation, Development (2004). A summary of the Frascati manual. *Main Definitions and Conventions for the Measurement of Research and Experimental Development*, **84**.

Ott, M., Choi, Y., Cardie, C., and Hancock, J. T. (2011). Finding deceptive opinion spam by any stretch of the imagination. In *Proceedings of the 49th Annual Meeting of the Association for Computational Linguistics: Human Language Technologies–Volume 1, HLT '11*, pages 309–319. Stroudsburg, PA. Association for Computational Linguistics.

Owen-Smith, J., and Powell, W. W. (2003). The expanding role of university patenting in the life sciences: Assessing the importance of experience and connectivity. *Research Policy*, **32**(9):1695–1711.

Owen-Smith, J., and Powell, W. W. (2004). Knowledge networks as channels and conduits: The effects of spillovers in the Boston biotechnology community. *Organization Science*, **15**(1):5–21.

Pan, I., Nolan, L. B., Brown, R. R., Khan, R., van der Boor, P., Harris, D. G., and Ghani, R. (2017). Machine learning for social services: a study of prenatal case management in illinois. *American journal of public health*, **107**(6):938–944.

Pang, B., and Lee, L. (2008). *Opinion Mining and Sentiment Analysis*. Now Publishers.

Park, H.-S., and Jun, C.-H. (2009). A simple and fast algorithm for k-medoids clustering. *Expert Systems with Applications*, **36**(2):3336–3341.

Paul, M., and Girju, R. (2010). A two-dimensional topic-aspect model for discovering multi-faceted topics. In *Association for the Advancement of Artificial Intelligence*.

Pennebaker, J. W., and Francis, M. E. (1999). *Linguistic Inquiry and Word Count*. Lawrence Erlbaum.

Pentland, A., Greenwood, D., Sweatt, B., Stopczynski, A., and de Montjoye, Y.-A. (2014). Institutional controls: The new deal on data. In Lane, J., Stodden, V., Bender, S. and Nissenbaum, H., editors, *Privacy, Big Data, and the Public Good: Frameworks for Engagement*, pages 98–112. Cambridge University Press.

Peters, M., Neumann, M., Iyyer, M., Gardner, M., Clark, C., Lee, K., and Zettlemoyer, L. (2018). Deep contextualized word representations. In *Conference of the North American Chapter of the Association for Computational Linguistics*.

Peterson, A., and Spirling, A. (2018). Classification accuracy as a substantive quantity of interest: Measuring polarization in westminster systems. *Political Analysis*, **26**(1):120–128.

Petrakos, G., Conversano, C., Farmakis, G., Mola, F., Siciliano, R., and Stavropoulos, P. (2004). New ways of specifying data edits. *Journal of the Royal Statistical Society, Series A*, **167**(2):249–274.

Plaisant, C., Grosjean, J., and Bederson, B. B. (2002). SpaceTree: Supporting exploration in large node link tree, design evolution and empirical evaluation. In *IEEE Symposium on Information Visualization*, pages 57–64. IEEE.

Plank, B., Søgaard, A., and Goldberg, Y. (2016). Multilingual part-of-speech tagging with bidirectional long short-term memory models and auxiliary loss. In *Proceedings of the 54th Annual Meeting of the Association for Computational Linguistics (Volume 2: Short Papers)*, pages 412–418. Berlin, Germany. Association for Computational Linguistics.

Plumb, G., Molitor, D., and Talwalkar, A. (2018). Model agnostic supervised local explanations. In *Technical report, Proceedings of the 32nd International Conference on Neural Information Processing Systems (NIPS 18)*, pages 2520–2529. Red Hook, NY, USA. Curran Associates Inc..

Potash, E., Brew, J., Loewi, A., Majumdar, S., Reece, A., Walsh, J., Rozier, E., Jorgenson, E., Mansour, R., and Ghani, R. (2015). Predictive modeling for public health: Preventing childhood lead poisoning. In *Proceedings of the 21th ACM SIGKDD International Conference on Knowledge Discovery and Data Mining, KDD '15*, pages 2039–2047. ACM.

Powell, W. W. (2003). Neither market nor hierarchy. *Sociology of Organizations: Classic, Contemporary, and Critical Readings*, **315**:104–117.

Powell, W. W., White, D. R., Koput, K. W., and Owen-Smith, J. (2005). Network dynamics and field evolution: The growth of interorganizational collaboration in the life sciences. *American Journal of Sociology*, **110**(4):1132–1205.

President's Council of Advisors on Science, Technology (2014). Big data and privacy: A technological perspective. Technical report, Executive Office of the President.

Provost, F., and Fawcett, T. (2013). *Data Science for Business: What You Need to Know About Data Mining and Data-analytic Thinking*. O'Reilly Media.

Puts, M., Daas, P., and de Waal, T. (2015). Finding errors in Big Data. *Significance*, **12**(3):26–29.

Rabiner, L. R. (1989). A tutorial on hidden Markov models and selected applications in speech recognition. *Proceedings of the IEEE*, **77**(2):257–286.

Ram, K. (2013). Git can facilitate greater reproducibility and increased transparency in science. *Source Code for Biology and Medicine*, **8**(1).

Ramage, D., Hall, D., Nallapati, R., and Manning, C. (2009). Labeled LDA: A supervised topic model for credit attribution in multi-labeled corpora. In *Proceedings of Empirical Methods in Natural Language Processing*.

Ramakrishnan, R., and Gehrke, J. (2002). *Database Management Systems*, 3rd Edition. McGraw-Hill.

R Core Team (2013). *R: A Language and Environment for Statistical Computing*. Vienna, Austria: R Foundation for Statistical Computing.

Reid, G., Zabala, F., and Holmberg, A. (2017). Extending tse to administrative data: A quality framework and case studies from stats nz. *Journal of Official Statistics*, **33**(2):477–511.

Reiter, J. P. (2012). Statistical approaches to protecting confidentiality for microdata and their effects on the quality of statistical inferences. *Public Opinion Quarterly*, **76**(1):163–181.

Ribeiro, M. T., Singh, S., and Guestrin, C. (2016). "why should I trust you?": Explaining the predictions of any classifier. In *Proceedings of the 22nd ACM SIGKDD International Conference on Knowledge Discovery and Data Mining*, pages 1135–1144. San Francisco, CA, USA.

Richardson, L. (n.d.). Soup. http://www.crummy.com/software/BeautifulSoup/. Accessed February 1, 2020.

Ritter, A., Mausam, Etzioni, O., Clark, S. (2012). Open domain event extraction from twitter. In *Proceedings of the 18th ACM SIGKDD International Conference on Knowledge Discovery and Data Mining*, KDD 12, pages 1104–1112. New York, NY, USA. Association for Computing Machinery.

Rodolfa, K., Salomon, E., Haynes, L., Mendieta, I., Larson, J., and Ghani, R. (2020). Predictive fairness to reduce misdemeanor recidivism through social service interventions. In *Proceedings of the ACM Conference on Fairness, Accountability, and Transparency (ACM FAT*) 2020*.

Rose, A. (2010). Are Face-Detection Cameras Racist? http://content.time.com/time/business/article/085991954643,00.html. Accessed February 12, 2020

Rubin, D. B. (1976). Inference and missing data. *Biometrika*, **63**:581–592.

Ruggles, S., Fitch, C., Magnuson, D., and Schroeder, J. (2019). Differential privacy and census data: Implications for social and economic research. In *Technical report, AEA Papers and Proceedings* (Vol. **109**, 403–08).

Sah, S., Shringi, A., Ptucha, R., Burry, A. M., and Loce, R. P. (2017). Video redaction: a survey and comparison of enabling technologies. *Journal of Electronic Imaging*, **26**(5):1–14.

Saket, B., Simonetto, P., Kobourov, S., and Börner, K. (2014). Node, node-link, and node-link-group diagrams: An evaluation. *IEEE Transactions on Visualization and Computer Graphics*, **20**(12):2231–2240.

Salton, G. (1968). *Automatic Information Organization and Retrieval*. McGraw-Hill.

Samuel, A. L. (1959). Some studies in machine learning using the game of Checkers. *IBM Journal of Research and Development*, **3**(3):210–229.

Sandhaus, E. (2008). The New York Times annotated corpus. *Linguistic Data Consortium*, http://www.ldc.upenn.edu/Catalog/CatalogEntry.jsp?catalogId=LDC2008T19.

Saraiya, P., North, C., and Duca, K. (2005). An insight-based methodology for evaluating bioinformatics visualizations. *IEEE Transactions on Visualization and Computer Graphics*, **11**(4):443–456.

Schafer, J. L. (1997). *Analysis of Incomplete Multivariate Data*. CRC Press.

Schafer, J. L., and Graham, J. W. (2002). Missing data: Our view of the state of the art. *Psychological Methods*, **7**(2):147.

Schermann, M., Hemsen, H., Buchmüller, C., Bitter, T., Krcmar, H., Markl, V., and Hoeren, T. (2014). Big data. *Business & Information Systems Engineering*, **6**(5):261–266.

Scheuren, F., and Winkler, W. E. (1993). Regression analysis of data files that are computer matched. *Survey Methodology*, **19**(1): 39–58.

Schierholz, M., Gensicke, M., Tschersich, N., and Kreuter, F. (2018). Occupation coding during the interview. *Journal of the Royal Statistical Society: Series A (Statistics in Society)*, **181**(2):379–407.

Schnell, R. (2014). An efficient privacy-preserving record linkage technique for administrative data and censuses. *Statistical Journal of the IAOS*, **30**:263–270.

Schnell, R. (2016). German Record Linkage Center.

Schnell, R., Bachteler, T., and Reiher, J. (2009). Privacy-preserving record linkage using Bloom filters. *BMC Medical Informatics and Decision Making*, **9**(1):41.

Schoenman, J. A. (2012). The concentration of health care spending. NIHCM foundation data brief, National Institute for Health Care Management. http://www.nihcm.org/pdf/DataBrief3%20Final.pdf.

Schölkopf, B., and Smola, A. J. (2001). *Learning with Kernels: Support Vector Machines, Regularization, Optimization, and Beyond.* MIT Press.

Schwartz, A. E., Leardo, M., Aneja, S., and Elbel, B. (2016). Effect of a school-based water intervention on child body mass index and obesity. *JAMA pediatrics*, **170**(3):220–226.

Scott, S. L., Blocker, A. W., Bonassi, F. V., Chipman, H., George, E., and McCulloch, R. (2013). Bayes and big data: The consensus Monte Carlo algorithm. In *In EFaBBayes 250 conference, volume 16. http://bit.ly/1wBqh4w, Accessed January 1, 2016*

Sethian, J. A., Brunet, J.-P., Greenberg, A., and Mesirov, J. P. (1991). Computing turbulent flow in complex geometries on a massively parallel processor. In *Proceedings of the 1991 ACM/IEEE Conference on Supercomputing*, pages 230–241. ACM.

Severance, C. (2013). Python for informatics: Exploring information. http://www.pythonlearn.com/book.php.

Shawe-Taylor, J., and Cristianini, N. (2004). *Kernel Methods for Pattern Analysis.* Cambridge University Press.

Shelton, T., Poorthuis, A., Graham, M., and Zook, M. (2014). Mapping the data shadows of Hurricane Sandy: Uncovering the sociospatial dimensions of 'big data'. *Geoforum*, **52**:167–179.

Shlomo, N. (2014). Probabilistic record linkage for disclosure risk assessment. In *International Conference on Privacy in Statistical Databases*, pages 269–282. Springer.

Shlomo, N. (2018). Statistical disclosure limitation: New directions and challenges. *Journal of Privacy and Confidentiality*, **8**(1).

Shneiderman, B. (1992). Tree visualization with tree-maps: 2-D space-filling approach. *ACM Transactions on Graphics*, **11**(1): 92–99.

Shneiderman, B. (2008). Extreme visualization: Squeezing a billion records into a million pixels. In *Proceedings of the 2008 ACM SIGMOD International Conference on Management of Data*, pages 3–12. ACM.

Shneiderman, B., and Plaisant, C. (2015). Sharpening analytic focus to cope with big data volume and variety. *Computer Graphics and Applications, IEEE*, **35**(3):10–14. See also http://www.cs.umd.edu/hcil/eventflow/Sharpening-Strategies-Help.pdf.

Sies, H. (1988). A new parameter for sex education. *Nature*, **332**(495).

Silberschatz, A., Korth, H. F., and Sudarshan, S. (2010). *Database System Concepts*, 6th Edition. McGraw-Hill.

Smalheiser, N. R., and Torvik, V. I. (2009). Author name disambiguation. *Annual Review of Information Science and Technology*, **43**(1):1–43.

Smola, A. J., and Schölkopf, B. (2004). A tutorial on support vector regression. *Statistics and Computing*, **14**(3):199–222.

Snow, J. (1855). *On the Mode of Communication of Cholera*. John Churchill.

Spielman, S. E., and Singleton, A. (2015). Studying neighborhoods using uncertain data from the american community survey: A contextual approach. *Annals of the Association of American Geographers*, **105**(5):1003–1025.

Squire, P. (1988). Why the 1936 Literary Digest poll failed. *Public Opinion Quarterly*, **52**(1):125–133.

Stanford Visualization Group (n.d.). Dorling cartograms in ProtoVis. http://mbostock.github.io/protovis/ex/cartogram.html. Accessed February 1, 2020.

Stanton, M. W., and Rutherford, M. (2006). *The High Concentration of US Health Care Expenditures*. Agency for Healthcare Research and Quality.

Stasko, J., Görg, C., and Liu, Z. (2008). Jigsaw: Supporting investigative analysis through interactive visualization. *Information Visualization*, **7**(2):118–132.

Steorts, R. C., Hall, R., and Fienberg, S. E. (2014). SMERED: a Bayesian approach to graphical record linkage and deduplication. https://arxiv.org/abs/1312.4645.

Stephens-Davidowitz, S., and Varian, H. (2015). A hands-on guide to Google data. http://people.ischool.berkeley.edu/hal/Papers/2015/primer.pdf.

Stock, J. H., and Watson, M. W. (2002). Forecasting using principal components from a large number of predictors. *Journal of the American Statistical Association*, **97**(460):1167–1179.

Stopczynski, A., Sekara, V., Sapiezynski, P., Cuttone, A., Madsen, M. M., Larsen, J. E., and Lehmann, S. (2014). Measuring large-scale social networks with high resolution. *PloS One*, **9**(4).

Strandburg, K. J. (2014). Monitoring, datafication and consent: Legal approaches to privacy in the big data context. In Lane, J., Stodden, V., Bender, S. and Nissenbaum, H., editors, *Privacy, Big Data, and the Public Good: Frameworks for Engagement*. Cambridge University Press.

Strasser, C. (May 5, 2014). Git/GitHub: A primer for researchers. http://datapub.cdlib.org/2014/05/05/github-a-primer-for-researchers/.

Strauch, C. (2009). Nosql databases. http://www.christof-strauch.de/nosqldbs.pdf.

Stuart, E. A. (2010). Matching methods for causal inference: A review and a look forward. *Statistical Science*, **25**(1):1.

Sutton, R. S., and Barto, A. G. (2018). *Reinforcement Learning. An Introduction*. Cambridge, MA: The MIT Press.

Sweeney, L. (2001). Computational disclosure control: A primer on data privacy protection. Technical report, MIT. http://groups.csail.mit.edu/mac/classes/6.805/articles/privacy/sweeney-thesis-draft.pdf.

Szalay, A. S., Gray, J., Thakar, A. R., Kunszt, P. Z., Malik, T., Raddick, J., Stoughton, C., and vandenBerg, J. (2002). The SDSS skyserver: Public access to the Sloan digital sky server data. In *Proceedings of the 2002 ACM SIGMOD International Conference on Management of Data*, pages 570–581. ACM.

Talley, E. M., Newman, D., Mimno, D., Herr, II, B. W., Wallach, H. M., Burns, G. A. P. C., Leenders, A. G. M., and McCallum, A. (2011). Database of NIH grants using machine-learned categories and graphical clustering. *Nature Methods*, **8**(6):443–444.

Tanner, A. (April 25, 2013). Harvard professor re-identifies anonymous volunteers in DNA study. *Forbes*, http://www.forbes.com/sites/adamtanner/2013/04/25/harvard-professor-re-identifies-anonymous-volunteers-in-dna-study/#6cc7f6b43e39.

Tennekes, M., and de Jonge, E. (2011). Top-down data analysis with treemaps. In *Proceedings of the International Conference on Imaging Theory and Applications and International Conference on Information Visualization Theory and Applications*, pages 236–241. SciTePress.

Tennekes, M., de Jonge, E., and Daas, P. (2012). Innovative visual tools for data editing. Presented at the United Nations Economic Commission for Europe Work Session on Statistical Data. Available online at http://www.pietdaas.nl/beta/pubs/pubs/30_Netherlands.pdf.

Tennekes, M., de Jonge, E., and Daas, P. J. H. (2013). Visualizing and inspecting large datasets with tableplots. *Journal of Data Science*, **11**(1):43–58.

The Northpointe Suite (2016). Response to ProPublica: Demonstrating accuracy equity and predictive parity. https://www.equivant.com/response-to-propublica-demonstrating-accuracy-equity-and-predictive-parity/. Accessed February 12, 2020.

Thompson, W. W., Comanor, L., and Shay, D. K. (2006). Epidemiology of seasonal influenza: Use of surveillance data and statistical models to estimate the burden of disease. *Journal of Infectious Diseases*, **194**(Supplement 2):S82–S91.

Tibshirani, R. (1996). Regression shrinkage and selection via the lasso. *Journal of the Royal Statistical Society, Series B*, **58**: 267–288.

Trewin, D., Andersen, A., Beridze, T., Biggeri, L., Fellegi, I., and Toczynski, T. (2007). Managing statistical confidentiality and microdata access: Principles and guidelines of good practice. *Technical report, Conference of European Statisticians, United Nations Economic Commision for Europe*.

Tuarob, S., Pouchard, L. C., and Giles, C. L. (2013). Automatic tag recommendation for metadata annotation using probabilistic topic modeling. In *Proceedings of the 13th ACM/IEEE-CS Joint Conference on Digital Libraries, JCDL '13*, pages 239–248. ACM.

Tufte, E. (2001). *The Visual Display of Quantitative information*, 2nd Edition. Graphics Press.

Tufte, E. (2006). *Beautiful Evidence*, 2nd Edition. Graphics Press.

Ugander, J., Karrer, B., Backstrom, L., and Marlow, C. (2011). The anatomy of the facebook social graph. Technical report, https://arxiv.org/abs/1111.4503.

University of Oxford (2006). British National Corpus. http://www.natcorp.ox.ac.uk/.

UnivTask Force on Differential Privacy for Census Data (2019). Implications of Differential Privacy for Census Bureau Data and Scientific Research.

Ustun, B., and Rudin, C. (2016). Supersparse linear integer models for optimized medical scoring systems. *Machine Learning*, **102**:349–391.

Ustun, B., and Rudin, C. (2019). Learning optimized risk scores. *Journal of Machine Learning Research*, **20**(150):1–75.

Valentino-DeVries, J., Singer, N., Keller, M., and Krolick, A. (2018). Your Apps Know Where You Were Last Night, and They're Not Keeping It Secret.

Valliant, R., Dever, J. A., and Kreuter, F. (2018). *Practical Tools for Designing and Weighting Survey Samples*. Springer, 2nd edition.

Varian, H. R. (2014). Big data: New tricks for econometrics. *Journal of Economic Perspectives*, **28**(2):3–28.

Ventura, S. L., Nugent, R., and Fuchs, E. R. H. (2015). Seeing the non-stars:(some) sources of bias in past disambiguation approaches and a new public tool leveraging labeled records. *Research Policy*.

Vigen, T. (2015). *Spurious Correlations*. Hachette Books.

Voigt, R., Camp, N. P., Prabhakaran, V., Hamilton, W. L., Hetey, R. C., Griffiths, C. M., Jurgens, D., Jurafsky, D., and Eberhardt, J. L. (2017). Language from police body camera footage shows racial disparities in officer respect. *Proceedings of the National Academy of Sciences*, **114**(25):6521–6526.

Wallach, H., Mimno, D., and McCallum, A. (2009). Rethinking LDA: Why priors matter. In *Advances in Neural Information Processing Systems*.

Wallgren, A., and Wallgren, B. (2007). *Register-Based Statistics: Administrative Data for Statistical Purposes*. John Wiley & Sons.

Wang, C., Blei, D., and Fei-Fei, L. (2009a). Simultaneous image classification and annotation. In *Computer Vision and Pattern Recognition*.

Wang, Y., Bai, H., Stanton, M., Chen, W.-Y., and Chang, E. Y. (2009b). PLDA: parallel latent Dirichlet allocation for large-scale applications. In *International Conference on Algorithmic Aspects in Information and Management*.

Ward, K. C. (2017). Word2vec. *Natural Language Engineering*, **23**(1):155–162.

Ward, M. O., Grinstein, G., and Keim, D. (2010). *Interactive Data Visualization: Foundations, Techniques, and Applications*. CRC Press.

Weinberg, B. A., Owen-Smith, J., Rosen, R. F., Schwarz, L., Allen, B. M., Weiss, R. E., and Lane, J. (2014). Science funding and short-term economic activity. *Science*, **344**(6179):41–43.

Wezerek, G., and Van Riper, D. (2020). Changes to the Census Could Make Small Towns Disappear.

Whang, S. E., Menestrina, D., Koutrika, G., Theobald, M., and Garcia-Molina, H. (2009). Entity resolution with iterative blocking. In *Proceedings of the 2009 ACM SIGMOD International Conference on Management of data*, pages 219–232. ACM.

White, H. C., Boorman, S. A., and Breiger, R. L. (1976). Social structure from multiple networks. I. Block models of roles and positions. *American Journal of Sociology*, **81**:730–780.

Wick, M., Singh, S., Pandya, H., and McCallum, A. (2013). A joint model for discovering and linking entities. In *Proceedings of the 2013 Workshop on Automated Knowledge Base Construction*, pages 67–72. ACM.

Wikipedia (n.d.). Representational state transfer. https://en.wikipedia.org/wiki/Representational_state_transfer. Accessed February 1, 2020.

Wilbanks, J. (2014). Portable approaches to informed consent and open data. In Lane, J., Stodden, V., Bender, S. and Nissenbaum, H., editors, *Privacy, Big Data, and the Public Good: Frameworks for Engagement*, pages 98–112. Cambridge University Press.

Winkler, W. E. (2009). Record linkage. In Pfeffermann, D. and Rao, C. R., editors, *Handbook of Statistics 29A, Sample Surveys: Design, Methods and Applications*, pages 351–380. Elsevier.

Winkler, W. E. (2014). Matching and record linkage. *Wiley Interdisciplinary Reviews: Computational Statistics*, **6**(5):313–325.

Wongsuphasawat, K., and Lin, J. (2014). Using visualizations to monitor changes and harvest insights from a global-scale logging infrastructure at twitter. In *Proceedings of the IEEE Conference on Visual Analytics Science and Technology*, pages 113–122. IEEE.

Wu, X., Kumar, V., Quinlan, J. R., Ghosh, J., Yang, Q., Motoda, H., McLachlan, G. J., Ng, A., Liu, B., Philip, S. Y., Zhou, Z.-H., Steinbach, M., Hand, D. J., and Steinberg, D. (2008). Top 10 algorithms in data mining. *Knowledge and Information Systems*, **14**(1):1–37.

Wuchty, S., Jones, B. F., and Uzzi, B. (2007). The increasing dominance of teams in production of knowledge. *Science*, **316**(5827):1036–1039.

Yarkoni, T., Eckles, D., Heathers, J., Levenstein, M., Smaldino, P., and Lane, J. I. (2019). Enhancing and accelerating social science via automation: Challenges and opportunities. Technical report, DARPA.

Yates, D., and Paquette, S. (2010). Emergency knowledge management and social media technologies: a case study of the 2010 haitian earthquake. In *Proceedings of the 73rd ASIS&T Annual Meeting on Navigating Streams in an Information Ecosystem, volume 47 of ASIS&T '10*, Silver Springs, MD. American Society for Information Science.

Yost, B., Haciahmetoglu, Y., and North, C. (2007). Beyond visual acuity: The perceptual scalability of information visualizations for large displays. In *Proceedings of the SIGCHI Conference on Human Factors in Computing Systems*, pages 101–110. ACM.

Zachary, W. W. (1977). An information flow model for conflict and fission in small groups. *Journal of Anthropological Research*, **33**(4):452–473.

Zafar, M. B., Valera, I., Rogriguez, M. G., and Gummadi, K. P. (2017). Fairness Constraints: Mechanisms for Fair Classification.

In Singh, A. and Zhu, J., editors, *Proceedings of the 20th International Conference on Artificial Intelligence and Statistics*, volume 54 of *Proceedings of Machine Learning Research*, pages 962–970, Fort Lauderdale, FL, USA: PMLR.

Zayatz, L. (2007). Disclosure avoidance practices and research at the US Census Bureau: An update. *Journal of Official Statistics*, **23**(2):253–265.

Zemel, R., Wu, Y., Swersky, K., Pitassi, T., and Dwork, C. (2013). Learning fair representations. In Dasgupta, S. and McAllester, D., editors, *Proceedings of the 30th International Conference on Machine Learning*, volume 28 of *Proceedings of Machine Learning Research*, pages 325–333, Atlanta, Georgia, USA: PMLR.

Zeng, Q. T., Redd, D., Rindflesch, T. C., and Nebeker, J. R. (2012). Synonym, topic model and predicate-based query expansion for retrieving clinical documents. In *American Medical Informatics Association Annual Symposium*, pages 1050–1059.

Zhang, L.-C. (2012). Topics of statistical theory for register-based statistics and data integration. *Statistica Neerlandica*, **66**(1): 41–63.

Zhu, J., Chen, N., Perkins, H., and Zhang, B. (2013). Gibbs max-margin topic models with fast sampling algorithms. In *Proceedings of the International Conference of Machine Learning*.

Zhu, X. (2008). Semi-supervised learning literature survey. http://pages.cs.wisc.edu/jerryzhu/pub/ssl_survey.pdf.

Zolas, N., Goldschlag, N., Jarmin, R., Stephan, P., Owen-Smith, J., Rosen, R. F., Allen, B. M., Weinberg, B. A., and Lane, J. (2015). Wrapping it up in a person: Examining employment and earnings outcomes for Ph.D. recipients. *Science*, **350**(6266):1367–1371.

Zygmunt, Z. (2013). Machine learning courses online. http://fastml.com/machine-learning-courses-online.

Index

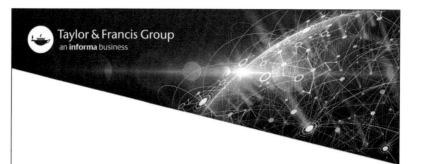

9780367568597